Springer Praxis Books
Popular Science

This book series presents the whole spectrum of Earth Sciences, Astronautics and Space Exploration. Practitioners will find exact science and complex engineering solutions explained scientifically correct but easy to understand. Various subseries help to differentiate between the scientific areas of Springer Praxis books and to make selected professional information accessible for you.

The Springer Praxis Popular Science series contains fascinating stories from around the world and across many different disciplines. The titles in this series are written with the educated lay reader in mind, approaching nitty-gritty science in an engaging, yet digestible way. Authored by active scholars, researchers, and industry professionals, the books herein offer far-ranging and unique perspectives, exploring realms as distant as Antarctica or as abstract as consciousness itself, as modern as the Information Age or as old our planet Earth. The books are illustrative in their approach and feature essential mathematics only where necessary. They are a perfect read for those with a curious mind who wish to expand their understanding of the vast world of science.

Serge Marguet

A Brief History of Nuclear Reactor Accidents

From Leipzig to Fukushima

Serge Marguet
SINETICS, Bureau O2 EF.05
Electricite De France, Recherche et Dév
Palaiseau, France

ISSN 2626-6113   ISSN 2626-6121   (electronic)
ISBN 978-3-031-10499-2   ISBN 978-3-031-10500-5   (eBook)
https://doi.org/10.1007/978-3-031-10500-5

This Springer imprint is published by the registered company Springer Nature Switzerland AG.
The registered company address is: Gewerbestrasse 11, 6330 Cham, Switzerland

Preface

It is undeniable that if nuclear energy fascinates, it also frightens and even scares. It is difficult to find reliable information on the subject because of the technical nature of the nuclear field where self-proclaimed experts assert without demonstrating, explain without showing, and conclude without arguing. The anti-nuclear debate is often bogged down by simplistic shortcuts often based on the terror that nuclear accidents inspire. Pro-nuclear people hide by saying: "*It's too complicated, I can't explain!*" The purpose of this book is to inform, explain, and conclude based on proven facts and my long experience with the subject. On reflection, my own expertise could be questioned by the reader. If my expertise is proven by my participation and appointment in numerous committees and expert bodies throughout my 35 years of career in the nuclear field, my impartiality is undoubtedly more difficult to assert. Indeed, how could an expert be totally impartial. How could a surgical expert be totally foreign to the hospital environment. It is the case with nuclear energy as with any other human action; the expert is necessarily a stakeholder, since it is obvious that one must be familiar with the subject to provide an expert opinion.

Nuclear accidents are unfortunately a vast subject. In this book, I have concentrated on reactors in which nuclear fissions are voluntarily produced while avoiding the important issue of irradiation accidents in hospitals or accelerators, contamination in waste storage sites, or criticality accidents in radioactive liquid solutions. I have reviewed not only the emblematic accidents such as Three-Mile-Island, Chernobyl, and Fukushima, but also accidents that are much less well known but just as rich in lessons. "*Those who cannot remember the past are condemned to repeat it,*" this quote from the philosopher George Santayana (1863–1952) in 1905 perfectly sums up the philosophy of this book. Whatever your initial point of view on nuclear energy, I hope that this book, which I wanted to be reader-friendly - judge by yourself!, will allow you to feel really informed, if not convinced!

Palaiseau, France   Serge Marguet

Introduction

Abstract Since the discovery of nuclear fission in 1939, physicists have postulated the possibility of using it for civilian energy production, but also for military applications. The German wartime tests to produce an atomic weapon caused the first nuclear reactor accident in Leipzig on June 23, 1942, destroying the experimental heavy water pile by fire. Other criticality accidents occurred in the USA during the Manhattan program and in the first reactors producing plutonium.

The First Ever Nuclear Reactor Accident

Few fields fascinate the public as much as the atomic adventure. This fascination has a double meaning: interest but also fear. According to some exegetes, this fascination stems from the original flaw of nuclear energy: the atomic bomb, the absolute weapon that was supposed to put an end to all wars because of its atrocious efficiency. Nuclear power was born in secrecy, and Enrico Fermi's first uranium atomic pile (Photo 1) began to generate heat in a sustained manner, hidden under the bleachers of a Chicago stadium on December 2, 1942.

Photo 1 Enrico Fermi. Designer of the first operational nuclear reactor in history. Fermi is considered by his peers as a giant of modern physics

Fermi[1] is the first to succeed, but not the first to try! As early as 1940, the Canadian physicist George Laurence[2] (Photo 2) tried in Ottawa

---

[1] Enrico Fermi (1901–1954). Italian physicist. After brilliant studies in mathematics, he studied physics at the University of Pisa. He published his first article in 1922 on general relativity, of which he was one of the main advocates in Italy. In 1926, he became professor of physics at the University of Rome. In 1934, he proposed a revolutionary theory of β^- decay by introducing a new particle: the neutrino. He then devoted himself to the creation of new radioactive isotopes and was awarded the Nobel Prize in 1938. Faced with the rise of totalitarianism in Italy and his wife being of Jewish origin, he immigrated in 1939 to the USA, where he built in Chicago the first operational reactor of Humanity, known as Chicago Pile-1 or CP-1, which diverged on December 2, 1942. He participated intensively in the Manhattan Project to build the American atomic bomb and became an American citizen in 1945. He died of stomach cancer, possibly due to his exposure to radiation from the reactors and the testing of the first aerial atomic bombs.

Fermi"s certificate of entry into the USA and the Italian stamp in honor of the first critical reaction in the CP-1 pile in Chicago. On the right Fermi (wearing glasses) visits the Italian motorcycle firm Guzzi in 1954 shortly before his death.

[2] George Craig Laurence (1905–1987). Canadian physicist. After a doctorate in physics at Cambridge under the direction of Ernest Rutherford, he worked from 1930 for the National Research Council of Canada. In 1940, he attempted to build a fission reactor, succeeding in inducing fissions in a subcritical device, but not in maintaining the reaction. After the war, he worked at the Chalk River nuclear center on the piles ZEEP, NRX, and NRU.

(Canada) to build a small critical reactor made of uranium oxide bags surrounded by coke that can sustain a fission chain reaction. The coke is a relatively pure form of coal to slow down the neutrons that become more able to produce fissions[3] (Fig. 1). The experiment failed because of the lack of purity (limited by short funds) of the materials used (the insufficiently purified coke containing traces of neutron absorbers). These materials turn out to be too absorbent for neutrons. Moreover, the choice of a rather homogeneous geometry hinders the establishment of fissions (the neutrons are captured by uranium 238 to the detriment of fissions) and especially the absence of enrichment in uranium 235, because of the use of natural uranium that contains only 0.711% uranium 235, prevented the establishment of a regime of self-sustaining fissions.

Photo 2   Georges Craig Laurence (1905–1987) was a Canadian pioneer in reactor physics who tried unsuccessfully in 1941 to build a small critical reactor

---

[3] This may seem counterintuitive, but the lower the velocity of neutrons, the greater their probability of producing fission in uranium 235. Even if the analogy is false, we can remember the idea of a soccer goalkeeper (^{235}U), who catches slow balls (slow or thermal neutrons) more easily than fast shots (fast neutrons).

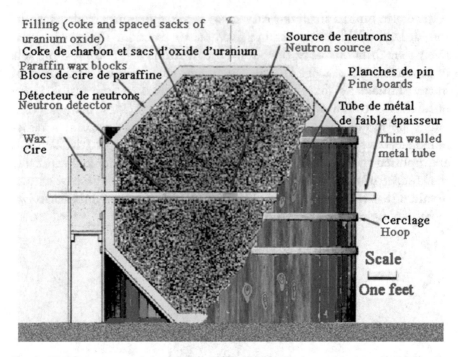

Filling (coke and spaced sacks of uranium oxide)
Coke de charbon et sacs d'oxide d'uranium

Paraffin wax blocks
Blocs de cire de paraffine

Détecteur de neutrons
Neutron detector

Wax
Cire

Source de neutrons
Neutron source

Planches de pin
Pine boards

Tube de métal
de faible épaisseur
Thin walled
metal tube

Cerclage
Hoop

Scale
One feet

Fig. 1 The Ottawa experiment (1941–1942). The device proved to be subcritical even though a neutron source was placed in the center of a tube embedded in the filling of coke (a usual form of coal) and uranium oxide multiplied by fissions, which was revealed by the neutron detector placed in the same tube. Carbon is a good slowing down agent for neutrons, which have a greater chance of causing fission if they have a slow speed. Actually, the probability of fission increases greatly when the neutrons are slow. Physicists therefore tried to slow them down efficiently without the moderator (retarder), the coke, capturing them too much. The experiment is therefore a half-failure because the device turns out to be subcritical and therefore incapable of sustaining a chain reaction, but fortunately, for the observers placed around the device! The paraffin wax is a hydrogenated material, which acts as a neutron reflector by reflecting neutrons toward the nuclear fuel and by limiting the neutron leakage of the device. The homogeneous mixture of coke and uranium oxide is not necessarily a good idea because it is better to separate the uranium and the carbon moderator in a heterogeneous geometry to favor reactivity (measure of the capacity of the device to establish a chain reaction). This technic limits the capture of neutrons by uranium 238, which is not fissile in the presence of slow neutrons (^{238}U can fission only with fast neutrons). The idea of uranium bags is already better than fully homogeneous mixture of uranium and graphite, but the whole thing remained too homogeneous

More worryingly, in June 1942, Werner Heisenberg[4], Nobel Prize in Physics in 1932, answered the question posed by the German general Erhard Milch[5] that a bomb the size of a pineapple would be enough to destroy a city like London, which is close to reality.

---

[4] Werner Heisenberg (1901–1976). Werner Heisenberg was a German physicist who won the Nobel Prize in Physics in 1932 for his work in quantum mechanics. After studying physics at the University of Munich, where he defended a thesis on fluid turbulence under the direction of Arnold Sommerfeld, he worked with the great physicists of his time: Max Born, Arnold Sommerfeld, and Niels Bohr. He introduced the use of matrices in quantum physics. At the age of 26, he was appointed professor of physics at the University of Leipzig, where he later launched the heavy water reactor experiments under the Nazi regime. In 1927, he formulated the uncertainty formalism that bears his name. Initially attacked by the Nazis who considered quantum physics to be ""Jewish,"" he was ""rehabilitated"" in 1939, mainly because his mother was a close friend of the mother of Heinrich Himmler, the supreme leader of the SS! Heisenberg directed the German nuclear program from 1942 to 1945, especially the experiments in Leipzig and Haigerloch. After the war, Heisenberg denied having wanted to develop an atomic bomb by voluntarily delaying the progress of the program (?). His ambiguous position toward the Third Reich earned him criticism, although he was not worried after the war.

From left to right: Enrico Fermi, Werner Heisenberg, and Wolfgang Pauli sitting on the shores of Lake Como (Italy). The first two worked on the atomic bomb in opposite camps.

[5] Erhard Milch (1892–1972). German air force general. *Generalfeldmarschall* Milch was charged by Hitler with the supervision of German aeronautical production, and specifically the special weapons (V1, V2, super-bombers). He was sentenced to life imprisonment for war crimes at Nuremberg but was released in 1954.

Milch still enjoys a certain popularity linked to the prestige of the Luftwaffe and special weapons.

As part of the German war effort, Heisenberg was part of the team attempting to develop a Nazi atomic bomb. As soon as fission was discovered by the German chemists Otto Hahn[6] and Fritz Strassmann[7] in 1938 (Photo 3), and that the chain reaction had been

---

[6] Otto Hahn (1879–1968) was a German chemist. After studying chemistry in Munich and Marburg, he was introduced to radioactivity in 1904 in the English laboratory of Sir William Ramsay. Back in Germany, he worked at the Berlin Institute of Chemistry where he met Lise Meitner. In 1907, he discovered radium 228, thorium 230, and protactinium, a new chemical element in 1917. He studied heavy transuranic nuclei with his assistant Fritz Strassmann from 1935. This work led him to discover uranium fission at the end of 1938 by detecting the presence of radioactive barium in a liquid solution of uranium irradiated by neutrons. Hahn remained in Germany throughout the Second World War, keeping his distance from the Nazi dictatorship, and trying to protect his Jewish collaborators such as Lise Meitner. He learned of the use of the atomic bomb on Japan while being held prisoner with other German scientists at Farm Hall (photo below) in England. He was so horrified that it was feared that his life would be at risk by committing suicide. He was awarded the Nobel Prize in Chemistry in 1944 for his discovery of fission. A chemical body bearing his name, hahnium, was proposed for element 105, but it was finally the name dubnium that was officially retained. The same misfortune happened to him for element 108, which was finally named hassium, and again for element 110, which became darmstadtium. It is reasonable to think that his name will be definitively retained for a super-heavy nucleus in the future.

Lise Meitner on the left and Otto Hahn, and the stamp of the late East Germany in honor of Otto Hahn.

Farm Hall near Cambridge where the German scientists were kept as prisoners from July 3, 1945, to January 3, 1946.

[7] Fritz Strassmann (1902–1980) was a German chemist who co-discovered the nuclear fission of uranium-235 with Otto Hahn after studying chemistry and completing a thesis at the University of Hanover. From 1929, he worked at the prestigious Kaiser Wilhelm Institute in Berlin, where he specialized in analytical chemistry and became a very close collaborator of Otto Hahn, as well as Lise Meitner. After the war, he worked at the Max Planck Institute in Mainz. He received the Enrico Fermi Prize in 1966 as well as Hahn and Meitner.

Photo 3 The wood table on which fission was discovered. The uranium liquid solution was contained in the beaker seen on the right of the photo. The source of neutrons comes from a mixture of radium and beryllium placed in the center of a yellowish cylindrical block of paraffin wax located next to the beaker. The paraffin has the function of slowing down the neutrons, which makes them more likely to produce fissions on uranium 235. This wood table is piously preserved at the Karlsruhe nuclear research center

proven in 1939, some German physicists embarked on a train of research whose unavowed goal was the production of a nuclear weapon. The German army supported the informal military program *Uranverein*", "*the uranium club*", which brought together a few hundred scientists

Fritz Strassman and the discovery report published in 1944.

concerned with the subject. First and foremost, Werner Heisenberg, technical leader of the German bomb research, Carl-Friedrich Von Weizsäcker who filed a patent in 1941 on the concept of an atomic weapon, son of the German diplomat and State Secretary for Foreign Affairs Ernst Von Weizsäcker from 1938 to 1943 under Hitler, but also Paul Hartek who worked on the enrichment of uranium and heavy water,[8] Walter Gerlach, Kurt Diebner[9] (member of the Nazi party and

---

[8] Heavy water D_2O has the same chemical properties as normal water H_2O, but the ^1H isotope of normal (light) water is replaced by the ^2H isotope in heavy water. To simplify the writing, the term deuterium D was invented to refer to ^2H. Heavy water is a very expensive and difficult to produce product that is used as a neutron moderator in an atomic pile. A pile operating with heavy water can produce plutonium that can be used for a bomb. In 1934, Norsk Hydro built the world"s first commercial heavy water production plant in Vemork, Norway (see photo below), with a capacity of 12 tons per year. During the Second World War, the Germans invaded Norway in order to dispose of the Norwegian ports, but with an ulterior motive related to heavy water production. The British decided to destroy the plant by several military actions (commandos) in order to prevent Germany from developing its nuclear weapons program. On November 16, 1943, the Allies dropped more than four hundred bombs on the site, prompting the Germans to move all production to Germany. On February 20, 1944, Knut Haukelid, a Norwegian partisan, sank the ferry carrying heavy water on Lake Tinn. Contrary to what the Allies have long claimed, the Germans would have known about this raid and would have deceived the Allies, as most of the heavy water was actually evacuated by truck and used for the *Uranverein* program. The story of the heavy water sabotage was the basis for the French film ""*La bataille de l"eau lourde*"" in 1947 and the American film ""*The heroes of Telemark*"" in 1965 with Kirk Douglas in the main role.

[9] Kurt Diebner (1904–1965) was a German physicist and scientific advisor to the *Heereswaffenamt*, the Armaments Office of the German Army. He organized a conference on September 16, 1939, between the various German physicists concerned with nuclear energy, followed by a new conference on September 26, which led to the launch of a research program on the atomic bomb, the construction of a pile, and enrichment in U235. At the suggestion of Paul Harteck, the choice of pile was a natural uranium reactor moderated with heavy water. Erich Bagge, Diebner's assistant and a former student of Heisenberg's, took care of the isotopic separation. On October 5, 1939, Diebner took over the effective direction of the Kaiser Wilhelm *Institut für Physik* (KWIP) from the Dutchman Peter Debye, who did not want to take German nationality and participate in the war effort (he left for Cornell University in the USA). Fortunately for the rest of history, Heisenberg never accepted the tutelage of Diebner, whom he did not consider a physicist (despite his doctorate in physics), and he quickly "went it alone," fuelled by his poor relations with Diebner. This difference sums up the general atmosphere of the *UranVerein*, where several

manager of a nuclear research group), Erich Bagge (member of the Nazi party), Walther Bothe, Klaus Clusius who worked at the University of Zurich on heavy water production, Karl Wirtz who worked on reactor physics during the war and then at the Karlsruhe research center after the war, and Robert Döpel experimental physicist who worked on the heavy water reactor program in Leipzig and who was captured by the Russians to work on the Soviet atomic weapon.

Theoretical work by Heisenberg, the tests (one cites the test named "L-IV", L for Leipzig), were carried out in the first half of 1942, and extrapolations (erroneous) by Heisenberg indicated that the spherical geometry, with five tons of heavy water and 10 tons of metallic uranium, could be critical. The calculations are simpler in spherical geometry because of the symmetry of revolution (a problem that can be reduced to a geometry with a single radial dimension of successive layers). The tests conducted by Robert Döpel showed indeed a production of neutrons, but still subcritical. An article by Klara Döpel, Robert Döpel's wife, and Werner Heisenberg was first published in the *Kernphysikalische Forschungsberichte* (Research Reports in Nuclear Physics), a classified journal of the *Uranverein*. The first L-I and L-II tests used uranium oxide and 164 kg of heavy water. The replacement in 1942 of uranium oxide by uranium metal plates increased the production of neutrons more than expected. The L-III reactor at Leipzig used 108 kg of metallic uranium (and still 164 kg of heavy water); then L-IV reached 750 kg of uranium (with still the same 164 kg of heavy water) in the spring of 1942. L-IV showed in April 1942 an increase of 13% in the neutron flux, "*the experimental proof of the effective multiplication of neutrons in a concentric sphere of D_2O and uranium*," as the Döpels wrote in July 1942. These results indicated that a self-sustaining reaction was within the realm of possibility, provided allegedly that 5 tons of

---

competing teams were working separately, the probable cause, along with the lack of means, of the German failure in the field. Kurt Diebner.

heavy water and about ten tons of uranium were available. Increasing the size of the pile reduces neutron leakage.

The Leipzig research group was led by Heisenberg until 1942. Heisenberg then withdrew from practical experiments and left the execution of the L-III and L-IV experiments (Fig. 2) mainly to his colleagues under the direction of

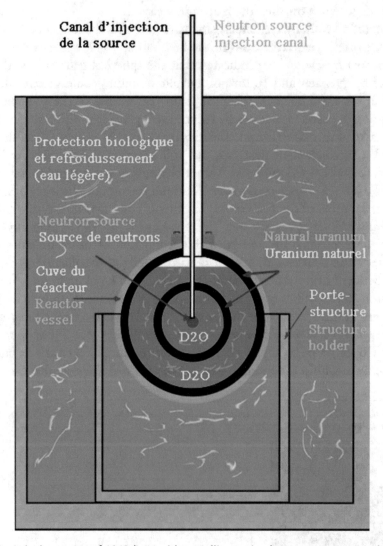

Fig. 2 Leipzig reactor of 1942 (L-IV with metallic uranium). It was necessary to leave a little vacuum in the vessel to take into account the thermal expansion of the heavy water during the heat-up. One of the realistic causes of the explosion of June 23, 1942, is that this void would have been filled with air at dismantling, reacting explosively with the uranium at high temperature, especially in the form of powder, and especially if the uranium had been hydrated during the 20 days of operation. It is known that uranium hydrides are particularly pyrophoric. It should be noted that no system for

Döpel. The theoretical calculation of a pile made of laminated materials would be the work of two young physicists because Heisenberg was rather uninterested in the practical aspects: Karl-Heinz Höcker (1915–1998), a former student of von Weizsäcker, and Paul O. Müller (1915–1942), a former student of Erwin Shrödinger (both were mobilized, Müller was killed on the Russian front, and Höcker was able to be reinstated at the KWIP in 1942 after strong and motivated pressure on the Army). The team, at the suggestion of Paul Harteck, quickly understood that it was better to separate (heterogeneous geometry) the fissile material from the moderator than to achieve an intimate mixture. In a homogeneous mixture, the neutrons do not have time to slow down because they are captured without fission by uranium 238, without being slowed down enough to induce fissions in uranium 235. However, the probability of fission is much greater with slowed neutrons than with fast neutrons. In a heterogeneous geometry, neutrons arriving in heavy water can slow down without risk of capture (heavy water is not very absorbent) and can return to the fuel to induce fissions by geometrically avoiding capture in ^{238}U, hence the need for technological ingenuity in the respective distribution of uranium and heavy water. This idea of heterogeneity is still used in present

---

controlling the chain reaction existed on all the types of piles built by the Germans, which highlights a flagrant incompetence in the kinetic aspects of the reactor. The consequences of uncontrolled over-criticality seemed to escape them (radiation protection of operators). A form of modern fantasy tends to say that the Germans would have developed a low-power atomic bomb, but their inability to enrich uranium in the isotope 235, let alone plutonium 239, which is an artificial isotope produced in a reactor (which they did not possess), makes this hypothesis totally unrealistic. Any scientific evidence other than conspiracy rumors or vague testimonies of unusual explosions (the Italian journalist Luigi Romersa or the airplane pilot Rudolf Zinsser in October 1944) that could have used conventional explosives does not support even the testing of a "dirty" bomb, i.e., a conventional bomb loaded with radioactive waste. Other testimonies relate to a very bright explosion at the Ohrdruf concentration camp on March 3, 1945. Even after so many years, residual radioactivity would be easily detected in case of success, as well as fission products such as technetium or promethium that do not exist in nature and a large quantity of unfissioned uranium, even if the atmospheric nuclear tests after 1945 tend to create background noise. Another argument is that the Germans never had uranium enriched to more than 0.8% U235, and even then, in ridiculous quantities, not to mention plutonium 239, the extraction technique for which was totally unknown to them, and which their experimental subcritical piles could not produce continuously. At most, one can imagine a test of compression of natural uranium by a conventional explosive, but the result could not be anything other than a dispersion of nuclear fuel without precise control of the compression zone. How could such a test have been unknown to the specialists at the head of the *Uranverein*? The astonishment of the most famous German physicists, held at Farm Hill, at the announcement of the explosion of the first American atomic bomb was not feigned: Heisenberg even thought that an entire atomic pile had been launched on Hiroshima! The German atomic bomb remains an anticipation-book uchrony that excites people who like to be scared.

reactors. Reactor physics calculations aim at calculating the pitch between the plates or the fuel cubes, since if the plates or the cubes are too close, the neutrons do not have time to thermalize (i.e., to slow down) before returning to the fuel, and if on the contrary the plates or the cubes are too far apart, the neutrons will be absorbed before returning to the fuel. There is therefore an optimum of moderation where the effective multiplication factor k_{eff} is maximum: we speak of the optimum moderation ratio, that is to say the ratio between the volume of moderator (heavy water) and the volume of fuel. To hope to be critical, the k_{eff} must be at least greater than 1 at the optimal moderation ratio; otherwise the pile can never hope to be critical whatever the arrangement of fissile materials. It should be noted that the Germans did not choose the simplest geometry. The principle of these experiments was to have the powder of metallic natural uranium and the moderator in the form of heavy water loaded in a device designed to slow down the neutrons produced by a radium-beryllium source. In the case of L-IV, the uranium was plated against the inner face of the spherical container and in a spherical inner shell (Fig. 2). The whole assembly swims in heavy water. The pile is submerged in light water that serves as a neutron reflector and biological protection outside the spherical shell.

One senses the desire to keep a spherical geometry, no doubt because the calculations were made in this geometry. However, plans show one of the geometries made up of a laminate of uranium (551 kg) and paraffin wax, an alkane derived from solid petroleum residues (Fig. 3, Photo 4). The paraffin C_nH_{2n+2} therefore contains carbon and hydrogen, which are excellent neutron moderators, although less effective than heavy water, which captures fewer neutrons, as the Germans must have realized. Such a geometry is simpler to realize and preserves the heterogeneous character of the pile. The Germans knew the neutrophageous character of uranium 238 (especially when the temperature increases because of the Doppler Effect), which makes homogeneous geometries particularly inefficient. Heisenberg even considered that the temperature increase was a stabilizing character of the pile to avoid a power excursion. However, he did not seem to differentiate between thermal and fast neutrons, being content to use rough estimates of cross-sections averaged over the entire energy spectrum, hence the confusion between a thermal spectrum pile and a fast spectrum bomb. On June 23, 1942, after 20 days of operation, Robert Döpel[10] noted the appearance of blisters at the level of the vessel seal, probably caused by a heat-up and a rise in pressure (dilatation of the heavy

---

[10] Georg Robert Döpel (1895–1982) was a German physicist who studied in Leipzig and Munich. He obtained his doctorate in 1924. He was a member of the *Uranverein* and worked with Werner Heisenberg at the University of Leipzig, where he directed the L heavy water reactor experiments. Captured by the Russians in 1945, he had to work on the atomic bomb project in the USSR. He married a Russian

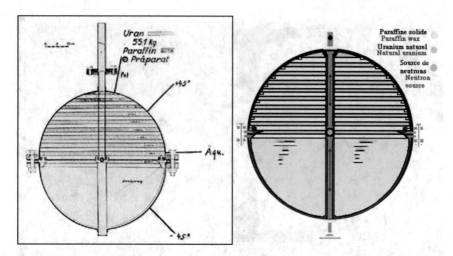

Fig. 3 L-I reactor with uranium oxide and paraffin layers (1940). Beginning in October 1940, Heisenberg and Karl Wirtz carried out a series of chain reaction experiments at the KWIP in Berlin using an arrangement of successive layers of natural uranium oxide and paraffin (used as a moderator), the whole immersed in light water (used as a neutron reflector, heat sink, and biological protection)

water, steam production?). It must be understood that the source of neutrons imposed by S *neutrons per second* can multiply by fission even in a subcritical environment and that the neutron level will stabilize at a level of $S/(1-k_{eff})$ neutrons per second, therefore higher than the neutron source as soon as the k_{eff} is non-zero. With a k_{eff}[11] of the order of 0.8, this is equivalent to multiplying the source by 5 and by 10 if the k_{eff} is worth 0.9. Note that the formula is not valid if $k_{eff} = 1$ since one would find an infinite result. This is because a much more complex calculation has to be performed when the reactor is

---

woman in 1954 and was not allowed to return to East Germany until 1957 in Thuringia to teach at the University.

Robert Döpel in 1935

[11] We will return in more detail to the concept of k_{eff}. For now, it is enough to understand that the k_{eff} is a multiplication coefficient of neutrons in the considered geometry and materials. Starting from n given neutrons, the next generation will count n times k_{eff} neutrons. This is the neutron equivalent of the famous R_0 coefficient used in pandemic epidemiology studies such as COVID-19.

Photo 4 Two pictures of L reactors from Leipzig. Vents can be seen on the pile on the right. The pile on the left should be L-II with a spherical geometry. L-II allowed an increase in neutron flux to be measured on October 28, 1941, but it contained only 142 kg of uranium oxide and 164 kg of heavy water, far from the critical mass necessary to reach sustainable criticality. The pile on the right is L-IV, the one that exploded on June 23, 1942

critical. Nevertheless, even in subcritical conditions, enough fissions can be produced to heat up the pile. The opening of the vessel by the operator lets in air, and uranium and especially uranium hydrides (which were created by direct contact with heavy water) are particularly pyrophoric, i.e., they ignite in air. This property is used in weapons with depleted uranium cores, such as certain tank shells: the uranium of the shell in contact with the steel armor of an enemy tank creates a low-melting point eutectic, resulting in a *"butter-like"* penetration, and then the depleted uranium core explodes mechanically inside, setting the tank ablaze (it is not a nuclear explosion at all!). The greater density of the uranium makes the projectile heavier at constant volume and increases the kinetic energy at constant speed. In the case of the L-IV pile, the ignited uranium caused the water to boil, generating enough steam pressure to dismantle the reactor. As it burned, the uranium powder dispersed throughout the laboratory, causing a larger fire in the facility. It was reported that glowing uranium powder had reached the ceiling of 6 meters high, spreading a severe fire, and that the device had heated up to 1000 °C. Leipzig L-IV can be considered as the first severe accident in history. This will not shut down the German research on the subject.

In parallel to Heisenberg's work, Kurt Diebner developed his own concepts in Berlin. After having returned the control of the KWIP to the *Reichsforschungsrat*, the Army had nevertheless retained a research center, directed by Diebner and located in Gottow, about 50 km from Berlin. Diebner's work followed more or less the same steps as Heisenberg's. The two scientists hated each other cordially, and it is doubtful that they would work together. His first spherical reactor, G-I—G for Gottow, used cubes of uranium oxide inserted in paraffin in the fall of 1942. Since he had no metallic uranium, neither in plates nor in powder, Diebner used the unused uranium oxide of the *UranVerein*. He first considered alternating layers of paraffin and uranium oxide, but finally opted for cubes. The size of the cube was chosen to be smaller than the mean free path of a neutron so that it would have a good chance of being slowed down in the paraffin before fissioning another uranium-235 nucleus or being captured by a uranium-238. The G-II reactor (Fig. 4) used heavy water[12] (in the form of ice) instead of paraffin. The idea of heavy water ice is rather curious in a device intended to heat up, hence the variations in internal density when the ice melts. Here again, the geometry of fuel cylinders, much simpler to realize and more efficient, escaped Diebner. The cubic distribution was nevertheless more interesting than the layered one advocated by Heisenberg for L-I, as the theoretician Karl-Heinz Höcker had calculated. Höcker, a former doctoral student of von Weizsäcker and his collaborator at the KWIP and then in occupied Strasbourg, collaborated with Diebner's team in 1943 after his brief incorporation into the army. The cubes were more favorable to a chain reaction than the alternating or concentric layers of the Heisenberg device because the risk of resonant capture of neutrons by uranium 238 was much lower. Moreover, the cubes were much easier to fabricate than the large plates required by Heisenberg. On the other hand, the orderly structuring of the lattice of cubes in a sphere remains technically difficult to achieve (positioning to respect the regular lattice).

From March 1945, Heisenberg and his team at Berlin-Dahlem attempted to create a heterogeneous critical device consisting of a lattice of uranium cubes attached to chains that were immersed in heavy water enriched in deuterium contained in a vessel (Pile B for Berlin? Fig. 5). Curiously, he did not think of the much simpler and more efficient solution of a vertical lattice of

---

[12] The natural hydrogen contained in water has two isotopes. The nucleus of the first, the most abundant, has a single proton; the second, 7000 times less abundant, is sometimes called deuterium and has a neutron and a proton. Deuterium is therefore heavier than the single-proton hydrogen. Because of its nuclear properties, deuterium is much less neutron absorbing than natural hydrogen, hence the idea of using water enriched in deuterium, so-called ""heavy water,"" to slow down without absorbing neutrons, which become more efficient for the chain reaction. Heavy water is 10% heavier than light water (its density compared to water is 1.1), hence its name.

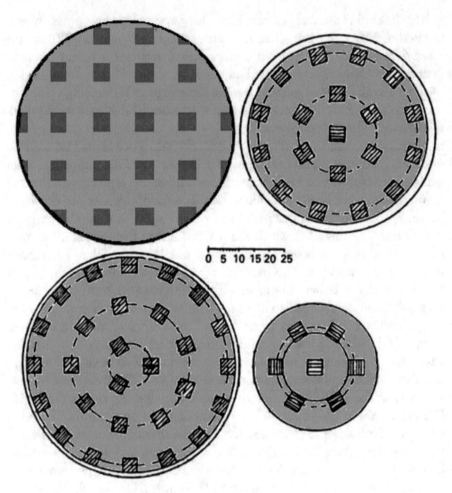

0 5 10 15 20 25

Fig. 4 The G-II pile (Gottow) designed by Diebner and his team

tubes containing uranium, a solution that would be adopted 20 years later in the pressurized water reactors (PWRs). This cube device, which in any case could not have been critical, was found by the American army in April 1945 in an underground brewery in Haigerloch[13] and dismantled for transfer to the USA by the ALSOS mission[14] organized to recover technology and German

---

[13] 60 km South of Stuttgart.

[14] The ALSOS mission was a secret mission created on April 4, 1944, by the Americans in order to gather information on the progress of the German nuclear program. It was composed of about 100 military personnel and scientists commanded by American Colonel Boris Pash, a former athletics professor at Hollywood College who was later charged with investigating the alleged anti-American activities of Robert Oppenheimer, and under the scientific direction of the Dutch-born physicist Samuel Goudsmit, nicknamed ""*Uncle Sam*." This mission first operated in Italy on the immediate rear of the advancing

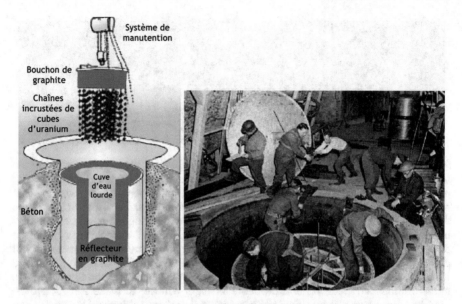

Fig. 5 The US Army dismantles the German "reactor" known as "B VIII" at Haigerloch (US Army photo, 1945). A military column including General Harrison's 1279th Engineer Battalion and led by Colonel Pash of ALSOS took the small town of Hechingen next to Haigerloch (operation "Humbug") with the objective of reaching it before the French. The French had the political intention of creating a vast zone of occupation east of the Rhine and were unaware of the scientific potential of Haigerloch. By trickery, the Americans succeeded in saying that the Heichingen area would be heavily bombed to frighten the French, and were the first to capture Carl Friedrich von Weizsäcker, Karl Wirtz, and Erich Bagge, the elite of reactor physics. Otto Hahn and Max von Laue were captured in Tailfingen. Werner Heisenberg, target number one, was caught at his chalet in Urfeld in the Bavarian Alps, where he had taken refuge. As for the pile, the very number of the name suggests preparatory tests in Berlin. The pile in Haigerloch had been moved from Berlin-Dahlem to avoid the bombing and to escape the dangerously approaching Russians. A posteriori, analysis showed that the reactor could not have reached criticality because of a lack of critical mass. The value of the k_{eff} multiplication coefficient is only 0.89, whereas it should be 1 to reach criticality. This value could have been reached with a much larger size (to reduce neutron leakage) or a slight enrichment in uranium 235 (1 to 2% instead of the 0.711% of natural uranium). The "pile" contained about 1500 kg of heavy water, 1500 kg of uranium metal, the refining process of which the Germans had mastered, in the form of cubes, 10,000 kg of graphite serving as a neutron reflector around the magnesium vessel (a metal that is a weak neutron absorber, unlike steel), and a source of initiating neutrons made up of a mixture of 500 mg of radium (radioactive α) and beryllium that produced neutrons by reaction (α, n). The cubes of natural uranium were fixed in a spaced-apart manner on chains and form a fuel lattice embedded in heavy water

---

troops, but with very inconclusive results, because the Italians, whether in Naples, Brindisi or Taranto, knew nothing of the secret German projects. ALSOS was redeployed to France at the German border and then to Germany itself, where it finally collected scientific reports, equipment, and fissile materials and recovered many scientists and specialized technicians. The insignia of the mission was a white alpha

Photo 5 Recovery of the Haigerloch uranium cubes by the ALSOS team of the American army led by Samuel Goudsmit, a scientist from MIT with knowledge of nuclear energy, who oversaw the task of scouring Germany behind the troops in order to recover researchers and expertise, especially in the nuclear field. The Russians, the British, and the French (in France alone, more than 1000 engineers and researchers between 1945 and 1950 in the field of submarines, missiles, aeronautics...!) did the same. One of the uranium cubes is shown in the Haigerloch museum. The total lack of precaution in the handling of the cubes by the soldiers suggests that it is not believed that the Germans could have operated the pile, Goudsmit having the knowledge to detect radioactive materials and probably a Geiger counter to make sure. A professional advice: never pile up cubes of fissile material as these soldiers do at the risk of a bad surprise! It should be remembered that even subcritical operation will produce fission products in quantities depending on the operating time and the level of neutron flux reached

scientists (Photo 5). This same ALSOS mission was able to establish, as soon as Strasbourg was taken at the end of November 1944, that the Germans had only just begun to build a pile in August 1944, when Samuel Goudsmit[15] and

---

struck by a red lightning bolt, rather indiscreet for such a secret mission. The mission was disbanded on October 15, 1945, after having scoured the western part of Germany, the rest being scoured by the Russians in a more aggressive way. An identical mission of smaller size was ordered to examine the state of Japanese science after Japan"s surrender.

[15] Samuel Abraham Goudsmit (1902–1978). Born in La Hague (Netherlands). While studying theoretical physics at the University of Leiden in 1925, Goudsmit discovered the phenomenon of electron spin with George E. Uhlenbeck. He was a student of Paul Ehrenfest at the University of Leiden (Netherlands), from which he received a doctorate in 1927. He was then a professor at the University of Michigan (USA) from 1927 to 1946. In 1941, Goudsmit joined the Radiation Lab at the Massachusetts Institute of Technology (MIT), where he conducted radar research and headed the laboratory"s document room, which contained invaluable information about German technical capacities. In May 1944, Goudsmit became scientific director of the Manhattan Project"s Alsos Mission, a top-secret operation responsible for gathering intelligence on Germany"s atomic program. He worked there with captain Reginald C. Augustine and Fred Wardenburg to build an efficient hunting team. The mission investigated German scientists" progress toward nuclear weapons as the Allies liberated the European continent. While in Europe, he traveled to his childhood home in The Hague, where he found that his parents had been killed in a concentration camp. Concerning the German bomb, Goudsmit concluded that the failure of the German project was attributable to a number of factors, including dictature bureaucracy, Allied bombing campaigns, the persecution of Jewish scientists, and Werner Heisenberg"s failed leadership. After the war, Goudsmit briefly taught at Northwestern University and then was chairman of the physics department at Brookhaven National Laboratory. He also edited the American Physical Society"s *Physical Review* for 25 years.

Fred Wardenburg analyzed Von Weisäcker's papers that had been left at the University of Strasbourg where he was working (Bar-Zohar, 1965, p. 64). The German nuclear program collapsed in 1945 at the same time as the Third Reich, Adolf Hitler having never shown a particular interest in these subjects, as he was more interested in super-tank programs, Messerschmitt 262 jet fighters, and long-range launchers such as the V2 (the famous "retaliation weapons"). However, it has been reported that Hitler himself wondered about an uncontrollable chain reaction leading to the extinction of all life on Earth! The Nazi nuclear program never monopolized more than a hundred people, compared to the 120,000 people affiliated with the American Manhattan Project. It was therefore an arms race won by the Allies that contributed to the beginnings of nuclear power, and the founding fathers of reactor physics were soon faced with difficult moral choices.

From 1942 onward, the American war effort was considerable, and the Manhattan Project produced most significant results: a highly enriched uranium bomb totally destroyed Hiroshima on August 6, 1945 (the Americans were so sure of its success that it was not even tested!); then a plutonium bomb (this bomb will be tested in the desert of Alamogordo) destroyed Nagasaki on August 9, 1945, with the human consequences that we know. The Americans preferred to sacrifice two Japanese cities rather than consider a terribly deadly landing on Japanese soil, widely predicted by the fierce resistance of the Japanese army on every island in the Pacific. A controversy about the real need to launch the second bomb erupted after the war, accusing some scientists of wanting to "experiment" with this new type of bomb. Accidents had already occurred during the preparation of the bomb. It was at Los Alamos (USA) on February 11, 1945, the first criticality incident in human history took place. An uncontrolled start of criticality took place on the Dragon reactor using a uranium compound, UH_3, compressed in Styrex.[16] As a notable effect, it is noted that one operator

---

Samuel "*Uncle Sam*" Goudsmit

[16] An extruded polystyrene insulation.

had a significant loss of hair, but without lethal effect. On August 21, 1945, the first criticality accident occurred at the Los Alamos research center, resulting in one identified victim, independently of the many victims of the two atomic bombs over Japan, which could not be called an accident. The accident occurred when a block of tungsten carbide used as a reflector slipped from the hands of an experienced operator, Harry K. Daghlian Jr. who positioned these blocks by stacking them around a fissile core (Photos 6 and 7). The block in question felt

Photo 6 Herbert Lehr (left) and Harry Daghlian, Jr. (right), loading the assembled tamper cap containing the plutonium "compartment" and initiator into a sedan for transport from McDonald Ranch to the firing tower on July 13, 1945. The photo on the right shows a reconstruction made in 1946 with an object of the same size

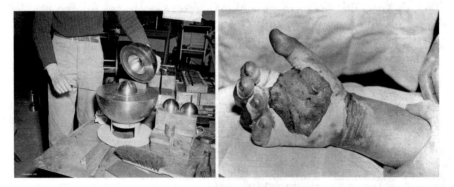

Photo 7 Reconstruction made in 1946 of the fit-up of the upper reflector of the "demon core" and photo of Daghlian's hand after this deadly irradiation

close to a subcritical sphere[17] of 6.2 kg of plutonium 239, making it over-critical by the reflector effect, the carbide blocks sending the neutrons back to the fissile core. The operator then suffered a lethal dose estimated (although he was not wearing a dosimeter) to 200 *rads*[18] on his body with peaks of 40,000 rads on his hands, and he died 26 days after his fatal exposure from radiation sickness[19] (destruction of the spinal marrow with the impossibility of producing red blood cells along decay of all the organs).

It is reported that Enrico Fermi had written a memorandum before this case to Robert Oppenheimer,[20] responsible for the Manhattan Project, to

---

[17] A reactor is said to be ""critical"" when the fission chain reaction is stabilized and continuous. A subcritical reactor sees a progressive smothering of the chain reaction when it has taken place, or the impossibility of the establishment of the critical reaction at start-up, whereas a super-critical reactor sees an exponential progression of the chain reaction. An atomic bomb is a particular highly over-critical core whose geometry and constitution (materials) are designed to voluntarily maintain the over-criticality as long as possible. A nuclear reactor has a core whose geometry and constitution are calculated to be just critical. We will return to this crucial concept later. It should be noted that the uranium 238 present in a civil pile prevents a total nuclear explosion as in the case of a bomb (because of the Doppler effect of uranium 238). A mechanical explosion by expansion of the components, thermochemical interaction, or hydrogen explosion is always possible.

[18] Remember that a rad is the old unit of dose absorbed by a gram of biological tissue subjected to 100 erg, a unit of energy that is no longer used today: 1 erg = 10^{-7} Joule. The official unit today is the Gray, which is one Joule deposited in 1 kg of material. The damage suffered by the tissues is proportional to the dose received.

[19] This event is particularly well portrayed in Roland Joffé's 1989 film *"Masters of the Dark"* with Paul Newman and John Cusack as the operator. The events described in the film relate rather to the death of Louis Slotin (1910–1946), fatally irradiated on May 21, 1946, while demonstrating reflector assembly around the same plutonium sphere that killed Daghlian. One can easily find photos of the reconstruction that took place afterward.

[20] Robert Oppenheimer (1904–1967). American physicist. After a degree in chemistry, he became an outstanding theorist (he was responsible for major theoretical advances on black holes) and obtained a doctorate at the age of 22 under the direction of Max Born in Göttingen. He then became a professor of physics at Berkeley. Despite his possibly sympathetic views toward communism, he was appointed to head the Manhattan Project to build the first atomic bomb. In 1947, he replaced Albert Einstein at Princeton.

Oppenheimer was the scientific director of the Manhattan project

point out the danger of performing these manipulations by hand rather than by remote control, but that Oppenheimer believed that such operations would delay the schedule for making the first atomic bomb, "*Little Boy.*" The same core, later nicknamed "*The demon core,*" was the cause of another fatal accident on May 21, 1946, under somewhat similar conditions, when Louis Slotin caused a screwdriver to shatter during a dangerous fit-up of the two half-spheres of the core, the half-spheres then closing by engaging the over-criticality. Richard Feynman,[21] future Nobel Prize winner in physics and a young scientist on the Manhattan team, used the term "*tickling the dragon's tail*" when referring to these risky experiments.

From the first Fermi pile in 1942 (Photo 8), there was concern about the safety of the reactor and a redundancy of reactor shutdown resources was planned. First of all, a scientist, Norman Hilberry, nicknamed afterward "*the Axe man*",[22] armed with an axe, stands over the pile, ready to cut the rope holding a cadmium-coated shutdown control rod, a powerful neutron

---

[21] Richard Feynman (1918–1988). American physicist. After brilliant studies at MIT, he introduced the concept of path integral in quantum physics and published many books on physics and popularization. He then taught at the Californian Institute of Technology (Caltech). He was awarded the Nobel Prize in Physics in 1965 for his work on quantum electrodynamics.

Richard Feynman

[22] I personally find it hard to see this great physicist holding an axe in the role he is given! This is perhaps an urban legend that is repeated over and over again. I understand that he has always denied this anecdote.

Norman Hilberry (1899–1986), director of *Argonne National Laboratory* from 1957 to 1961.

Photo 8 The famous CP-1 pile of Fermi (*pile* in English means stack. The word is also used in French, although the usual meaning is *battery*). As there are practically no photos of the pile itself and of this historical event, only this painting allows us to understand the excitement preceding the divergence of December 2, 1942. At the top of the pile is the wooden frame under which the shutdown control rod is suspended (Painting by Gary Sheahan (Joseph "Gary" Sheahan (1893–1978) was born in Winnetka, Illinois. He studied at the University of Notre Dame and the Chicago School of Art before joining the *Chicago Tribune* as an illustrator in 1922. He is best known for his World War II paintings, having participated in the D-Day landings.))

absorber, which will then fall by gravity into the core. The image is so striking that at the end, the word SCRAM, widely used in the nuclear industry to signify the shutdown, has been associated a posteriori[23] with the significance *Safety Cute Rope Axe Man* or *Safety Control Rod Axe Man*. A second operator, the physicist Wallace Koehler, is always standing over the pile, armed with a tub full of a cadmium sulfate solution, ready to spray it in the reactor. Cadmium is indeed one of the most powerful absorbers of slow neutrons. We find here the current principles of redundancy in safety systems. The pile is controlled by a horizontal cadmium control rod handled by hand by the operator George Weil, and Enrico Fermi supervises the neutron flux measuring devices.

In the 1950s, it was civil nuclear power that gave the atom its moral support. If nuclear energy can kill, it can also produce heat. Numerous more or less realistic projects flourished: the use of small atomic bombs to pierce

---

[23] Volney Wilson, head of instrumentation, is also credited with answering the electrician who was wiring the red shutdown button and asking what would happen if that button were to be pressed, so that he could title it correctly on the control panel: Wilson answered "*You scram out of here as fast as you can!*"

mountains or canals (the technique was often used in the USSR!), airplane reactors, submarine reactors (since atomic fission does not require air, the submarine could remain under the sea for several months, if the air needed by the crew can be recycled), transportable terrestrial reactors for the Arctic zones... The production of electricity by reactors vaporizing water to turn a turbine is a concept that will find a great industrial echo. On December 20, 1951, in Idaho Falls (USA), the fast neutron reactor EBR-1, designed by Walter Zinn[24] and cooled by liquid sodium, produced enough electricity to light the building that contained the reactor. The power, of 300 kW electric,

Photo 9 Historic graffiti. The EBR-1 team led by Walter Zinn wrote this historical phrase on the reactor wall (adapted from photo): *"Electricity was first generated from Atomic Energy on December 20, 1951. On December 21, 1951, all the electricity in this building was supplied from atomic energy. Those present..."*

---

[24] Walter Zinn (1906–2000). American engineer and physicist. He participated in the Chicago team led by Enrico Fermi, and then in the Manhattan project to build the atomic bomb. He became the first director of the famous Argonne National Laboratory from 1946 to 1956, where he oversaw the construction of several experimental reactors. In particular, he designed the EBR-1 fast neutron reactor. He was the first president of the American Nuclear Society.

Walter Zinn

Photo 10 The building containing the Obninsk reactor (August 1955), in the Kaluga region. The first power reactor in the world. The construction of the plant began on January 1, 1951, and the reactor diverged on June 27, 1954, to shut down on April 29, 2002

remained certainly modest but opened the door to an industrial production of electricity (Photo 9).

From this pioneering work, the development of nuclear reactor technology was to be rapid and awareness of the risk of accidents was to develop at the same time as physicists' understanding of reactor behavior. The first industrial-scale piles were built at Hanford (USA) to produce plutonium for the American military program. On June 27, 1954, the Russian plant in Obninsk (5 Mega Watt electric or MWe, 30 MWthermal or MWth) produced the first industrial production of nuclear electricity on the city's grid. This reactor (Fig. 6, Photos 10 and 11) was conceived by the very discreet *"Laboratory n°2,"* in fact a secret institute in charge of conceiving the atomic bomb, and which will become in 1960 the Kurchatov Institute, named after its first director Igor Kurchatov.[25] The reactor was permanently shut down on April 29, 2002 to become an Atom Museum.

---

[25] Igor Vassilievich Kurchatov (1903–1960) was the father of the Soviet atomic bomb. After studying physics at the University of Crimea and shipbuilding at the University of Petrograd, he joined the Physical-Technical Institute in 1925 where he worked on radioactivity under the direction of Abraham Ioffé. During the war, he developed a system for demagnetizing the shells of ships, which proved to be effective. He then worked on the Soviet atomic bomb, which was fueled by scientific leaks from the USA (Klaus Fuchs, the Rosenbergs, etc.). The bomb was developed in Laboratory n°2, which later became the

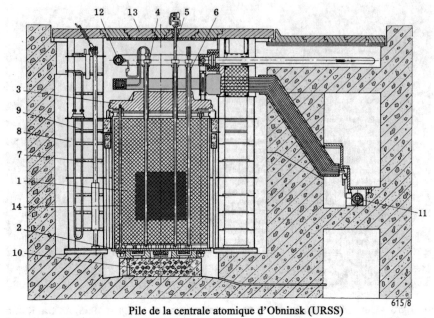

Pile de la centrale atomique d'Obninsk (URSS)

1. Empilement de graphite; 2. Plaque inférieure; 3. Plaque supérieure; 4. Canal de chargement; 5. Canal pour barre de sécurité; 6. Canal pour barre de réglage automatique; 7. Canal pour chambre d'ionisation; 8. Protection latérale (eau); 9 et 10. Réfrigérateurs; 11. Collecteur de distribution; 12. Collecteur d'évacuation; 13. Bouclier de protection supérieur (fonte); 14. Colonne refroidie du réflecteur.

Fig. 6 Section of the Obninsk reactor. The reactor (in red) in the center of the pile, operated with uranium enriched to 5%, moderated by graphite blocks also serving as reflectors, and cooled by water. This concept is called RBMK and Obninsk is the ancestor of Chernobyl

Kurchatov Institute. This plutonium bomb exploded on August 29, 1949. He wore a large beard following the vow not to cut his beard until the bomb worked. At the end of his life, he worked on civil applications of the WWER type reactor.

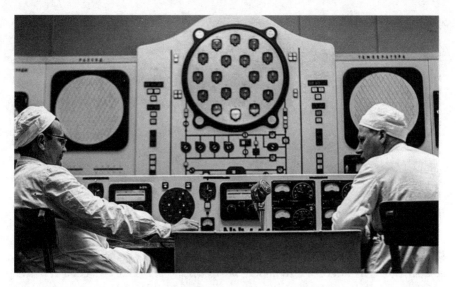

Photo 11 Control console of the Obninsk reactor in 1964 (source https://fr.rbth.com/tech/83027-russie-institut-kourtchatov-premiere-bombe-atomique, Lev Nosov, Sputnik). The panel on the right indicates "temperature" probably at the outlet of the reactor. One does not see recorders with graph paper, making it possible to follow temporally the behavior of the reactor

Photo 12 Pierre Ailleret

In France, Pierre Ailleret[26] (Photo 12), the first director of the EDF Research and Development Division and member of the scientific committee of the CEA since 1950, closely follows the implementation of the French military industrial program to produce plutonium. His brother was the general who directed the

---

[26] Pierre Ailleret (1900–1996). After graduating from Polytechnique in 1918, then the *Ecole de Ponts et Chaussées*, he worked in Indochina on the construction of an electric grid and then in the private electric industry in France. He was appointed EDF"s first Director of Studies and Research from the time of privatization in 1946 until 1958, as he was considered ""*the most learned electrician in the company*"" by his peers. At that time, EDF was mainly focused on hydraulic and oil-fired plants. We can legitimately call him the father of nuclear power at EDF because of his decisive action in the 1950s.

Photo 13   Fuel of the G1 pile after the 1956 accident. The uranium disintegrated into powder under the effect of a 20-minute fire in which the reactor remained at power

French nuclear tests in Algeria. Pierre Ailleret lobbied for the installation of the G1 military pile in Marcoule of an electricity production system that recovers the air heated by the reactor. At the divergence[27] in 1956, the demonstrator produces a mere 5 MWe when 6 MWe were needed to turn the blowers that injected air into the reactor, so a negative balance! But the technical feasibility has been demonstrated, and a power reactor, EDF-1, was built in Chinon. In October 1956, a fuel cartridge[28] of G1 badly positioned in its channel and thus badly cooled heats up and catches fire. 7 kg of nuclear fuel melt (Photo 13). Thanks to the cladding rupture detection system, the pile was shut down, but the quasi-manual extraction of the incriminated cartridge proved to be complex, obliging the operators to handle pieces of spent fuel with tongs, as there were no handling systems adapted to the problem at the time. This was the first accident of this nature in France, and it was largely unknown to the public.

Despite this hazard, the French reactor type Natural Uranium Graphite Gas (UNGG) designed by the CEA was launched. Following the various oil crises, France chose the American pressurized water reactor type, more powerful and economically more profitable, whose first prototype (1967) was the Chooz-A reactor at the border with Belgium (Photo 14). This reactor has the unique feature of having been built in a cavern entirely dug into the mountain. This cavern replaces advantageously the reactor building, at the expense of a reduced accessibility.

The movement will grow until it reaches a very high level of energy independence: nearly 80% of the energy produced comes from a Fleet of 56

---

[27] The divergence is the start-up operation that makes a core critical. This operation is naturally followed with great attention by the operators.

[28] The term ""*cartridge*"" comes from the analogy of the loading of a bullet into a gun barrel, with the placement in its housing (or channel) of a fuel assembly.

Photo 14 The impressive reactor cavern of the Ardennes Nuclear Power Plant (Chooz-A) under construction. The location of the reactor vessel pit is clearly visible

pressurized water reactors[29] (REPs). In this concept initially developed by Westinghouse in the USA and then in France by Framatome under a license, water is heated to approximately 326 °C in the reactor (Fig. 7) under high

---

[29] 56 nowadays after the shutdown of the two Fessenheim reactors on the Rhine in 2020. The Flamanville-3 EPR will become the 57th reactor as soon as it starts up.

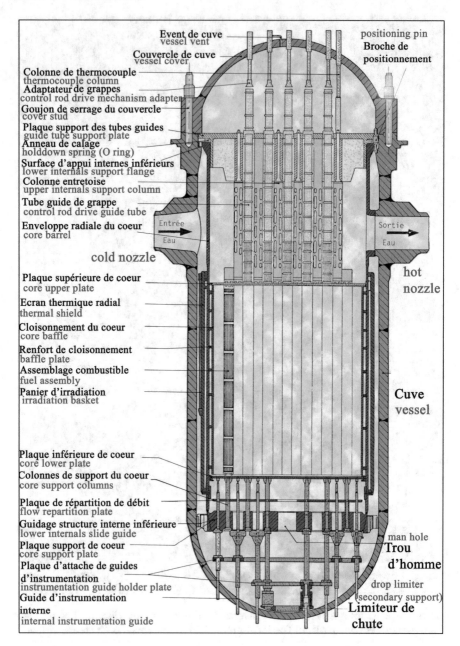

Fig. 7 Description of the vessel of a 900 MWe pressurized water reactor (PWR of the type used by *Electricité De France*) and its contents: the structures that penetrate the top vessel cover are the control rod clusters that are connected to the so-called RGL cluster control system. The cluster rod is hooked under the vessel cover to a bundle of absorbing rods by a system called "spider" (because of its shape) and the cluster slides in guide tubes that are part of the assembly. The clusters fall by gravity in case of an emergency shutdown in less than two seconds. The bottom vessel penetrations are used to insert mobile fission chamber inside the active core for measurement purpose

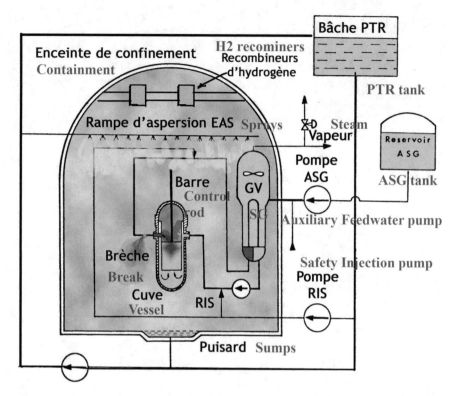

Fig. 8   The main emergency circuits of a pressurized water reactor

pressure (155 bars) in such a way that it remains liquid without passing into the steam phase (the saturation temperature of water at 155 bars is approximately 345 °C).

Once heated, the water circulates in a primary circuit and transfers its heat to steam generators (SGs). From these SGs, steam is released and, via a secondary circuit, it turns a turbine whose rotating shaft drives an alternator that produces electricity. The interest of this concept is to physically separate the reactor core from the turbine to avoid contamination of the latter in the event of a rupture of the fuel cladding, i.e., the hermetic cladding that contains the nuclear fuel. The risk of accident has been taken into account to some extent in the design of PWRs, in particular the possibility of a pipe rupture in the primary circuit in the containment (*Accident de Perte de Réfrigérant Primaire* (APRP) in French, *LOss of Coolant Accident* (LOCA) in English), and safety systems have been designed to counteract (Fig. 8). France, through its national electricity company EDF, has nearly 60 power reactors. Therefore, nuclear risk is a particularly crucial issue for France, and the operator utility EDF must be irreproachable in terms of nuclear safety.

Contents

List of Figures

List of Photos

List of Tables

1

The Physics of Nuclear Accidents

Abstract This chapter presents the physics necessary to understand a severe accident. First, the kinetic behavior of neutrons in a reactor is recalled, then the degradation of the reactor core is considered, either in the context of uncontrolled overpowering, or in the context of heat up due to lack of cooling. Neutronic, thermal-hydraulic, thermal, and chemical aspects are addressed with the appearance of corium, this mixture of molten nuclear fuel and structural materials, but also the production of hydrogen by oxidation of metals in the core. These complex subjects are presented in a sufficiently simplified way to be accessible to a non-specialist.

The Physics of Nuclear Fission

It would be a challenge to talk about accidents in reactors without a minimum understanding of the physical phenomena that come into play on this occasion. The principle of a nuclear power reactor is to heat a coolant which will store and transport the heat of the nuclear fuel,[1] place of the fissions induced by neutrons. Nuclear fission produces heat, by

---

[1] The term nuclear fuel is used by analogy with fossil fuel in conventional plants, but the term is misleading because there is no combustion in the conventional sense. Fission is not a chemical reaction like combustion and the loss of mass by nuclear fission in a reactor is negligible.

fragmentation of fissile isotopes (mainly uranium 235 and plutonium 239). This process induces the production of fission products, generally radioactive, and new neutrons (on average about 2.5 per fission) which, when released in a judiciously arranged geometry, allow fissions to take place again. This is called a "*chain reaction.*" There are many popular books that describe this phenomenon in more detail,[2] therefore we will only dwell here on the aspects that are important for reactor safety. As early as 1939, the *Collège de France* team, including Frédéric Joliot[3], Lew Kowarski[4] and Hans

[2] Let us quote in the collection « *Que sais-je?* » n°3307: *La neutronique* et n°317: *L'énergie nucléaire* de Paul Reuss, n°2032: Sûreté de l'énergie électronucléaire; n°2243: Les réacteurs atomiques de Daniel Blanc, n°1037: *Les centrales nucléaires* de Georges Parreins et n°2362: *Les surgénérateurs* de Georges Vendryès, (Chelet 2006) on radioactivity. These small books have a content accessible to most people. For readers who are interested in theories on reactor physics, I recommend more difficult books: (Ash 1979; Barjon 1993; Glasstone and Sesonske 1994; Hetrick 1993; Keepin 1965; Lamarsh and Barrata 2001; Marguet 2017; Reuss 2003; Rozon 1992).

[3] Frédéric Joliot known as Joliot-Curie (1900–1958). French physicist. Awarded the Nobel Prize in Chemistry in 1935 with his wife Irène Curie (daughter of Marie Curie) for the discovery of artificial radioactivity. After studying engineering at the *Ecole Supérieure de Physique et de Chimie Industrielles de Paris*, he became a chemistry assistant to Marie Curie. He then climbed the ladder at the predecessor of the CNRS and at the Collège de France. His work focused on radioactivity and fission. After the war, he was appointed director of the CEA, where he supervised the construction of the first French reactor, Zoé, which diverge in 1948. However, his proven communist views (he was a member of the Central Committee of the French Communist Party) made him gradually withdraw from French projects concerning the atomic bomb, to which he was very hostile.

Two unknown photos of Frédéric Joliot-Curie at the Jiu-jitsu club of France in 1936, a Judo club created by a collaborator of the time: Moshe Feldenkrais, who will become his assistant in 1938. In the group photo, we find Irène, his wife, seated (second from the left).-->

[4] Lew Kowarski (1907–1979). A naturalized French physicist and engineer of Russian origin. After studying in Lyon, and then earning a doctorate in Paris on the metrology of neutron counting, he worked with Joliot at the *Collège de France* on the fission of uranium. In 1940, at the request of Joliot-Curie, he fled to England with the French stock of heavy water from Bordeaux, then to Canada when England was threatened with invasion by the Germans. There he directed the construction of the first Canadian experimental heavy water reactor and participated in the Manhattan project as part of the British team. At the end of the war, he returned to France to the CEA where he was the designer of the Zoé reactor, thanks to his precious notebooks brought back from Canada. He joined CERN in Geneva in 1953 where he worked on the construction of particle accelerators.

Photo 1.1 Frédéric Joliot-Curie, Lew Kowarski and Hans Von Halban, the fission pioneers in France in 1939

Von Halban[5] (Photo 1.1), shows that the fission of an uranium 235 atom produces on average about 2.46 fission neutrons (some fissions produce 2 neutrons, others 3 neutrons, or even more, the exact average figure of 2.46 called $\bar{v}$ is of course an average, symbolized by the bar symbol). A very simple reasoning is to assume that each of these $\bar{v}$ fission neutrons could produce still $\bar{v}$ new neutrons and so on. We would then obtain, from an initial neutron, a geometric progression of the number of neutrons at each generation: $\bar{v}$ neutrons at the first generation, $\bar{v}^2 = \bar{v} \times \bar{v}$ at the second generation, and $\bar{v}^n$ at the nth generation.

In a nuclear reactor, the *generation time ℓ*, which corresponds to the lifetime of the neutron between its "birth" by fission and its "*death*" by absorption to eventually produce another fission, is of the order of 10^{-7} second in a sodium- or liquid-lead-cooled reactor (called fast neutron reactors because the neutrons do not have time to be slowed down) to 10^{-3} second in a reactor where neutrons are slowed down by shocks on graphite. This is called a *moderator*. for graphite and thermal reactors because the neutrons can be slowed down enough to be in thermal equilibrium with the reactor environment, i.e., to acquire the same velocity distribution as the vibration of the moderator atoms. In pressurized water reactors, which are the backbone of EDF's Fleet of reactors in France, this prompt neutron lifetime is about 2,5 10^{-5} second, or 40,000 generations of neutrons per second. If we base ourselves on this last figure, we will obtain after a single second the staggering figure of:

[5] Hans Von Halban (1908–1964). French physicist of Austrian origin. After a doctorate in physics at the University of Zurich, he worked with Otto Frisch on the slowing down of neutrons in heavy water. He fled with Kowarski to Canada, where he worked on the Manhattan project. After the war, he worked at Oxford, then became the director of the Orsay Linear Accelerator Laboratory in 1958.

$$\bar{v}^{40000} \approx 10^{15637} \text{ neutrons}$$

As an order of magnitude and for comparison, there would be 10^{80} protons in the visible Universe. One does not need to be a physicist to understand that such a reactor would be absolutely ungovernable. In reality, the previous reasoning is fallacious. Many neutrons, although absorbed by fissile nuclei, will not produce fission but neutron capture, transforming, for example, uranium 235 into uranium 236 according to the reaction that produces γ-photons:

$$_{0}^{1}n + _{92}^{235}U \rightarrow _{92}^{236}U + \gamma$$

Some neutrons are even absorbed by non-fissile nuclei, for example, in the structures of the reactor or the coolant. Finally, other neutrons will physically leave the reactor (we speak of leakage), all the more easily as the reactor is small. If we consider all these neutron losses, the real number of neutrons produced per generation, which is called the *effective multiplication coefficient* k_{eff} is much weaker than initially expected and we will try, by an adequate geometry and an adjusted isotopic composition, to make it equal to 1 in an industrial reactor (unlike an atomic bomb where we will try to make it as close as possible to $\bar{v}$). The situation where $k_{eff} = 1$ is called *critical*, a term that is not very reassuring for the general public, but which just means that the neutron population is stable since each generation of neutrons will produce the same quantity of neutrons as the one that disappeared.

Technical Insert: "Multiplication Coefficient and Reactivity"

The technical inserts in this book allow the avid reader to go deeper into the subject. They can be skipped without damage by the reader who wants to get to the point.

The multiplication factor k_{eff} has a very important role in reactor physics. Let us assume now that the value of k_{eff} is not strictly equal to 1 but slightly higher, i.e., 1.00100 (we speak of a supercritical reactor), i.e., an error of 1 per 1000, which is already a very difficult precision to reach whatever the technology considered, the variation of the neutron population n will be governed by the equation:

$$\Delta n = -n\frac{\Delta t}{\ell} + k_{eff}n\frac{\Delta t}{\ell}$$

(continued)

(continued)

The negative term corresponds to the disappearance by absorption of neutrons during the time period Δt, disappearance that causes the birth of *keff neutrons for each neutron that disappears. Each generation lives a time ℓ in the* reactor. In this expression, ℓ is the lifetime of the prompt neutrons. Turning to the derivative, we find that:

$$\frac{dn}{dt} = \left(k_{eff} - 1\right)\frac{n}{\ell} \quad \text{i.e., after integration}: n(t) = n_0 e^{\frac{\left(k_{eff}-1\right)}{\ell}t}$$

Taking k_{eff} = 1.00100 and ℓ=2.5 10^{-5} second, characteristic value of a Pressurized Water Reactor, we still find that after one second, the neutron population is multiplied by 10^{17}, figure much lower than the previous one based on the use of the average number of fission neutrons $\overline{\nu}$, but just as technologically unreasonable. These calculations would be hopeless since they would only allow the construction of bombs if there were not the existence of *delayed neutrons*. It appears that some neutrons are not emitted instantaneously after fission, but after a time that can be long, several tens of seconds. These neutrons are emitted by fission products that are so excessively rich in neutrons that they can emit neutrons by radioactivity. These fission products accumulate in the reactor in such quantities that they mainly drive the reactor behavior.

We owe to Eugène Wigner[6] the representative diagram of the fission (Fig. 1.1) where time appears on the abscissa and space on the ordinate. Just after fission,

---

[6] Eugene Paul (actually Jenö, his real Christian name) Wigner (1902–1995) was a Hungarian physicist who later emigrated to the United States. He joined the faculty of Princeton University in 1930 and became an American citizen in 1937. That same year, he introduced the total isospin vector of a nucleon system, which allows the classification of nuclear states into isotopic spin multiplets, for the understanding of nuclear reactions. He was one of five scientists to inform President Franklin D. Roosevelt in 1939 about the possible military use of atomic energy by the Germans. During World War II, he participated in the design of plutonium reactors and worked on the Manhattan Project. Nobel Prize in Physics jointly in 1963 for the discovery of the principle of symmetry. He was one of the first to propose the use of water for its slowing down and cooling function in reactors.

(continued)

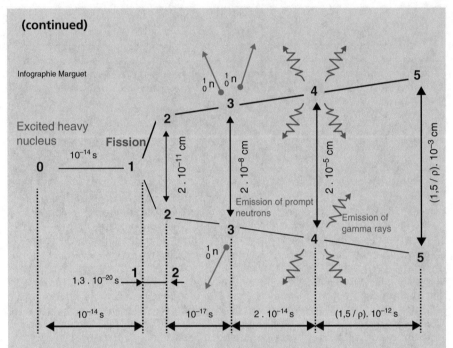

Fig. 1.1 Wigner's representation of fission with delayed neutrons. **0** A neutron, usually thermalized in a water reactor, has collided with a heavy nucleus. **1** The heavy nucleus is excited and fragments like a drop of water that is cut in two (water drop model). **2** The two fragments move away from each other carrying a large part of the fission energy released as kinetic energy (about 17 MeV). **3** There is emission of prompt neutrons (≈ 99% of neutrons are emitted at less than 10^{-14} second after the fission). Between two and three fast prompt neutrons are emitted. **4** Prompt γ emission (about 7 γ-rays emitted). **5** The medium slows down the fission fragments very strongly, especially as the density ρ of the medium is important. All the kinetic energy is then transformed into heat. **6** One of the fission products has a large excess of neutrons and decays to β^-, n (emission of an electron and a neutron). In our diagram, the other fission product is assumed to be stable

prompt neutrons are emitted (step 3), then in the last step, some fission products emit delayed neutrons by radioactivity. The fraction of delayed neutrons, that we classically note β is very low: about 650 neutrons per 100,000 neutrons produced, or 650 pcm[7] according to the commonly used reactivity unit, for uranium 235 fuel . The fraction of delayed neutrons is an average intrinsic characteristic of a fissile nucleus. For uranium 235, this fraction is 650 pcm, while for plutonium 239, it is 210 pcm. One immediately deduces that fuels containing plutonium, as

---

[7] pcm = pour cent mille (= per hundred thousand)

(continued)

(continued)

is generally the case for fast neutron reactors, have a lower fraction of delayed neutrons, which is translated by the fact that the reactor will be more "nervous," a term that is not very scientific but that translates well the behavior of the reactor to a reactivity request. *reactivity* ρ is a term with multiple meanings in the dictionary, but in reactor physics, it is defined as the criticality relative discrepancy:

$$\rho \equiv \frac{k_{eff} - 1}{k_{eff}}$$

It is customary to use the pcm (pour cent mille) as a unit of reactivity, which is a dimensionless quantity, but it is also sometimes counted in multiples of β when quantifying the ratio ρ/β which is then expressed as Dollars ($), an unusual conversion of the US currency unit!

The neutron behavior of a reactor is intimately regulated by the reactivity of the system. Lothar Nordheim[8] (Photo 1.2) showed that the average behavior of the neutron population *n(t)* in the reactor could be modeled according to a system of differential equations

$$\begin{cases} \dfrac{\partial n(t)}{\partial t} = \dfrac{(\rho - \beta)}{\ell} n(t) + \sum_{i=1}^{6} \lambda_i C_i(t) \\[2ex] \dfrac{\partial C_i(t)}{\partial t} = -\lambda_i C_i(t) + \dfrac{\beta_i}{\ell} n(t) \end{cases}$$

In these equations, we recognize the reactivity ρ, the fraction of delayed neutrons β, and the lifetime of prompt neutrons ℓ. The index *i* that appears qualifies the concentrations C_i of the groups of delayed neutrons produced by the fractions β_i, because it is customary to differentiate six families of delayed neutron precursors i.e., families of fission products that are grouped by similar half-life.

---

[8] Lothar Nordheim (1899–1985), a physicist of German origin of Jewish faith, emigrated to the United States in 1934 because of the rise of Nazism, like many German physicists. He began teaching at Duke University in 1937, where he spent most of his career. Appointed head of the theoretical physics group at Clinton Laboratories, the forerunner of Oak Ridge National Laboratory, during the times of the Manhattan Project, he worked on problems of kinetics and explosions in the development of the hydrogen bomb. He also contributed extensively to the theory of the layered nucleus. The Fowler-Nordheim model establishes that the emission of electrons following a Fermi-Dirac statistic on the surface of metals under the effect of an electric field obeys a tunneling effect. This effect has many practical applications.

(continued)

(continued)

Photo 1.2   Lothar Nordheim en 1954 (Duke University)

We will not dwell on the complex mathematical treatment of these seven differential equations, noting only that the number of neutrons in the reactor can be put in the form of 7 exponentials:

$$n(t) = \sum_{k=1}^{7} n_k e^{\omega_k t} \quad C_i(t) = \sum_{k=1}^{7} C_{ik} e^{\omega_k t}$$

We can also demonstrate that the principal *pulsation* ω_1 (as the term is used) can be put in the form:

$$\omega_1 \approx \frac{\rho - \beta}{\ell}$$

The other pulses are all negative. It is the comparison between the reactivity and the fraction of delayed neutrons that is crucial with respect to the kinetic behavior of the reactor. If $\rho > \beta$, the pulsation will be strongly positive and the exponential term that contains it will rapidly diverge with time. The population increases as the reactivity increases, as does the power and, without automatic action or operator intervention, the reactor is doomed.

Reactivity is the technical term that characterizes the ability of a fissile system to multiply its neutrons. The higher the reactivity, the more neutrons will be produced in the chain reaction. Note that a reactivity can be negative, in which case the reactor's neutron population will decrease. In layman's terms, we can say that the reactor "shuts down." The causes of reactivity variation are diverse. By design, the reactor design requires a sufficient reactivity reserve, insofar as the appearance of fission products during burn-up tends to reduce the reactivity of the core during the cycle, and insofar as these fission products are very often neutron absorbers. To be able to operate for a sufficiently long time to guarantee an economic interest, it is necessary to be able to propose an initial excess of reactivity which will be compensated at each moment by the control means. These means are of several kinds. Control rods are widely used and are made of neutron-absorbing materials (cadmium, hafnium, boron materials, steels, etc.) that are inserted into the core and that can be removed during the cycle. Some types of reactors, such as the pressurized water reactors (PWRs)), use an absorbent diluted in the reactor water: boric acid. Finally, so-called "burnable poisons" are used, often rare earth oxides such as gadolinium or erbium, which are mixed with the fuel and whose absorbing action decreases with time. The leakage can also be controlled by modifying the geometry of the neutron reflectors around the core. For a given reactor geometry, one can then calculate the quantity of fissile material to be placed in the reactor: the critical mass, of which Francis Perrin has established the theory.

Some Basic Technology

Some basic notions of reactor technology are necessary to further understand the accidental behavior of reactors. The production of electric current by a nuclear reactor is the result of the passage of a hot gas that is expanded to turn a conventional turbine. This gas can be steam produced by heat exchangers that extract heat from the coolant cooling the reactor (such as carbon dioxide for graphite-moderated natural uranium reactors cooled with carbon dioxide—UNGG (Photo 1.3), helium for very high-temperature reactors—HTR), either steam is produced directly in the reactor (boiling water reactors—BWRs) or vaporized water in steam generators (Pressurized Water Reactor—PWR, Fast Breeder Reactors—FBR).

Let us consider the case of an EDF 900 MWe pressurized water reactor : the principle is to heat water from 286 °C at the core inlet to 326 °C at the core outlet. The water progresses from the bottom to the top of the core. In the example of the Yankee Row plant (USA, 1961–1992, Fig. 1.2), the cooling of the core is ensured by 3 primary loops and the water heated in the core

Photo 1.3 Natural uranium graphite gas reactor (first French reactor type). The Saint Laurent-A loading platform and the Main Fuel Handling Device. The rails of the fuel loading machine and the tulip-shape plug closed next to a CO_2 channel can be seen) (photo: Bouchacourt-Foissote-Valdenaire)

is sent by the hot legs in 3 Steam Generators (SGs)) where water passes through inverted U-shaped tubes. The heat is transferred through the wall of the steam generator tubes (SGTs) to heat the secondary circuit, where the secondary water is vaporized under a pressure of approximately 55 bars. The steam is then sent to the turbine, which turns an alternator. Once cooled, the residual steam is condensed into liquid form in a condenser which acts as a heat exchanger with the cold source (river or sea). There is a complete physical

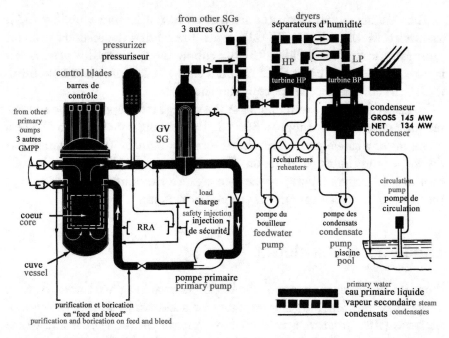

Fig. 1.2 Primary and secondary circuits of the Yankee Rowe pressurized water plant—USA (134 MWe)

separation between the primary circuit where the water is more or less radioactive (in fact the particles it carries, i.e., dissolved metal oxides or impurities), and the secondary circuit free of radioactivity except in case of leaks . Primary circuit water is activated under neutron flux by cooling the zirconium[9] cladding, which contains the nuclear fuel pellets. These claddings, called rods because of their geometry, constitute the first barrier. The primary circuit and the SGs form the second barrier. The primary and secondary circuits are placed inside a prestressed concrete containment to resist external aggression[10]: the reactor building containment (BR).), which constitutes the third barrier. Therefore, the image of Russian dolls is commonly used to conceptualize the principle of barriers that separate highly radioactive fission products from the environment.

---

[9] Zirconium is a metal (residue of the glass industry) widely used in the nuclear industry because it absorbs very few neutrons (almost transparent to the path of neutrons).

[10] Handling accidents, small aircraft crashes, external fire, …

The turbine and its alternator are located in the Turbine Building (BT), also called Machine Room (SDM), which means that the secondary circuit must exit the Reactor Building (BR) to supply the turbine. This penetration is potentially a weak point since it would allow, in case of Steam Line Break (SLB), to short-circuit the BR containment.

To keep the primary water liquid at the average rated operating point (vessel water temperature of approx. 306 °C), the pressure is increased to 155 bars by means of a single pressurizer placed on a hot branch. This pressurizer has the shape (and almost the function) of a cumulus, which would contain at its base electric heaters vaporizing water to create a steam mattress at the top of the pressurizer, which compresses the primary circuit.

The Reactor Accidents

From the design of the reactors, the engineers have imagined several types of *design basis accidents* . A first category of accidents is *Reactivity Initiated Accidents* (RIA[11]), where a neutron flux[12] prevails in the reactor and where over-reactivity is accidentally introduced into the core. The reactivity of a reactor varies due to the variation of some physical parameters (Table 1.1):

– The increase in the quantity of fissile materials in the core. One thinks in particular of the loading phase when fuel assemblies are introduced into the core. The cause can be a loading error which can lead to the constitution of a critical mass of assemblies before the core is completely filled. These are known as criticality accidents. The accident at the Tokai Mura uranium conversion plant in Japan in 1999, when the operators exceeded the critical mass of uranium in a dissolution tank, is an illustration of this type of problem applied to a geometry other than that of a reactor. In the case of fast reactors, there is a very strong sensitivity to the geometry of the lattice: if the fuel rods were to be compacted (mechanical buckling of the rod bundle), a gain in reactivity would be obtained. It is therefore necessary to ensure that the array of rods is rigid, hence the presence of hexagonal

---

[11] The term is also used in French. As reactor technology is often of American origin, the use of acronyms and English words is (unfortunately) very common in the French nuclear industry.

[12] The neutron flux is a complex notion that we will not discuss here. For our purposes, it is sufficient to imagine a set of neutrons, invisible to the naked eye of course, moving in a straight line between each collision with matter, in a disorderly fashion like a swarm of mosquitoes.

Table 1.1 Main causes of reactivity variation in Pressurized Water Reactors

	Reactivity insertion speed	Amplitude of the phenomenon taken into account in the design studies	Parades
Dilution of boric acid	1 pcm/s	2000 pcm	Design of the REA and RCV borication circuits to limit the rate of boron dilution Protection system and alarms that warn the operator that an action must be taken *Note*: The frequency of occurrence is much higher during shutdown states
Extraction of control rod groups in normal sequence	20 pcm/s	2000 pcm	Emergency scram on high power threshold or low DNBR[a]
Extraction of control rod groups in abnormal sequence	100 pcm/s	2000 pcm	Doppler Effect efficient before scram
Cooling by power extraction through the secondary circuit (steam line break scenario)	100 pcm/s	6000 pcm	Safety injection switched on Emergency shutdown with possible aggravation (stuck control rod cluster) Necessary operator action (isolation of the failed steam generator)
Rod ejection (at zero power, which is penalizing because the doppler effect is delayed)	900 pcm during 1/10 s Thus 9000 pcm/s	900 pcm, (weight of the most anti-reactive control rod cluster	Effective doppler effect before the scram

[a]DNBR: *Departure from Nucleate Boiling Ratio*, characterizes the point at which the fuel cladding is dried out at the surface and is no longer in contact with the coolant. From this point on, the temperature of the fuel, which is no longer cooled, increases sharply. The French term for this is Minimum Critical Heat Ratio (REC_{min})

tubes surrounding the assemblies and often of helical spacer wires that wrap around the rods to prevent them from coming too close together.
- The increase in neutron moderation, for example, by an increase in the density of water (cold shock, for example, as in the case of a steam line break in the SG, whose rapid depressurization creates an uncontrolled

demand for power in the secondary side, resulting in a cold shock on the primary that is more or less asymmetrical depending on the position of the breach in relation to the steam barrel header which collects the steam from all the steam loops before sending it to the turbine). This densification of the water improves the thermalization of neutrons. If the neutrons are better slowed down (they are said to be *thermalized*), they are more efficient for the chain reaction because the probability of fission is inversely proportional to the velocity v of neutrons (we speak of a law in $1/v$).

- The draining of the reactor: the neutrons are no longer slowed down, nor absorbed by the coolant, the neutron velocity spectrum approaches a pure fission spectrum [13] (thus with more fast neutrons, we speak of "*hardening*" of the spectrum) and, under certain conditions, particularly in fast reactors where the lattice of rods is very tight and the inventory of fissile nuclei is large, this favors the fission of isotopes such as uranium 238, which is not very fissile in thermalized spectrum.

- The decrease of neutron capture due to the fuel, for example, by a decrease of the fuel temperature, hence a decrease of the Doppler effect . The Doppler effect[14] is the name of the physical effect that increases the capture of neutrons by nuclei with capture resonances such as uranium 238, i.e., very high probabilities of interaction between neutrons and nuclei. We will come back to this physical effect in more detail later.

- The reduction of neutron captures in the core due to materials other than fuel, for example, due to the ejection of an absorbing rod or by dilution of boric acid inadvertently in the water of the PWRs.

The second category of accidents is represented by cooling accidents, i.e., situations where the core is insufficiently cooled. Nuclear reactors are unique in that their core must be constantly cooled even after fission has stopped. Unlike combustion engines, which only continue to generate heat after shutdown through energy storage in the engine mass (while the combustion heat source is stopped), nuclear fuel continues to generate heat intrinsically through

---

[13] The notion of spectrum describes the distribution of the neutron population as a function of their speed (and therefore their energy).

[14] Christian Andreas Doppler (1803–1853). Austrian mathematician and physicist. After studying physics in Vienna, he taught in various institutions in Vienna and Prague. He discovered the optical Doppler effect of frequency shift of a moving light source. He founded the Institute of Physics at the University of Vienna in 1850. The Nuclear Doppler effect has little to do with the optical effect except for a vague analogy of relative speed. Nevertheless, it is universally known by this name in the nuclear world.

radioactivity of the fission products for very long periods after the fission process has been stopped by *scram*. Because it is radioactive, there is no way to stop this major heat source even after the fission reactions have died down. This is called residual power. The order of magnitude of the residual power is about 7% at the onset of the scram (about 200 Mega Watt thermal, MWth for an EDF-900 MWe reactor[15]) of the nominal thermal power of the reactor one second after rod drop, but it is still about 1% (28 MWth) after one hour and 1‰ (3 MWth) after one year. This thermal power must imperatively be evacuated by pumping cold water into the core for several months if the core remains loaded with fuel (thanks to the shutdown core cooling system aka RRA).

A loss of reactor cooling can have several causes:

− The rupture of a primary water line or a leak on a component. This is called a Lost Of Coolant Accident (LOCA)).
− Loss of the heat transfer pumps (simultaneous failure of the pumps or loss of the electrical source).
− The loss of the cold source that cools the secondary, and by extension the loss of the secondary circuit. We can still run primary water in the core but we can no longer cool it.

Unlike reactivity accidents which have rather fast kinetics (a few hundredths of a second to a few minutes), cooling accidents have longer kinetics (a few tens of minutes to several days). The only rapid case of a cooling accident would be a complete cut-off of the hemispherical vessel bottom of the reactor, which would dry out the reactor core almost instantaneously. This accident is unanimously considered highly unlikely by the scientific community. The solution would be to completely drown the vessel pit with borated water to prevent any criticality return.

The safety of a nuclear reactor has been summarized by the guarantee of the three *safety functions*:

Control the reactivity of the reactor: By using various means of control (control rods which can drop by gravity in the absence of electrical power, injection of diluted absorbent into the coolant...).

---

[15] A 900 MWe reactor delivers a real thermal power of 2775 MWthermal because the efficiency of the installation is only 32.5%.

Cooling the reactor in all circumstances: A redundancy of circuits is always necessary to prevent the failure of one of the circuits. The application of defense in depth, which consists in analyzing scenarios such as *"What if such and such a system fails?"*, allows to accumulate failures by thinking and to analyze their consequences. The aim of Probabilistic Safety Assessment (PSA) is to quantify the risk of a cumulative failure scenario.

Containment of radioactivity: This is the purpose of the three barriers, namely the fuel cladding, the primary circuit for the coolant and the containment building. Some reactor types, such as the soviet RBMK reactor type of the Chernobyl type, do not have a containment building as such. Their exterior building has a structure closer to that of a factory hangar (cheap to build but?).

What would be the consequences of an accident? The understanding of the accident scenario is vital for the establishment of countermeasures. In the case of reactivity accidents, it is considered that the accident unfolds too quickly for the operator to have time to act. Therefore, automatons are the first to act. Several measurement systems monitor the state of the reactor: measurement of the temperature at the reactor outlet, measurement of the neutron flux outside the reactor by measurement chambers, measurement of the ambient radioactivity in the reactor building...The case of the rod ejection accident sheds light on the behavior of the reactor. On July 22, 1954, American scientists voluntarily provoked a rod ejection on the BORAX reactor (4.16 kg of 93% enriched uranium moderated by heavy water, Idaho, USA). The reactor was at the end of its operation, and it was wanted to use it one last time to observe the effects of a power excursion accident.[16] In less than a tenth of a second, the power of the reactor rose to 10 MWth, and the resulting mechanical explosion blew pieces of the reactor a hundred meters away. It is now known that the post-operational calculations had largely underestimated the power developed during the accident, necessitating a re-analysis of the kinetic coefficients of the reactor.

Let us imagine on a PWR the situation where control rods would be inserted in the reactor (operation at intermediate power, for example, or shutdown in hot state but under pressure). The pressure of the PWR primary circuit is 155 bars. If the control rod cluster housing, which is connected to the primary circuit above the vessel cover, becomes depressurized for any reason (cracking of the housing, break due to an impact...), a break appears,

---

[16] The term excursion should be understood here as a violent power peak.

which will violently depressurize the primary circuit. As the pressure is 155 bars in the primary circuit and 1 bar (atmospheric pressure) outside the vessel cover, the control rods are expelled from their housing like a bullet. This effect can be maximized by considering that the ejection is instantaneous. One thus ejects anti-reactivity since the rod had the function of absorbing neutrons. If the neutron "weight" of the control rod is important (higher than the fraction of delayed neutrons), one obtains an exponential power excursion which can be very violent (several hundred times the power of the reactor!), but which will prove to be very short. It is the very high level of neutron flux that will trigger the shutdown and drop the rods that are not already at the bottom of the reactor. When the power increases in a very localized way (at the location of the ejected rod), the temperature of the fuel soars to more than 3000 °C. At these temperatures, a saving effect will appear in the fuel: the *Doppler Effect*.

Indeed, the agitation of uranium 238 atoms at high temperature is such that they have a very strong affinity for capturing neutrons. To help us understand, we can use the image of a soccer goalkeeper who would move frantically to intercept a shot on goal, with the ball playing the role of the neutron. Although the scientific reality is of course more complex (quantum interaction between the incident neutron wave and the potential pit of the target nucleus), this image is sufficient to understand that all uranium 238 atoms will behave like frantic goalkeepers whose action of capturing neutron balls will crush the neutron multiplication almost as quickly as it had appeared. The Doppler effect is the prerogative of isotopes that have capture resonances and that do not fission in the thermal spectrum (uranium 238, plutonium 240...). This is one of the essential differences with atomic bombs, which contain only uranium 235 or plutonium 239. Since the power pulse is all the finer in time that the amplitude of the power peak and the temperature of the fuel are high, the energy delivered, which is nothing other than the integral of the power over time, can remain relatively low and heat will be stored in the fuel (note that the uranium oxides used in PWRs are poor conductors of heat, which favors the rise in temperature). The fuel will then expand as the temperature rises, and the cladding will explode mechanically. Solid fuel, but also liquefied in the form of liquid droplets because of the very high local temperature (Fig. 1.3 for the melting temperatures of the core materials), will then be sprayed into the coolant (water for PWRs and BWRs, liquid metal in FBRs). The risk is then that the heat transfer is very violent towards the coolant, which strongly vaporizes creating a detonation wave. This wave fragments the other droplets of fuel, increasing the exchange surface and amplifying the phenomenon: we speak of a steam explosion. This term is not really adapted

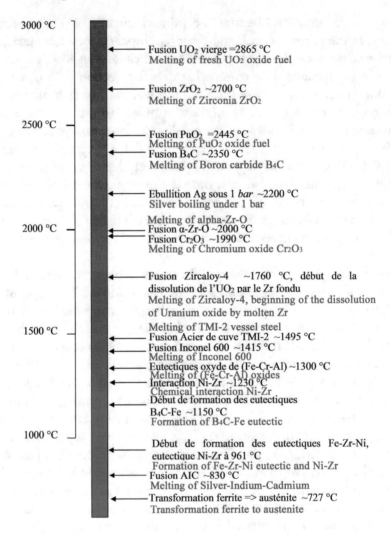

Fig. 1.3 Scale of melting temperatures in an accident situation

because there is no chemical explosion involving a fuel and an oxidizer as in the case of gasoline and air. We should rather speak of interaction with water), but the term explosion is still adapted because of its consequences, the pressure wave being able to destroy the structures it encounters before being attenuated. We will illustrate this accident with the case of the SL-1 reactor (USA) in the following chapter. Let us insist on the fact that such an accident is not an atomic explosion since the loss of geometry is very rapid, and the power excursion ceases as soon as the fuel is dispersed. In the case of an atomic bomb, one will try by construction to contain the over-critical geometry as long as possible in order to favor nuclear fissions.

Cooling accidents are the result of a thermal-hydraulic analysis. A priori, the shutdown rods will drop into the reactor due to some signal (low water level in the pressurizer, very low pressure...). The problem is therefore rather to be able to inject water into the core. Very prosaically, it is easier to inject the coolant if the pressure of the primary circuit is low, which is why it is wise to depressurize it. In a PWR, several systems are available to inject water into the primary circuit. Each loop has an accumulator tank, a kind of large cumulus pressurized to 40 bars, which can passively fill the primary circuit as soon as its pressure falls below the accumulator setting pressure. These accumulator tanks are sized to quench the core in the event of a large break on a hot leg, for example. The accumulator tanks give the necessary respite to engage the safeguard injections (RIS system), which will inject borated water into the reactor by taking it from a large reserve of borated water, the PTR tank. This reserve, although substantial (approximately 1500 m^3), is not infinite. It was therefore imagined that the water injected into the primary circuit would leak through the break wherever it was positioned in the reactor building, and that the water would end up in the lower parts of the reactor where it would be collected in sumps. This water is then returned to the core after cooling through heat exchangers. This is called recirculation. This recirculation must not be hindered by accidental blockage of the sumps.

If one cannot inject enough water, the core, by vaporizing the residual water that remains in the vessel, will denature, and the temperature of the fuel will increase. After a few hours, the core materials will start to melt. (Fig. 1.3), starting with silver-indium-cadmium[17] control rods. Uranium oxide has a high melting point, but this can be lowered when dissolved with the zirconium that constitutes the cladding (formation of eutectics[18]).

If the core still cannot be cooled, the core materials gradually melt and form a liquid magma called *corium*. If it consists only of liquid metals, it remains very fluid (even more fluid than water!) insofar as there is no silica as in the case of volcano lava which can be very pasty. This corium will progress downwards by gravity and reach the vessel bottom. If the corium falls into the water and fragments, a steam explosion could be feared, Figure 1.4, as in the

---

[17] These metallic materials are chosen because they are strong neutron absorbers.

[18] An eutectic is a mixture of simple bodies at a given stoichiometry. Some eutectics have a significantly lower melting point than the initial materials used in their composition.

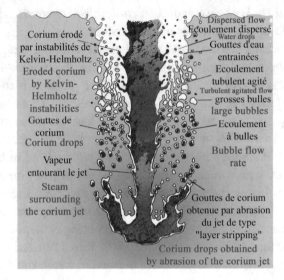

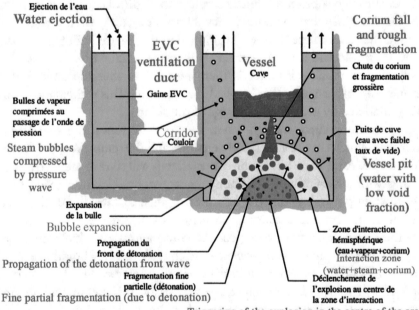

Fig. 1.4 Fragmentation of a jet of corium falling into water and risk of steam explosion which sets in motion by piston effect the columns of water around the vessel and in the EVC duct of access to the pit, hence a possible water hammer

case of reactivity accidents. But it seems from experience that the corium, massively made up of uranium and zirconium partially oxidized is finally little reactive with water (contrary to alumina, for example, which causes powerful explosions) for reasons which are not perfectly elucidated by the scientists in spite of many experimental programs.

In any case, the steam explosion in the vessel would have little energy and would not threaten the integrity of the vessel. When enough corium is collected in the vessel bottom, and in the absence of a protective crust, it attacks by melting the thickness of the vessel until it pierces it in the worst case, the corium then spreading in the vessel pit. As the internal source of heat by radioactivity is always present (some fission products degas from the corium, but others remain), one can imagine that the corium attacks the concrete raft of the reactor and pierces it. The final containment barrier is then lost in a scenario called "*The China Syndrome*" after the homonymous film (see chapter on TMI-2). If the concrete raft is breached, the radioactive material would then find its way into the substrate and could contaminate runoff water or groundwater.

Numerous solutions have been devised to stop this disaster scenario. Two separate, fully redundant safety injection trains (RIS) are used to inject borated water into the core. The power supply to the steam generators, which are the primary circuit cold source, is backed up by an ASG auxiliary feedwater emergency circuit. In case of lack of electrical power, the control rods of the PWRs automatically drop by gravity into the core to stop the chain reaction. An EAS spray circuit, very similar in function to a shower head, allows cold water to be sprayed inside the reactor building, just like a fire spray. The purpose of this action is to lower the pressure and to condense the water steam coming from the break in the primary circuit, because if the reactor building resists well in compression in case of external aggression, it will not hold at more than 5 bars of internal pressure. However, the heat released in a closed space slowly increases the pressure according to the classic equation of perfect gases ($pV = nRT$). This EAS system avoids losing the ultimate barrier. In the same way, the French utility EDF introduced passive auto-catalytic recombiners of hydrogen, an explosive gas[19], which comes from the oxidation of zirconium cladding at high temperature by water vapor, according to the reaction:

$$Zr_{metal} + 2H_2O_{steam} \rightarrow ZrO_2 + 2H_{2\,gas} + \approx 600 \quad kJ\,/\,mol\ of\ Zr$$

This chemical reaction is very exothermic and accelerates at high temperature, which means that the zirconium in the cladding will "burn" in the superheated water vapor. Similarly, all the metal structures in the vessel will oxidize to produce a total of nearly 1000 kg of highly flammable hydrogen gas, the rapid combustion of which can create a detonation that endangers the containment. The function of the recombiners is to eliminate the hydrogen by transformation into less dangerous water steam. This chemical reaction occurs passively by catalysis on platinum catalyst plates.

---

[19] The explosion of the German airship Zeppelin *Hindenburg* on May 6, 1937 upon its arrival in New York remains a grim memory and sounded the death knell for the use of hydrogen in aviation.

In the case of fast neutron breeder reactors, the same type of accident is found with the particularity that the primary circuit is not under pressure, which is a definite advantage for the design of components and piping. On the other hand, liquid sodium, widely used as a coolant because of its excellent thermo-physical properties (high boiling temperature of 880 °C, high thermal capacity) and neutron properties (low neutron absorption), has the disadvantage of being very reactive with water, and even with air, which means that there is a significant risk of a sodium fire that would be very difficult to clear out. A generalized flashover of the sodium in the primary circuit is such a catastrophic scenario that water intrusion in the vessel has been eliminated by design. In the case of the pool concept, where the primary pumps are immersed in the vessel, an intermediate circuit is available, consisting of heat exchangers which are immersed in the vessel (IHX). This is the case of the French reactor Superphénix (Figs. 1.5 and 1.6).

These immersed sodium-sodium exchangers will heat secondary sodium, which will further heat water in other exchangers located far from the vessel. The design of the sodium-water exchangers is such that the slightest leak causes a rapid discharge which drains the exchanger to limit the sodium–water interaction. There are fast neutron reactors (Russian) that use other less water-reactive coolants such as a lead-bismuth mixture, but corrosion by lead must be carefully controlled. The second function of the intermediate circuit is the containment of the radioactivity of the primary circuit because it is activated according to the reaction:

$$^{23}Na + {}_0^1 n \rightarrow {}^{24}Na$$

Sodium 24 is radioactive with a half-life of 15 hours (γ rays at 1.37 MeV and 2.78 MeV). This would contaminate the turbine in case of leakage if the intermediate circuit did not ensure the physical separation of the core and the turbine. Fast neutron reactors are very sensitive to geometry and it must be guaranteed that it will not move during an accident. The large inventory of plutonium, the fuel of choice for this type of reactor, means that the fraction of delayed neutrons is much smaller, resulting in very rapid kinetic behavior in the event of the introduction of reactivity.

The Radioactive Releases

A nuclear plant emits radioactive releases in quantities precisely limited by law. Each site must take into account specific release authorization decrees. These decrees are issued by the government with the support of scientists (engineers,

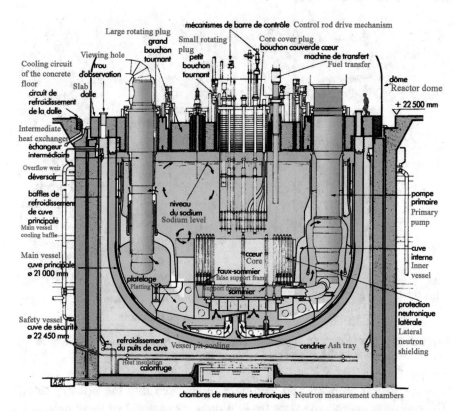

Large rotating plug
grand bouchon tournant

Viewing hole
trou d'observation

Cooling circuit of the concrete floor
circuit de refroidissement de la dalle

Slab
dalle

mécanismes de barre de contrôle Control rod drive mechanism

Small rotating plug
petit bouchon tournant

Core cover plug
bouchon couverde cœur
machine de transfert
Fuel transfer

dôme
Reactor dome

+ 22 500 mm

Intermediate heat exchanger
échangeur intermédiaire

Overflow weir
déversoir

baffles de refroidissement de cuve principale
Main vessel cooling baffle

niveau du sodium
Sodium level

Main vessel
cuve principale
⌀ 21 000 mm

platelage
Platting

cœur
Core

faux-sommier
false support frame

Support frame
sommier

Safety vessel
cuve de sécurité
⌀ 22 450 mm

refroidissement du puits de cuve Vessel pit cooling

cendrier Ash tray

Heat insulation
calorifuge

pompe primaire
Primary pump

cuve interne
Inner vessel

protection neutronique latérale
Lateral neutron shielding

chambres de mesures neutroniques Neutron measurement chambers

Fig. 1.5 View of the SUPERPHENIX reactor vessel: the sodium-sodium heat exchangers extract heat from the reactor. A safety vessel encloses the primary circuit to collect any leakage. A rotating plug allows the assemblies to be handled without having to open the reactor, whose sodium remains protected from the air by a mattress of argon gas

doctors, etc.) whose role is to define the limits that are dangerous for humans and the biotope, as well as the significant margins that are imposed in relation to these limits. In an accident situation, the evaluation of accidental releases is extremely decisive in the decision to evacuate the population. In the worst case, one can imagine the meltdown of all or part of the reactor due to lack of cooling. Taking the example of a Pressurized Water Reactor (PWR), the fuel (uranium or plutonium oxide) contains radioactive fission products if it has been burned up in the reactor during operation. If the fuel melts at high temperature, the molten part attacks the zircaloy cladding and corium is created.

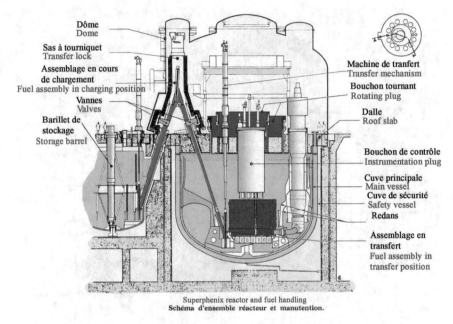

Dôme / Dome

Sas à tourniquet / Transfer lock

Assemblage en cours de chargement / Fuel assembly in charging position

Vannes / Valves

Barillet de stockage / Storage barrel

Machine de tranfert / Transfer mechanism

Bouchon tournant / Rotating plug

Dalle / Roof slab

Bouchon de contrôle / Instrumentation plug

Cuve principale / Main vessel

Cuve de sécurité / Safety vessel

Redans

Assemblage en transfert / Fuel assembly in transfer position

Superphenix reactor and fuel handling
Schéma d'ensemble réacteur et manutention.

Fig. 1.6 The SUPERPHENIX reactor and the famous fuel storage barrel whose leak made it impossible to refuel the core without opening the vessel

Gaseous fission products and aerosols[20] are released into the primary circuit. The composition and the state (solid, liquid, or gaseous) of these radioactive releases (most of the fission products are radioactive β^-, most heavy nuclei are radioactive α) depend on the burn-up of the fuel, the temperature and the chemical properties of the constituents and the environment (acidity pH, temperature, viscosity). Once in the primary circuit, the fission products react chemically with each other and mechanically with their environment during transport. The list of these interactions is long: condensation/evaporation, sedimentation by gravity, thermophoresis, diffusiophoresis, resuspension by flows of water, steam, or air, deposition in pipe elbows by impaction in particular in the U-tubes of the steam generators, degradation of complex molecules under the effect of radiation (absorbed dose). After the transport phase, the fission products are found at the break in the primary circuit, which caused the

---

[20] The word aerosol was coined by the German physicist and meteorologist August Schmauss (1877–1952) in 1920. It refers to the suspension of solid or liquid particles in a gaseous medium. These particles are small enough (dimension <100 μm) to exhibit negligible gravity fall velocity. In 1929, a Norwegian inventor, Erik Rotheim, patented a device consisting of a pressurized container equipped with a valve for the diffusion of liquid and/or gaseous products. The first insecticide was born in 1941. The aerosol can was born. Aerosols are of interest in the fields of filtration, pollution, insecticides, perfumery, cleanliness of clean rooms for the manufacture of electronic chips, space, oceanography, chemistry... and severe accidents. Renoux and Boulaud (1998) intelligently presents both the physics and the metrology of aerosols.

accident and spread in the containment (when there is one). After a more or less long time where the same phenomena occur in the containment, one can imagine a loss of containment by rupture linked to the increase of pressure in the containment. Let us recall that water at high temperature (326 °C) and high pressure (155 bars) has been released due to the break in the case of a PWR LOCA, and that this water has been instantaneously vaporized by the pressure drop in the containment (about 1 bar). Given the great variety of fission products released in the primary circuit, it is usual to group fission products and structural materials into families of similar properties to deal with chemical transport.[21] A distinction will be made between the vapor, aerosol in suspension and deposit phases (on walls or in water points (pools, sumps), and their granularity. We will group them in: (1) the noble gases called "rare" (xenon, krypton), which will be aggregated with the non-radioactive aerosols (in suspension or deposited), the structural materials (Cd, In, Ag, Sn, Mn) and the aerosols generated by the degradation of the concrete in case of vessel breakthrough; (2) the very present CsI and RbI; (3) TeO_2; (4) SrO; (5) MoO_2 mainly produced during the attack of the concrete by the corium; (6) CsOH and RbOH to account for Cs and Rb that did not alloy with iodine; (7) BaO; (8) All rare earth oxides La_2O_3, Pr_2O_3, Nd_2O_3, Sm_2O_3, Y_2O_3 which are rather non-volatile; (9) CeO_2; (10) Sb; (11) Te_2; (12) the heavy nuclei oxides UO_2, NpO_2, PuO_2. Each physical model is described by a rate of production or disappearance from separate tests, which is agglomerated in a calculation code whose objective is to describe the quantity of radioactive products in the containment: this is called the *source term*. This source term will be the input value for new calculation codes that will carry it into the atmosphere, depending on climatic conditions (wind, rain), to predict the deposits in the surrounding nature of the site concerned. Well qualified, these deposits will allow decisions to be made about evacuating the population, cordoning off areas and preventive treatment with stable iodine tablets (iodine 127). Iodine is indeed a chemical body of capital importance because it is preferentially fixed in the thyroid. The human thyroid gland, a butterfly-shaped gland at the base of the neck, produces two hormones: thyroxine, which activates metabolism, and calcitonin, which reduces the concentration of calcium in the blood. The thyroid absorbs the natural iodine found in the human diet. An insufficiency or an overdose of iodine produces hormonal and metabolic disorders of which goiter is a classic form. This "greed" of the thyroid for iodine is very dangerous in case of radioactive release of iodine because the radioactive isotopes of iodine 131 (yield 3%, half-life 8 days) and the couple tellurium 132 (half-life 3 days)/iodine 132

---

[21] We present here the groupings chosen in the MAAP calculation code, one of the main severe accident codes in the world.

(half-life 2 h 30) are produced in large quantity in a reactor. They will present a risk of contamination of the thyroid for about 2 months, considerably increasing the risk of thyroid cancer.

From a chemical point of view, iodine will come in the form of different iodine species: I_2 a metallic slate-gray solid at room temperature (melting temperature 114 °C, boiling 184.3 °C); CsI (melting 626 °C, boiling 1280 °C); CH_3I a colorless liquid whose carbon comes from boron carbide control rods or else from the moderator of reactors that contain graphite (melting –66 °C, boiling 42 °C) and AgI whose silver comes from the control rods of PWRs (558 °C, 1504 °C). During a severe accident, iodine will first reach the gaseous phase, then react in the primary circuit or in the containment with aerosols, including cesium and silver. These compounds can condense, and the iodine then joins the gaseous phase. It is in this form that it is most dangerous because gaseous iodine, unlike aerosol iodine (in the CsI molecule, for example), is not easily trapped by the filters and is not retained by the sand filter in case of voluntary depressurization. This sand filter is installed on the roof of French PWRs.

Technical Insert: "The Chemistry of Iodine"

Independently of the iodine molecules, which can pass from the liquid state to the gaseous state and vice versa according to the thermodynamic conditions, and of the reactions on the painted surfaces and the concrete, the reactions of iodine with its environment are very complex. Note:

Production of silver iodide in aqueous and gaseous phase: $2Ag + I_2 \Leftrightarrow 2AgI$. The silver comes from the fusion of the "black" control rods in Ag-In-Cd.

Oxidation of iodine by ozone in the gas phase: $4O_3 + 2I_2 \Leftrightarrow 4IO_3^-$. The ozone comes from the radiolysis of oxygen in the containment.

The formation of organic iodine gas: $2CH_3R + I_2 \Leftrightarrow 2CH_3I + 2R$. R is an organic radical.

CH_3I radiolysis under radiation: $2CH_3I + \gamma \Leftrightarrow 2CH_3 + I_2$ in dry atmosphere only.

Hydrolysis of iodine in aqueous phase: $H_2O + I_2 \Leftrightarrow I^- + HOI + H^+$.

The dismutation of HOI : $3HOI \Leftrightarrow IO_3^- + 2I^- + 3H^+$.

Oxidation of I^- by dissolved oxygen: $2I^- + 1/2\ O_2 + 2H^+ \Leftrightarrow H_2O + I_2$.

The formation of I_3^- : $I^- + I_2 \Leftrightarrow I_3^-$.

Radiolysis of I^- : $2I^- + \gamma \Leftrightarrow I_2 + 2e^-$.

Reduction of IO_3^- under radiation: $IO_3^- + \gamma \Leftrightarrow I^- + 3/2\ O_2$.

Formation and decomposition of organic iodine:

$2R \Leftrightarrow 2RI$ $R + OH \Leftrightarrow RI + OH$

Hydrolysis and radiolysis of CH_3I : $CH_3I + H_2O \Leftrightarrow I^- + CH_3OH + H^+$

$CH_3I + OH^- \Leftrightarrow I^- + CH_3OH$ $2CH_3I + \gamma \Leftrightarrow I_2 + 2CH_3$

This relatively tedious list shows the complexity of the calculation of the gaseous phases of iodine, those most likely to escape in case of containment rupture. Each reaction has its own reaction constant, and the equilibrium conditions between all these phases require a complete calculation code.[22]

---

[22] As the DECIDERA code in EDF.

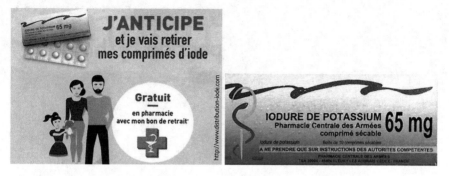

Photo 1.4 Iodine tablets distribution cycle and packaging box

There is an effective way to avoid radioactive iodine contamination independently of the evacuation. That is to use the thyroid's appetite for iodine by administering a stable iodine tablet to the subject. In the event of an accident, radioactive iodine is inhaled or ingested and will concentrate in the thyroid. In case of inhalation, part of it is expectorated during exhalation, the other part enters the lungs and the blood, and finally is absorbed by the body. If ingested, iodine enters the bloodstream in about 2 hours, but some is eliminated through the natural way. In an adult, it is estimated that about 30% of this iodine incorporated in the blood is fixed by the thyroid. The radioactive iodine fixed by the thyroid will then kill cells or damage their DNA by the emission of electrons due to the radioactivity, causing cancers in the long term. The idea of the stable iodine tablet is to saturate the thyroid with stable iodine before the subject is exposed to radioactive iodine, as the thyroid can only absorb a certain amount, after which the thyroid inactivates specific iodine transporters in the same way that a full stomach signals satiety to the brain. Taking stable iodine protects for at least 24 hours, providing a respite for the evacuation of people. Iodine 127 comes in the form of potassium iodide KI tablets produced in France by the *Pharmacie Centrale des Armées*. These tablets will be distributed to the populations concerned in the event of an accident. People who live within 10 to 20 km of a French nuclear plant can have pre-positioned tablets at home by picking them up free of charge at their nearest pharmacy. This provision allows them to be taken as soon as an accident involving a radioactive release is announced. The tablets are dosed at 65 mg and can be broken into four parts to respect the dose to be taken according to age (two tablets above 12 years, one tablet between 3 and 12 years, half a tablet from 1 month to 3 years, and a quarter tablet below 1 month for babies). The protection acts from 24 to 48 hours. The IRSN (the French Technical Support of the Safety Authorities) is considering the possibility of treatments including several doses over 7 days, except for pregnant women because fetuses are very sensitive to iodine. Consideration is being given to the possibility of overdosing, as it is to be expected that some people, too curious or too anxious, will take all the tablets at once (Photo 1.4).

2

Reactor Accidents in the Early Days of Nuclear Power

Abstract This chapter covers reactor accidents from the beginning of nuclear power until the 1980s. The accidents at Windscale (Great Britain), Vandellos (Spain), Vinča (Yugoslavia), SL-1 (USA), Santa Susana (USA), K-19 (USSR), Lagoona Beach (USA), Chapelcross (England), Grenoble (France), Lucens (Switzerland), Saint-Laurent-des-Eaux (France), Leningrad (Russia), Bohunice (Slovakia), and Constituyentens (Argentina) are analyzed in detail. Other less spectacular accidents are mentioned. The whole is abundantly illustrated with photos and diagrams.

The history of reactor technology is studded with accidents, like all other major human technologies. Those of the early days of nuclear power shed light more specifically on the engineers' understanding of physical phenomena. Improving reactor safety requires an understanding of past errors.

Windscale, a Fire in the Reactor (England, 1957)

The first major reactor accident occurred in 1957 on the first British military reactor at Windscale (Photo 2.1). Beginning in 1947, the British built a set of two graphite-moderated, air-cooled piles to produce weapons-grade plutonium. Windscale was located in the county of Cumberland, in the northwest of England, on the coast of the Irish Sea. This region was rather poor and sparsely populated, with barely 70 inhabitants per square kilometer at the time. Population therefore welcomed the arrival of an industrial complex that

Photo 2.1 An ominous-looking photo (cloudy sky? low angle shot?) of the two military reactors at Windscale. One notices the filters placed abnormally at the top of the chimneys

would bring more than 3,000 jobs to the region. The plant was built on the site of one of the British government's weapons factories that manufactured TNT during World War II. The purpose of the Windscale piles was to produce plutonium by burning-up uranium. These two nuclear piles allowed the British to produce some 80 kg of plutonium per year, the equivalent of about ten atomic bombs. This plutonium was used for Operation HURRICANE, the code name for the first British atomic test, carried out off the coast of Australia on 3 October 1952. In addition, next to the piles, there was a separation plant (1951–1964) for separating plutonium from spent fuel (Butex process). An advanced Gas-cooled Reactor AGR (Fig. 2.1), prototype of the British AGR reactor type, was built on the site to replace the accidented reactor. The AGR is a much more modern reactor with a containment vessel for radioactive products that contains the reactor vessel and the steam generator. This reactor, unlike the accidented reactor, produced electricity. This complex is now called Sellafield, extends over 10 km², includes a spent fuel processing plant, the four Magnox reactors at Calder Hall, a MOX fuel plant, and employs over 10,000 people.

The choice of air as a heat cooling fluid, rather than water used in the American Hanford piles, is the choice of simplicity: no problems of oxidation

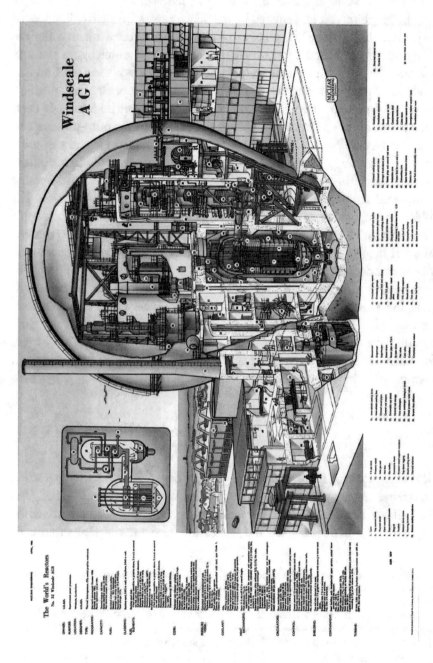

Fig. 2.1 Windscale AGR (1963-1981) cooled with carbon dioxide. Not to be confused with the accidented reactor. (Courtesy NEI)

or neutron effect due to reactor draining. The disadvantage of air is the risk of fire of the graphite, risk exacerbated by the *Wigner effect*. This effect had indeed been predicted by Eugène Wigner as early as 1942 during the work on American military reactors during the war. The ejection of carbon atoms by the neutron shock can take place from a neutron energy of 25 eV.[1] In fact, a fast neutron that thermalizes[2] in carbon atoms can displace them many times, especially during a burn-up[3] at low temperature, which creates vacancies in the lattice. If these ejected atoms do not recombine, which is the case when the burn-up temperature is low (around 115 °C), energy accumulates in the lattice and can reach 2000 J/g. This energy is released spontaneously and dramatically (temperature excursion higher than 1200 °C) when the reactor rises in temperature, because the displaced atoms find their place by increasing the thermal agitation. For graphite, to activate the Wigner Effect, the graphite burn-up temperature must be lower than 115 °C and the integrated fluence, which characterizes the damage of the material following the neutron shocks, must be higher than 0.1 displacement per atom (dpa[4]). Above 170 °C, the Wigner effect "disappears," or at least its harmful effect of heat up by restructuring, because the defects recombine at the same time as they are created. The Wigner effect can eventually generate graphite fires, especially if the air is present. The solution is to voluntarily restructure the graphite by long-term slow thermal annealing.[5] The Wigner Effect becomes really dangerous when the rate of energy release in relation to the temperature in the graphite exceeds its thermal capacity:

$$\frac{dE}{dT_{[\text{calorie/g/°C}]}} > C_p \left(T \right)_{[\text{calorie/g/°C}]}$$

Thermal capacity C_p of a material, which depends on the temperature, characterizes its capacity to store energy. This is called thermal inertia. The rate of release as a function of temperature depends on the dose received by the graphite (it increases with the dose), but also on the temperature (it

---

[1] The electron-Volt (eV) is the energy acquired by the accelerated charged electron in a potential difference of one Volt. It is a unit of energy more convenient to handle than the *Joule* in the context of particle physics.

[2] A neutron that loses its energy as a result of shocks in matter is said to *thermalize* when it reaches the average energy of the medium in which it evolves.

[3] Burn-up is the accepted term for a material subjected to a neutron flux.

[4] For a neutron fluence of one dpa, each atom of the structure concerned undergoes on average one displacement.

[5] We voluntarily heat the fuel to restructure it.

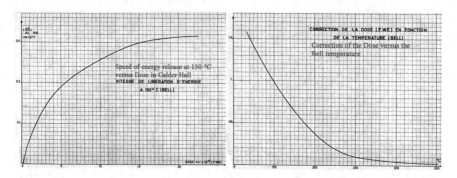

Fig. 2.2 Energy release rate in graphite at 150 °C as a function of the Wigner neutron flux (called dose at the time) and corrective factor of the dose as a function of temperature for neutron spectra encountered in the Calder Hall (Great Britain) or G2 (France) reactors. When graphite in a reactor is hotter than 150 °C, the corrective factor is less than 1. because there is a constant rearrangement of the carbon atoms of the graphite by the "annealing effect." The higher the temperature, the less the Wigner Effect. It is considered that the risk becomes negligible from 300 °C

decreases with temperature) (Fig. 2.2). As the temperature increases, the rate of relaxation decreases because the crystal lattice is constantly rearranged at high temperature. This is called annealing. This temperature annealing technique is sometimes applied to steel that contains cracks and defects caused by neutron impact (i.e., a reactor vessel). This type of annealing should not be confused with melting the steel to fill the cracks. The annealing temperature is in fact much lower than the melting temperature of the material to be restructured, and the annealing must last for long periods (several weeks) to be effective. If the material releases energy faster than it can store it, a limit is reached where the temperature goes out of control, and in the worst-case scenario, we speak of a *Wigner fire*. The first graphite reactors had low operating temperatures (below 200 °C), which justified a Wigner risk analysis (Fig. 2.3).

Nevertheless, at Windscale, the risk of an air-fed fire was judged, hastily at the time, to be negligible compared to the water loss accident. In anticipation of the risk of release of fission products from a possible defective fuel cartridge, filters were installed at the outlet of the 410-foot high stacks (123 m, 14 m in diameter, 50,800 tons of reinforced concrete). These chimneys, highly visible from the surrounding area, were called Cockcroft's follies, after the

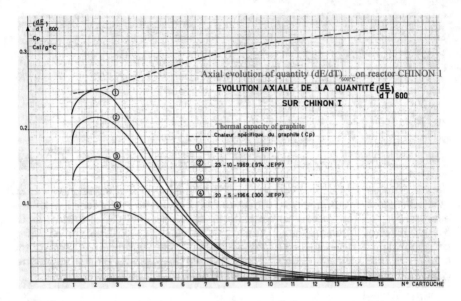

Fig. 2.3 Wigner effect predicted in the reactor of Chinon-A1 (nicknamed "the Bowl"). These predictive calculations at the end of 1969 show that the Wigner risk appears at the level of the first three cartridges (the most burned-up) as early as the summer of 1971. This problem can be treated by annealing at higher temperatures to restructure the graphite

physicist John Cockcroft[6] suggested the introduction of these filters after visiting the Oak Ridge site, which was experiencing problems with the unexpected release of uranium particles.

The filters, made of glass wool, would have been more effective if they had been placed at the base of the stacks, but the system was added after the stacks were built, so they could only be placed at the top. The purpose of these filters

------

[6] Sir John Douglas Cockcroft (1897–1967). British physicist. Nobel Prize in Physics with Ernest Walton on the transmutation of atomic nuclei by proton acceleration. After studying mathematics at the University of Manchester, he worked at the famous Cavendish laboratory, then became a professor at Cambridge. During the war, he became a member of the Maud Committee on the atomic bomb, then was sent to safety in Canada where he directed the Chalk River laboratories and participated in the Manhattan Project. After the war, he became the director of the British atomic center at Harwell.

Cockcroft and George Gamow at the Cavendish laboratory in 1931.

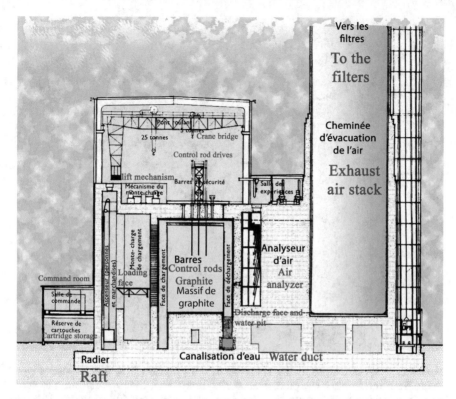

Fig. 2.4 Cross-section of the Windscale reactor building

was to retain solid particles during normal operation of the pile, and not to deal with a massive release in the event of an accident, but they were nevertheless very useful in mitigating the consequences of the accident.

After solving many technical problems, the British diverged the first pile in October 1950. The two piles, each located in a reinforced concrete containment to ensure biological protection against burn-up, are made up of a cylindrical graphite block with a horizontal base weighing 2030 tons, build from 50,000 graphite blocks, the whole being 15 m in diameter and 7.6 m long (Fig. 2.4). Each block is in the form of octagonal logs 25 feet long and 50 feet in equivalent diameter. The graphite assembly is drilled parallel to its axis to form 3440 horizontal channels (Photo 2.2). Inside the latter are natural uranium rods 2.5 cm in diameter, enriched to about 0.7% in uranium 235, the fissile isotope of uranium. The fuel is cladded in aluminum and provided with cooling blades to improve heat exchange (Photo 2.3). Each of the 3440 channels contained 21 *"fuel cartridges"* horizontally (Photo 2.4). There was a total of 72,240 fuel cartridges. The aluminum of the cladding strongly limits the

Photo 2.2 Fuel cartridges in the channels in the graphite. The fuel elements were loaded from the front of the piles, then during unloading, they were pushed towards the exit, from the back, where they fell into a compartment full of water. The sole purpose (not electricity production) of this reactor being to produce plutonium 239, the fuel elements were recovered after a short burn-up (so that the plutonium would be as rich as possible in plutonium 239), and then sent to a separation plant located on the Windscale site in order to limit the transport of dangerous materials. Within this plant, they were stored in a large pool to reduce their activity and their temperature because of the residual power

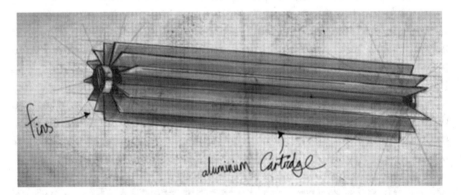

Photo 2.3 Drawing of a fuel cartridge with a uranium rod clad in aluminum and cooled by blades. Heated aluminum ignites easily in air, as does magnesium

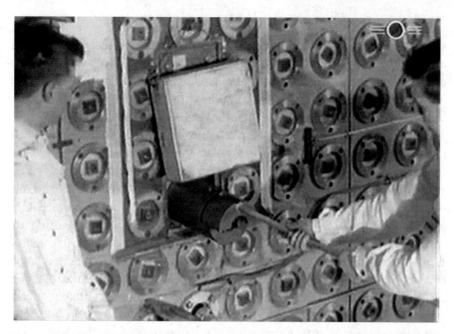

Photo 2.4 Handling in normal situation of a cartridge in front of the loading face of Windscale. The white parallelepiped is probably a biological protection. During the accident, it is with long steel rods that the courageous operators will push the cartridges towards the front face of the reactor. We notice that the operator on the right does not even have gloves, perhaps because the reactor is shutdown at the time of the photo

admissible temperatures due to the relatively low melting point of aluminum (660.3 °C). The load can be carried out continuously, pile in operation, thanks to a platform-hopper which positions the cartridges horizontally in the core by the loading face. The cartridges are loaded by pushing them through a push rod handled by manual operators. When a cartridge is loaded, the last cartridge is ejected from the channel into a recovery compartment filled with cooling water located below the other side of the reactor (Fig. 2.4). The (vertical) columns of graphite are pierced with horizontal fuel channels. The power of the pile is regulated by 12 horizontal control rods inserted on each side (24 rods in total). A set of 16 vertical shutdown control rods could drop vertically by gravity into the core in case of emergency. A group of 8 blowers was used to cool the core with air. A detection system made it possible to alert if a fuel channel released fission products. On October 3, 1952, the British detonated their first atomic bomb using plutonium extracted from Windscale, on an uninhabited atoll off Australia.

The phenomenon of Wigner energy storage was unknown when Windscale started. Lorna Arnold (Arnold 2007) reported that a first incident had occurred in May 1952 in pile n°2, where an unexplained increase in temperature was observed, which could be controlled by increasing the flow of the blowers. An identical phenomenon appeared in the pile n°1, which caused a light fire of lubricating oil of the plant blowers, which had escaped in the core. The understanding of the physical phenomenon made it possible to attempt a voluntary annealing in pile n°2 in January 1953. The operation was successful, and a rapid increase in temperature was observed in the lower part of the pile, after having operated the pile at reduced power for a certain time. From this point on, many voluntary anneals were successfully performed. The annealing procedure became standard and consisted of instrumenting the core with 66 thermocouples to monitor the annealing, which were removed during normal production period. Unfortunately, only one of these thermocouples was continuously readable and allowed to visualize the dynamic behavior of the heat up. In fact, the behavior of the pile was different at each annealing, which was attributed to "pockets" of graphite that had not properly released their Wigner energy, without being noticed by the thermocouples, especially in the areas near the load face where the Wigner energy was maximum because the graphite temperature was lower. This zone, difficult to access for the instrumentation, was not investigated in the end, so that the operator did not have access to the hottest point of the reactor during the annealing process.

On October 10, 1957, at 4:30 p.m., during the ninth annealing of pile n°1 begun on October 7, a fire broke out in the center of the reactor. Following a first low-power nuclear heating (2 MWth i.e., about 1% of the power of the pile), the temperature had risen to a little more than 200 °C, which made it possible to hope that the beginning of the release of Wigner energy would be sufficient to initiate the total annealing of the pile, making it possible to shut down the thermal chain reaction, which was actually done. But the temperatures seemed to stabilize and even decrease, suggesting that the annealing was incomplete and weakening. The reactor was diverged again, allowing the temperature to rise to 330 °C, and then the reactor was shut down again. On October 9, the temperature rose rapidly to over 400 °C. The fan doors were opened to air cool the pile according to official procedure. On October 10, radioactivity was detected in the stack of pile 1, an unusual occurrence since the pile was shut down at this stage of annealing. From noon onwards, the radioactivity increased at the chimney outlet. The temperature continued to rise, so that the staff started the blowers again to cool the reactor, which was like blowing air on a fire! One think then of a failed cartridge and not yet of a fire. It is then decided to open a channel to check in visu the suspect channel. The four visible channels were red hot and so distorted that it could not be possible to eject them! With long steel rods, the operators ejected the surrounding channels to prevent the fire from spreading to the rest of the pile.[7] The temperature of the pile measured now exceeds 1200 °C to the great horror of the physicists present in the control room.

In the end, 120 channels were on fire. Personnel will perform heroically by relentlessly pushing the partially burning fuel cartridges toward the back side of the pile with all available steel rods, protected only by portable respirators and conventional protective suits. On October 11, an attempt was made to inject carbon dioxide from the Calder Hall plant to try to smother the fire, but without any noticeable effect, because the quantities of gas were too small. It should be noted that tests carried out in France afterwards showed that it was very difficult to cool down a graphite fire even with argon. One can imagine that the graphite burns at least during a certain time thanks to the oxygen trapped in the carbon matrix, which degasses. Water was then brought in with the means at hand because no connection was foreseen (in particular, no water was to be present in the reactor building to avoid any criticality risk). Despite the risk of a steam explosion, the personnel sprayed with great apprehension (because of the criticality risk!) the pile on top in the hope of extinguishing the fire, initially with a minimum flow. After an hour of injection and the shutdown of the fans that kept a breathable atmosphere in front of the loading face, the situation improved. The paradox was that the cooling of the pile by air inevitably maintained the fire. For 30 hours, the pile was flooded by pumping water that had become highly radioactive after its passage through the core from the pit below the core to tanks. The situation was deemed to be under control on 12 October, the pile having become cold again. No special measures were taken regarding the population, and the local police were only notified one day after the first detection of radioactivity. Later, the government bought back contaminated milk from local producers at a generous price to avoid any local discontent (Fig. 2.5). two million liters of milk will finally be pierced into the Irish Sea.

The authorities communicated rather evasively on the affair under cover of defense secrecy, and a commission of inquiry was set up in October 1957, the Penney Commission , to draw the first conclusions of the accident. The main conclusion is that it was the second nuclear heating that was too fast and too close from the first, which must have produced ruptures of the cartridges, the

---

[7] We are still amazed by the "radiation protection" aspects of this operation, as the operators are almost in contact with the spent fuel.

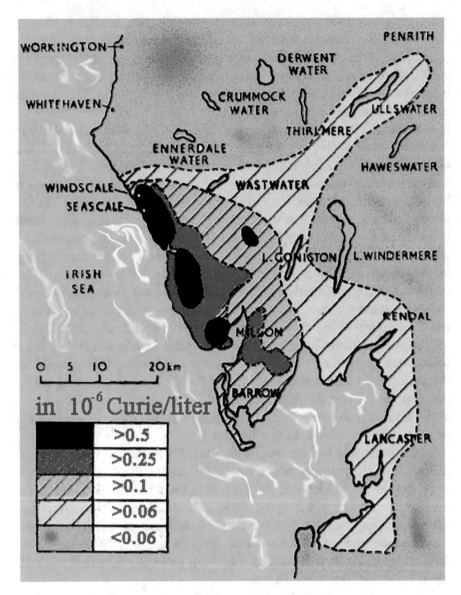

Fig. 2.5 Iodine activity in milk as of October 13, 1957

oxidation of the uranium then adding to the temperature excursion. The possible oxidation of the magnesium in the cartridges containing lithium placed there to make military tritium is another scenario that has been mentioned. The inadequacy of the location of the thermocouples is also widely criticized, as well as the absence of clear written operating procedures for annealing. The absence of what is now called an Internal Emergency Plan (IEP) was also pointed out. To be honest, nothing had been planned! The Penney report, which was very factual, was not made public in the end, under the pretext of defense secrecy. A watered-down version of the Penney Report was finally published in the *White Paper* on the Windscale accident. While the Penney report exonerated the operators, the White Paper seems to point the finger at the failures of individuals, presenting annealing as a routine procedure poorly managed by the operators, and insisting on the absence of risk in the case of the British Magnox reactor type used for energy production (cooled with carbon dioxide). In the end, it is especially the lack of knowledge on the behavior of burn-up graphite, in a context of all-out development of reactors in England, that raises questions. It was not until 1958 that these graphite problems were studied in detail. Reactor 1 was definitively shut down and Reactor 2 was also shut down shortly afterwards, because the cost of upgrading the instrumentation was considered unreasonable in relation to the life expectancy of the reactor. This expectation was reduced by the fact that the analysis of some graphite samples from reactor 2 showed oxidation rates 3,000 times higher than the expected average! As the excursion temperature of oxidation of carbon in the air is of the order of 320 °C (for oxidation in the mass of graphite), and as the release of Wigner energy raises the air temperature to at least 250 °C, this leaves a very small margin of barely 70 °C between the two thresholds for an annealing that does not massively oxidize the graphite. It is this technical observation that will finally sound the death knell for Reactor n°2. The plutonium will then be produced in the more powerful Calder Hall power reactors.

A significant amount of radioactivity was finally released during the accident, estimated today at 740 *TBq*[8] (20,000 *Ci*[9]) of ^{131}I, 22 *TBq* (600 *Ci*) de ^{137}Cs, 3 *TBq* (80 *Ci*) of ^{88}Sr and 330 *GBq* (9 *Ci*) of ^{90}Sr. At noon on October 10, the wind was light with a tendency to blow from the southwest. But at the start of the accident, the winds strengthened as they turned north and then northwest on the morning of October 11, sending easily detectable releases as far south as Yorkshire, largely to the southeast. The main iodine deposition was over Lancashire and Cumberland. By the end of the 11th, the plume reached Belgium, Frankfurt in Germany by the end of October 12, and even Norway on the 15th. France was largely spared because the wind flux was along the northern border with Belgium. The initial plume, oriented from the

---

[8] 1 Tera Becquerel (TBq) = 10^{12} Bq.

[9] One *Curie* equals 3.7 10^{10} *Becquerels*. 1 *Becquerel* corresponds to one disintegration per *second*. One *Curie* corresponds to the activity of one gram of radium 226 (3.7 10^{10} *Becquerels*), discovered by Marie Curie.

plant towards the southeast and running roughly along the coast, very quickly deposited radioactivity that significantly contaminated the soil over a distance of about 10 kilometers, hence the contamination of cow's milk in this farming region, since no containment measures had been taken. As for the operators, the thyroid dose measurement on 96 persons indicates a maximum dose of 9.5 rads,[10] the second highest being 2.1 rads and an average of 0.4 rad by inhalation of iodine 131. Outside the building, the maximum dose equivalent recorded over 13 weeks was 4.7 rems, well below the 12 rems recommended at the time.

It was not until the 1990s that epidemiological studies were published to try to determine the real impact of the accident. A controversy occurred in the 1980s when a librarian from the University of Newcastle-upon-Tyne, John Urquhart, contested the official figures of the low number of radiation-induced cancers by calculating the dose induced by polonium 210, isotope produced by the burn-up of bismuth 209 for military purposes[11] (polonium is used as an α-emitter to initiate fission in bombs though reaction (α,n) on beryllium). Polonium 210 is extremely radiotoxic, as the case of the poisoning of Alexander Litvinenko proved to the public in 2006 in England.

Vandellos, a Fire of Turbo-Blowers (Spain, 1989)

In terms of fire, we should mention the accident at the Natural Uranium-Graphite-Gas (UNGG) reactor No. 1 in Vandellos near Tarragona, Spain. This 480 MWe reactor is the twin of the French reactor of Saint Laurent de Eaux-A2, since it was built by a Franco-Spanish "joint venture" with HIFRENSA (*HIspano-FRancesa de Energia Nuclear*) in November 1966. It brought together EDF and Catalan producers, of which EDF held 25%. This was an attempt to export French nuclear know-how at the initiative of General de Gaulle and General Franco. Work began in 1968 (Photos 2.5 and 2.6), and the reactor was put into operation in 1972 (Photo 2.7). The reactor being of continuous fuel load, General Franco did not hide his ambitions of a Spanish

---

[10] The rad is the old unit of dose and corresponds to 0.01 Gray = 0.01 J/kg.

[11] This bismuth irradiation would have been largely hidden because the British government did not want it to be known that the bomb starters were still manufactured at that time by such an obsolete means. The bomb primers were not made like that already at that time by the other countries and the British showed a certain delay in the matter.

Photo 2.5 Construction work: Siding, pouring the concrete of the caisson and loading face (municipal archives of Vandellos)

Photo 2.6 Assembly of the inner containment and the lower compartments and support of the graphite block (municipal archives of Vandellos)

atomic bomb[12] based on plutonium 239, perhaps with the help of France (it has been said that General de Gaulle would not have been opposed to help (?), De Gaulle was known to be hostile to American domination within NATO).

---

[12] This is the *Islero* project, which began secretly in 1963 and was led by José María Otero de Navascués, director of the equivalent of the French Atomic Energy Commission (*Junta de Energía Nuclear* or JEN). The project relies on the production of plutonium 239 by Vandellos-1, which Franco intends to keep, on the model of what France did at the end of the 1950s with the G reactors at Marcoule, and on French assistance for a plutonium separation process. The pressure of the Americans, allies of Spain during the Cold War, and who feared scientific dissemination, put an end to this dream of greatness.

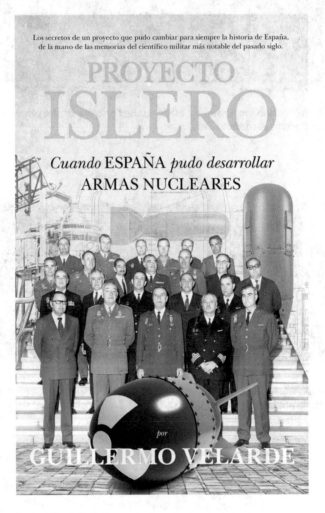

The book of Guillermo Velarde well documented on the question.

Photo 2.7 The Vandellos-1 plant in Spain

On the night of October 19, 1989, alarms began to sound in the control room of the Vandellos-1 nuclear plant. The first alarm announced the strong vibration of one of the shafts of the generator turbine. Several alarms went off when suddenly the operators heard explosions. A fire broke out from 9:39 p.m. in the generator in the turbine building. In fact, a shear crack in the shaft of turbine n° 2 led to a clean fracture, destroying 37 of the 92 blades of the turbine, causing rapid decompensation of the turbine. The rapid braking of this 5-tons turbine ignited the lubricating oil of the shaft bearings by friction. The explosion was amplified by the destruction of a hydrogen outlet terminal (cooling of the alternator). The flames spread at high speed, causing severe damage to the reactor cooling systems, and the fire was visible for miles around. This fire spread to the electrical circuits. Two of the four turbo-blowers that circulate the carbon dioxide coolant in the reactor were destroyed. The other two were accidentally drowned by firefighters in an attempt to reduce the fire. Josep Pino, chief of the Amposta fire station called to the rescue, will say "*The technicians fled the affected premises, and we were left alone; some technicians were taking water samples and others were calling France,*[13] *while we were shouting "the reactor is running away, the reactor is running away!"*" (quoted by Mr. P. Pons in an article in El País). "*Along the way, I heard "if the*

---

[13] France sold the reactor in the early 1970s.

alternator burns out, does it affect the reactor?" This only made me more worried. When I arrived, the access barrier was up, and people were fleeing in a hurry because at first you have to evacuate those who are not essential. But of course, I did not know it at the time", recalls Fèlix González, head of the emergency region of *Tierras del Ebro*, who was at the time in charge of the *Reus* fire station. Those in charge of the plant immediately called the employees who were on call, such as Carlos Arriola, who worked on the mechanical maintenance of the plant. "*There was a lot of smoke, the priority was to get the water out. I was one of the first to go down the reactor pit. There was almost no lighting, the sound of alarms, drums floating, a meter and a half of water deep...*" he recounts. "*One firefighter kept saying to me, 'But are we safe here?' We were up to our necks in water, and we didn't know if it was contaminated, until I tasted it, and luckily it was salty.*[14]" Most of the plant's staff came to help out with the problems. "*We were the only ones who knew about the plant and could solve the situation.*"

With difficulty, the operators finally managed to cool the reactor with the secondary cooling circuit, to prevent a general flashover of the graphite block, the situation that had occurred at Windscale. The accident will be classified afterwards as 3 on the INES scale and it is considered that only the firemen, who intervened without much preparation, were exposed to ionizing radiation. Repairs proved too costly (Photos 2.8, 2.9 and 2.10), the reactor was finally definitively shut, then progressively dismantled, and a sarcophagus was built around the reactor (Photo 2.11).

A week after the accident, a new failure due to a short circuit in an auxiliary transformer caused a small fire and a plume of smoke that panicked the surrounding population for no reason, causing a spontaneous evacuation.

Vinča, a Serious Criticality Accident (Винча, 1958)

After the Second World War, Eastern countries also embarked on the race to the atom under the impetus of the Russian big brother. However, Yugoslavia was a special case because Marshal Tito did not align himself strictly with the USSR, adopting a more open policy with the West. In 1958, a nuclear program was launched at the Institute of Nuclear Sciences « *Boris Kidrič* » of

---

[14] Thus, coming from the sea and not from the reactor.

Photo 2.8 Photo taken outside (shadow of the photographer) probably showing a fan motor cowling damaged by the fire (photo J.L. Sellart)

Photo 2.9 Firefighters and technicians after the fire in an unidentified area of the reactor

Photo 2.10 Journalists and photographers visit the degraded installation without special protection

Photo 2.11 Hexagonal sarcophagus from Vandellos-1 in 2014

Vinča, 15 km from Belgrade (Институт за нуклеарне науке Винча) and 2 km from the Danube (Photo 2.12). The institute was founded on January 21, 1948, and named after the leader of the Slovenian Liberation Front against the Nazi occupiers during World War II. The institute was placed under the

Photo 2.12 The site in Vinča in 1952. Building 2 is the physics department. Building 4 is the department of physical chemistry, building 5 is biology, building 6 is the particle accelerator V15, building 7 is the library, building 8 is radiobiology

authority of Professor Pavle Savič,[15] specialist in physical chemistry, trained at the Radium Institute in France, then former collaborator of the great Soviet

---

[15] Pavle Savič (Павле Савић, 1909–1994) is a Serbian physicist and chemist. He graduated in 1932. In 1936, he received a six-month scholarship from the French government to study at the Radium Institute in Paris; instead of 6 months, Savič stayed in France for 4 years. In 1937 and 1938, he worked with Irène and Frédéric Joliot-Curie on research relative to the action of neutrons on heavy elements. Together with the Joliot-Curie couple, Savić was nominated for the Nobel Prize in Physics. Savić returned to Yugoslavia to fight as a partisan against the German occupation. After the war, he was one of the first promoters of the idea to build the Institute of Nuclear Sciences in Vinča. He was the director of the Institute from 1960 to 1961. In 1966, he returned to his position at the University of Belgrade. He was elected president of the Serbian Academy of Sciences and Arts from 1971 to 1981.

Savič in Paris-1937. Serbian stamp in honour of Savič.

physicist Piotr Kapitsa. Although the idea of producing a Yugoslavian atomic bomb was evoked at the beginning, the Institute quickly turned to more peaceful and more affordable applications. A group of reactor physics was constituted in 1955 whose first task was to produce heavy water, an expensive liquid because its production requires much electric energy.

Two nuclear research reactors were built there, the RA and RB reactors, the largest of which, the RA reactor, had a power of 6.5 MWth and was fueled by 80% enriched uranium from the Soviet Union. The RB reactor was what is called a zero-energy reactor, in fact of very low energy, from a few 10 mW to about 50 Watts, to carry out critical experiments first with a natural uranium (metal) assembly moderated[16] by heavy water. Heavy water is a compound whose hydrogen is composed almost entirely of deuterium atoms ^{2}H while ordinary water is made of ^{1}H. To indicate deuterium, physicists usually use the symbol D and note the heavy water D_2O, while ordinary (or light) water is noted H_2O. Heavy water is a better neutron moderator than light water because of its near absence of neutron capture. The objective was to measure precisely the height of heavy water in the vessel of the small reactor and the bulge of the neutron flux by the method of measuring the critical buckling.[17] To do this, a 10 mm aluminum vessel (a light metal with a density of 2.7 compared to water) is mounted on a platform more than 4 m away from any surface that could reflect neutrons (this is called a reflector). The vessel can contain 6.36 m^3 of heavy water. The vessel is closed by a 7 cm aluminum cover with two small inspection windows. The presence of reflectors would reduce the critical size of the reactor and thus the mass of fissile material required (this is called the critical mass). The supporting structure (Figs. 2.6 and 2.7, Photo 2.13) is made of aluminum and can support a weight of 15 tons. Two working platforms allow the operators to control the pile. It is placed in the center of a pond 8 × 8 m wide and 1.5 m deep, which serves as a backup receptacle for heavy water in case of an incident.

The reactor (Fig. 2.8) is presented as a lattice of cylindrical fuel rods made of metallic uranium, 2.10 m high, 2.5 cm in diameter, and with a square pitch of the rod positions of 12 cm. The total weight of uranium is 3995 kg. The cladding of the rods is made of 1 mm thick aluminum. The rods are separated

---

[16] Moderation represents the capacity to slow down neutrons by successive shocks. The more the neutron is slowed down, the better its capacity to fission uranium, because the probability of fission of heavy nuclei increases when the speed of the neutron decreases.

[17] The method of the critical buckling consists in measuring the radius of curvature of the 3D neutron flux shape in the core. Without going into detail, this radius of curvature is related in critical situation (stable reactor) to the neutron properties of the fissile material of the reactor and to the geometry of the pile in what is called the *"fundamental mode."*

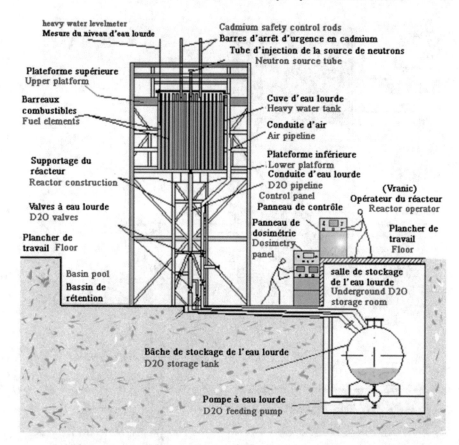

Fig. 2.6 Position of the most affected operators during the Vinča accident in Yugoslavia (15 October 1958), adapted from M. Pesič: *Some examples of accident analyses for RB reactor*, IAEA Technical meeting on Safety Analysis for Research reactors, Vienna, Austia, 5–7 June 2002)

by two grids at the top and bottom of the vessel. The absence of power simplifies the cooling of the reactor (Fig. 2.9).

On October 15, 1958, 6 months after the start-up of the first core, during a criticality experiment on the RB reactor, a bad evaluation of the height of heavy water necessary to make the device critical, led to a power excursion of the heavy water research reactor, following a bad adjustment of the heavy water level. The rate of rise of the heavy water in the vessel was rapid: 2.5 cm/ min. With the water level at 175 cm, 3.5 cm below the expected critical level, the operating team was distracted by non-team personnel entering the hall. The crew intended to stabilize the reactor at 177 cm just below the critical level, but the booster pump was allowed to run due to distraction, and the

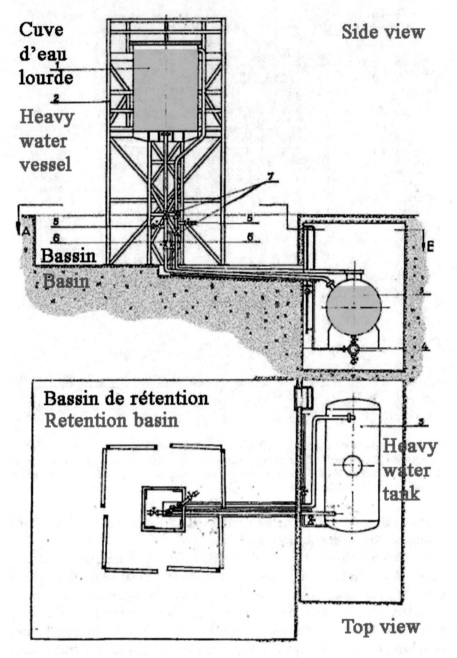

Fig. 2.7 View of the RB installation. 1: Reactor vessel, 2: Supporting structure, 3: Heavy water filling tank, adapted from (D. Popovič, S. Takač, H. Markovič, N. Raisič, Z. Zdravkovič, j. Radanovič: *Zero Energy Reactor « RB »*, Bulletin of the Institute of Nuclear Sciences *"Boris Kidrič,"* Vol. 9, N°168, March 1959, Laboratory of Physics)

Photo 2.13   Photo of the RB reactor in 1958. One can recognize the vessel placed on its support, itself placed in the pond with white walls in unevenness compared to the service desk. The control consoles are visible on the right of the picture, a few meters away from the building, without any particular biological protection

level continued to rise. The instrumentation used for dosimetry and the alarm systems were disconnected or partially removed, a serious mistake with serious consequences. 84 s after reaching the 175 cm level, the 178.5 cm level was reached, and the pump, still operating, raised the level to 4.5 cm above the critical level! The reactivity and power of the reactor then began to increase. Two BF_3 neutron radiation counters saturated during the power excursion, still without worrying the operators. A third counter, suspected to be out of service, was turned off. Yet another automatic recorder, located 540 m outside the hall and responsible for measuring air activity and possible radioactive deposits, did measure this power and gamma radiation increase for about 10 min. It is estimated that the heavy water level remained too high for 433 s.

The term *criticality* can mislead the reader. Indeed, for the reactor to remain in a stable operating condition, it must be critical, whereas the word in its common meaning rather raises concern. To reach this state, heavy water is slowly raised in the vessel. As long as the water level does not reach a "critical

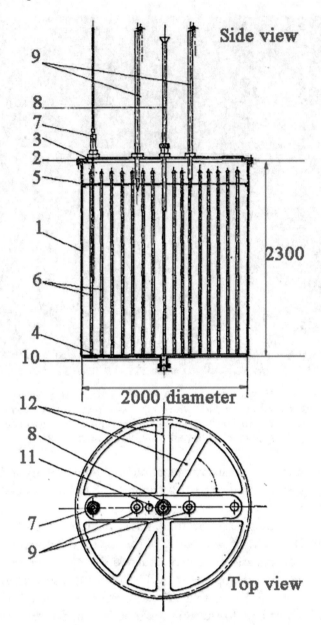

Fig. 2.8 Schematic of the RB reactor. 1: Aluminum reactor vessel (10 mm), 2: Aluminum vessel cover (7 cm), 3: Instrumentation channel cover, 4: Lower fuel rod grid, 5: Upper fuel rod grid, 6: Uranium rods, 7: Heavy water level measurement, 8: System for injecting the 500 *milliCurie* (Radium-Beryllium) neutron source per reaction (α, *n*), 9: Two neutron safety absorber rods, 10: Bottom of the vessel with the heavy water inlet and outlet, 11: Two sight glasses, 12: Radial ribs as stiffener

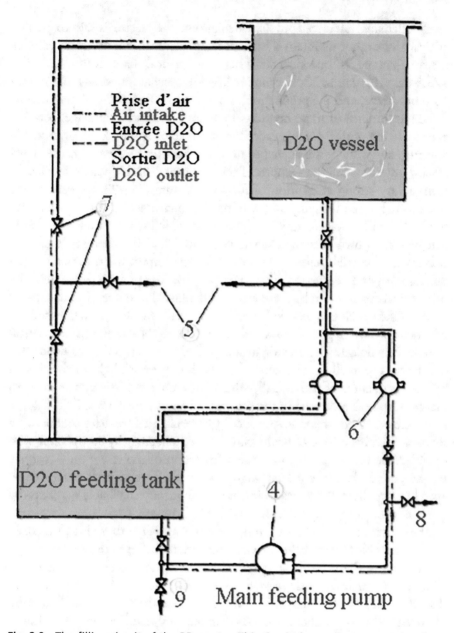

Fig. 2.9 The filling circuit of the RB reactor. This circuit does not even contain a heat exchanger since the pile is not supposed to produce energy. As heavy water is very expensive, it is carefully collected in a tank when the reactor vessel is emptied. The circulation of dry air prevents the heavy water from becoming loaded with moisture, which would lower its deuterium content

level" calculated by clever physics calculations, the reactor is said to be sub-critical and cannot maintain a stable level of neutron flux unless an external neutron source is introduced. When the critical level is reached, i.e., 178.5 cm ± 0.1 cm at 22 °C estimated by Yugoslavian physicists, the reactor becomes stable, and the production of neutrons by fission is compensated by the disappearance of these neutrons by absorption and by leakage from the reactor. If the critical level is exceeded, the reactor is said to be over-critical and runs away. Its power increases until the heat up of the fuel causes what reactor physicists call the Doppler effect to appear. This Doppler effect results from a very strong absorption of neutrons by uranium 238 present in the nuclear fuel, when the temperature of the fuel increases. This absorption leads to a very rapid power decrease of the reactor, which will re-diverge when the reactor cools down if the geometric conditions and the chemical composi-tions of the materials remain unchanged. If the temperature has risen sharply during the power excursion, there is a possible loss of critical geometry by mechanical explosion or by evaporation of the liquid in the reactor. In the case of the Vinča accident, no explosion but a relatively slow power excursion producing a flash of gamma rays and neutron flux. The overflow of the critical water level engaged the reactor in a so-called "over-critical" behavior. This excursion is generally accompanied by a flash of greenish light in the air and by the production of ozone O_3[18], which has a characteristic odor similar to bleach. Ozone is produced in the presence of an intense electric field (e.g., as in transformers), in this case, produced by the charged particles produced by the fissions. This release of ozone was detected olfactory by an operator who operated the reactor shutdown system (insertion of the safety rods), but six people close to the vessel were strongly irradiated. The core itself was not dam-aged because there was no explosion as such (contrary to what is suggested by the comics strip Figs. 2.10, 2.11, 2.12, 2.13, 2.14 and 2.15).

The subsequent heat up of the heavy water probably caused it to expand, and it is possible that some heavy water was discharged from the vessel through the air line at the top of the vessel, which was placed there to evacuate air when the water level in the closed vessel rose. Since the fuel rods were not degraded, this heavy water should not have been heavily contaminated by radioactive fission products. It must be understood that this type of criticality accident generally lasts only a few tens of milliseconds to a few seconds for

---

[18] Ozone is an allotropic variety of oxygen, less stable than the oxygen gas O_2. Ozone is detectable by the human sense of smell up to 0.01 ppm (parts per million). Ozone is known to the public through the ozone layer that surrounds the Earth between 13 and 40 km in altitude and which intercepts nearly 97% of ultraviolet rays. The hole in the ozone layer which is constantly growing at the North Pole worries scientists because too many ultraviolet rays cause skin cancers.

Fig. 2.10 The comic strip transcription of the human adventure of the rescue of the Vinča accident in the children's magazine *Okapi* No. 40 of July 1, 1973.The death of Albert Biron

Fig. 2.11 The Yugoslavian team

Fig. 2.12 Due to lack of information, the artists, although talented, describe rather the explosion of a power reactor than a modest experimental reactor. A fireball (!) surrounds the operator Vranic

Fig. 2.13 The French medical team

Fig. 2.14   The D Day

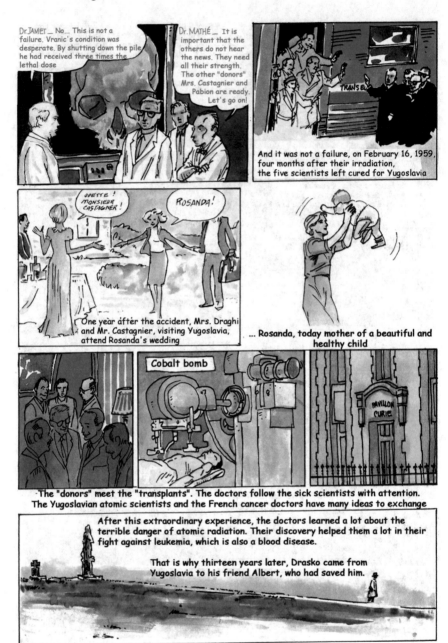

Fig. 2.15 Life wins over Death!

large over-criticality, and that only the fuel has time to heat up. In the present case, the power excursion linked to a weak over-criticality led the reactor into an overpowered state for about 400 s, which must have allowed the heavy water to heat up and expand thermally. The fact that there was a partial rupture of the vessel is not mentioned in the most serious references. The presence of contaminated water sometimes reported must rather refer to the badly managed draining of the air line. Neutron physics confirms that it is the Doppler effect that shuts down the power excursion, the emergency rod drop is only effective to ensure a subcritical geometry at the end of the accident (the power excursion is often faster than the rod drop). In the Vinča accident, recent calculations showed that the excess reactivity for a 4.5 cm heavy water surge was about +0.305 β_{eff}[19], i.e., a relatively moderate overactivity. Physicists know that rapid power excursions occur when the excess reactivity is of the order of or greater than β_{eff}. This means that the power excursion was finally slower and therefore longer than in the very fast accidents that we will describe in the case of the SL-1 reactor. The period of the reactor , i.e., the time for which the power is multiplied by the Neper constant (aka Euler constant) $e = 2.718$, is estimated at 12.3 s, leading to a power of 2.5 Mega Watt thermal with a total energy released during the excursion of 80 MegaJoules (Fig. 2.16), approximately 2.8 10^{18} fissions.

Six physicists and operators were standing near the reactor at the time of the accident: Radojko Maksič, Roksanda Dangubič, Draško Grujič, Živorad Bogojevič, Stjepan Hajdukovič and Života Vranič. Maksič and Vranič activated the shutdown via a control panel located very close to the vessel. It is estimated that Vranič, the closest to the reactor, experienced an irradiation of 433 *rem*[20] (4.33 *Sievert*), and the five other people were irradiated at

---

[19] The fraction of delayed neutrons β_{eff} expressed in pcm is used as a reference for whether the reactivity ρ is strong or not. When the ratio β_{eff}/ρ is small in front of 1, the overactivity is small, and the power excursion kinetics is relatively slow. This is the case for the Vinča accident, which will last on the order of 400 s. If this ratio approaches or exceeds 1, the kinetics become increasingly violent and the power peak will be much stronger, but the accident time much shorter. For the most violent peaks, the fuel core temperature will exceed the fuel melting temperature and the fuel rod will burst with dreadful consequences, releasing molten fuel into the medium surrounding the rods, heavy water in the case of Vinča (which did not happen because the supercriticality was low), light water in the case of Pressurized Water Reactors, or the pressure tubes containing light water in the case of Chernobyl. Such a release causes a steam explosion and the dissemination of highly radioactive fission products. In the case of Vinča, it was rather a flash of neutrons and photons that irradiated the operators.

[20] the rem or « *röntgen equivalent man* » is an old unit of measurement for equivalent dose. The unit now official since 1979 is the *Sievert* (symbole Sv). 1 rem = 10 milliSv. The rem is still widely used in industry. The equivalent dose takes into account the damage done to human tissues according to the type of particle (dose equivalent) whereas the dose in Gray is a unit of energy (Joule/kg). Above 4 Sv, it is estimated that 50% of those affected will die. Above 10 Sv, death is almost certain.

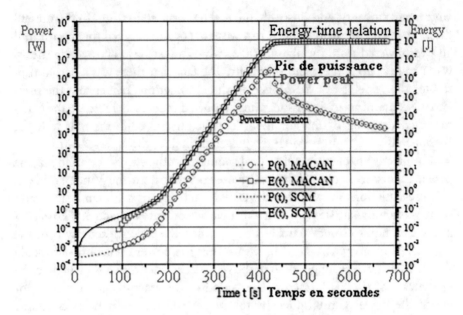

Fig. 2.16 Power excursion calculated by the MACAN and SCM calculation codes in the 1990s. It should be noted that the ordinate scales are logarithmic, i.e., each main scale is ten times the previous one. Paradoxically for the uninitiated, the power excursions are all the more violent as the initial power level is low, but the peak lasts less time because a high power leads to a higher temperature in the fuel, thus a stronger Doppler effect. A stronger Doppler effect will increase the absorption of neutrons and "crush" the power peak more quickly. We note a good match between the two calculation codes. Adapted from M. Pešič: *Some examples of accident analyses for RB reactor*, IAEA Technical meeting on Safety Analysis for Research reactors, Vienna, Austria, 5-7 June 2002)

205 - 320 - 410 - 415 and 422 *rem*. The day after the accident, the six irradiated were transferred to the hospital in Belgrade, but the Serbian doctors were baffled by this atomic disease described in the Japanese survivors of the atomic bombs of Hiroshima and Nagazaki, and on which the known medicines seemed to have no effect. Director Pavle Savič, a former student of Irene and Frédéric Joliot-Curie, called the Curie Institute in Paris for help. Savič learnt from professor B. Pendic of the Curie Foundation in Paris that the oncology

Photo 2.14   The Professor Georges Mathé

professor Georges Mathé[21] (Photo 2.14) experimented with a bone marrow transplant technique with his team. The French immediately agreed to treat the Serbian irradiated patients who were transferred to France as a matter of

---

[21] Georges Mathé (1922–2010) is a French oncologist. He was awarded a doctorate in medicine in 1950 (gold medal from the Paris Hospitals) and participated in the development of exanguino-transfusion, the first extra-renal purification procedure in 1948. He was introduced to immunology with Baruj Benacerraf in Bernard Halpern's laboratory in 1950, then to oncology with Joseph Burchenal at the Memorial Sloan-Kettering Cancer Center in New York in 1951. In 1953, he was appointed Chief of Clinic at the Faculty of Medicine in Paris, with Professor Paul Chevallier in Hematology at the Broussais Hospital. In 1954, he became assistant physician at the Paris hospitals, Deputy Director of the Research Center for Leukemia and Blood Diseases directed by Professor Jean Bernard at the Saint-Louis Hospital. The same year, he was appointed Associate Professor of Oncology at the University of Paris. In 1961, he became head of the hematology department at the Gustave-Roussy Institute in Villejuif, before founding the Institute of Cancerology and Immunogenetics (INSERM-CNRS). In 1963, he cured his first leukemia with a bone marrow transplant preceded by a irradiation. In the 1970s and 1980s, Georges Mathé participated in the development of poly-chemotherapy, cooperating in the development of several important molecules. When the AIDS epidemic appeared, he became interested in it as an immunotherapist and hematologist. In 1989, he designed a quintuple therapy that limited the side effects. He died on October 15, 2010, the anniversary of Vinča's accident, in the department he had created, at the Paul-Brousse Hospital in Villejuif. His research work resulted in the publication of more than 1000 articles and numerous books (adapted from Wikipédia and the Inserm website https://presse.inserm.fr/deces-du-professeur-georges-mathe/14728/).

Georges Mathé is honored worldwide as a pioneer in cancer research. On the left of the poster, the daughter of G. Mathé.

urgency on October 16, 1958 (Figs. 2.10, 2.11, 2.12, 2.13, 2.14 and 2.15). Mathé looked for donors in the Paris area to try to save their lives. It is important to understand that the technique was totally experimental and has never been applied to humans. The risk for both donors and recipients was significant.

Despite this, five Frenchmen agreed to donate their bone marrow for this last chance operation: the doctor and future professor Léon Schwartzenberg (member of the team of professor Mathé), Marcel Pabion, Albert Biron, Raymond Castanier and Odette Draghi, to whom we pay tribute here. The latter, although herself a mother of 4 children and informed of the risks of the operation, nevertheless insisted on helping by giving her marrow to Roksanda Dangubič. The operations took place from November 11 to 16, 1958. All the transplanted will survive, except the young Zivota Vranič (Photo 2.15), the most affected, who will die shortly after his transplant. Roksanda Dangubič will get married in the presence of Odette Draghi, and she will give birth to a perfectly healthy child. In the winter of 1972, Draško Grujič will come to the bedside of Albert Biron, who was very ill and who had given him his bone marrow, during the 3 weeks before his death. These bone marrow transplants gave great hope in the treatment of cancers, in particular leukemia. Professor Mathé kept all his life close links with Serbia by going regularly and free of charge to give treatments at the hospital of Belgrade (Photo 2.16).

Photo 2.15   Zivota Vranič was the young operator (24 years old at the time of the accident) who did not survive despite the bone marrow transplant given by Raymond Castanier. Vranič did not flee at the alert but helped bring the reactor back to subcritical, which ultimately cost him his life

Photo 2.16   Professor Mathé (left) in 2007 with Radojko Maksič at the opening of the cancer unit named after him at the *"Bežanijska kosa"* clinical center in Belgrade

Beginning in 1962, the RB reactor was modified several times, in particular by the introduction of uranium metal enriched to 2% and uranium oxide enriched to 80%. In January 1961, a French team from the CEA specialized in instrumentation (Jacky Weil, J. Furet…) contributed to an international IAEA dosimetry experiment by being in charge of monitoring and safety. Weil was the technician who had spotted the divergence criticality in Zoé, the first French reactor, on the millimeter paper of the neutron flux level recorder in 1948. The work of the French showed that the weight of the two cadmium control rod i.e.,–1300 pcm of reactivity, was still modest compared to the 1200 pcm of over-reactivity that could be reached in the event of total untimely filling of the vessel. This is why it was decided to add a third safety rod, making it possible to raise the anti-reactivity of the three rods to –2500 pcm. This additional rod will act as a water level control rod, its position being directly linked to the water level by a contact point. The control rod has been deliberately slowed down to a speed of 4 pcm/s to avoid any problem of untimely withdrawal (Fig. 2.17). The control room, which had

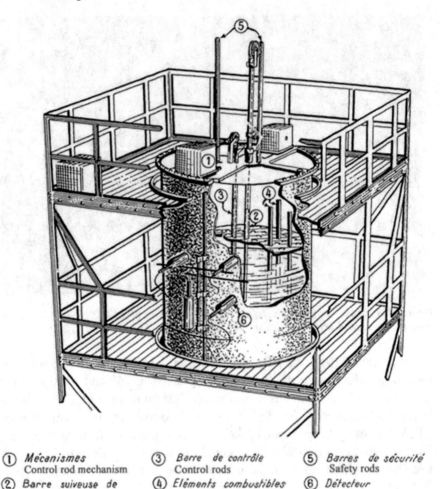

① *Mécanismes*
   Control rod mechanism

② *Barre suiveuse de*
   niveau d'eau
   Heavy water level following rod

③ *Barre de contrôle*
   Control rods

④ *Eléments combustibles*
   Fuel rods

⑤ *Barres de sécurité*
   Safety rods

⑥ *Détecteur*
   Detector

Fig. 2.17 Detail of the Vinča vessel after the French modifications

been moved 7 m from the reactor without a direct view of it after the accident, had its protection against radiation reinforced. The French reinforced this protection against radiation by bringing from Saclay protective concrete bricks and strips of cadmium, a powerful neutron absorber, and by building a baffle of concrete bricks in front of the entrance to the control room (Photo 2.17).

Photo 2.17 Improvement of the biological protection of the entrance door of the reactor building (left) and of the control room (right) protected by bricks of absorbing material (1961). These old photos, unfortunately of poor quality, give an idea of the improvements made in 1961 by the French in the field of radiation protection. A first in the collaboration between East and West in the middle of the cold war!

Zoé, a Near Criticality Accident (France, Circa 1948)

The Vinča accident is strangely reminiscent of a little-known incident that occurred around the end of 1948/beginning of 1949 on the first French reactor: Zoé . The Pile Zoé (Photos 2.18, 2.19 and 2.20) consists of an aluminum vessel containing heavy water D_2O (the moderator, 5 tons) and uranium oxide rods (1950 kg), surrounded by a 90 cm thick graphite reflector, all placed in a hollow concrete block used for the radiation protection of personnel (Photo 2.21). The primary pump, which circulates heavy water for cooling the Pile, is external to the reactor block, the logic being to have easy access for maintenance. There is a strong similarity to the Vinča device except for the strong concrete biological shield surrounding the vessel and the graphite reflector that saves fissile mass. Zoe diverged by going critical on December 15, 1948, at 12:12 pm. A nice Christmas present for the team (Photo 2.22) led by its creators Frédéric Joliot-Curie and Lew Kowarski .

The listing of the detector (Photo 2.23), which traces this feat, signed by Jacky Weil , who will later go to Vinča in 1961, is pictured in the museum that became the Zoé building on the CEA site in Fontenay aux Roses (France).

In all fairness, there was no accident in Zoe, but the similarity of the near accident, which we will describe, with what happened in Vinča is striking. The filling of the vessel in Zoé is done by a small booster pump, which is shown on Fig. 2.18 (number 11). In order to protect against the risk of untimely

Photo 2.18 The control room of Zoé in 1948. Note the "head up" unrecorded measurement dials on the vertical panel above. Two scrolling graph paper recorders are placed on the sides of the cabinet, hardly visible to the operator (CEA photo)

criticality, its operation is automatically limited in time by a protection that shuts down the pump after a programmed time (of the order of one minute). At the end of this time, the protection triggers the power supply to the pump, which shuts down. But this pump having a small flow, it appeared that it was extremely fastidious for the operators to constantly reset the pump during a complete filling of the vessel, the volume of the vessel being very important. The low flow rate of the pump was of course intended by the designer for safety reasons, but the impatient nature of humans being what it is, it did not take long for an excited operator to remove the time protection and let the pump run continuously. The reader who has followed my comments on Vinča will of course have understood what happened next. The critical level of heavy water was almost reached because of the forgotten disconnection of the booster pump protection or because the operator reacted too late to the pump shutdown. Fortunately, the uncertainties of the calculations of the time, all

Photo 2.19 The same control room of Zoé (France) renovated at the end of the 60s. Many "head-up" recorders were installed. We can see the "Human Factors" progress brought to the control console. ZOE's core shutdown on April 6, 1976 at 11:51 a.m. after 28 years of good and loyal service, and above all without accident! (CEA photo)

done by hand with the poor knowledge of the properties of fissile materials at the time, led to an underestimation of the real critical level, introducing a happy conservatism into this type of situation. Following this near miss, the safety of the pile was of course improved by physically preventing the booster pump from operating without its protection.

Zoé has rendered invaluable services in the acquisition of knowledge in reactor physics, in the irradiation of materials of all kinds, in the production of radioactive isotopes useful to industry and medicine (Photo 2.24). Nowadays, Zoé has become a museum that can be visited during open days or by contacting the CEA in Fontenay aux Roses (Photo 2.25). Some memories recall the importance of this reactor in the history of French nuclear power (Photo 2.26).

In both the Zoe and Vinča situations, human error is glaring: distraction in the case of Vinča, whose operators were unfortunately punished in their flesh;

1. Eau lourde. - 2. Barres d'oxyde d'uranium. − 3. Réflecteur en graphite.
− 4. Protection en béton. − 5. Colonne diffusante. − 6. Protection de la
colonne diffusante constituée par une porte en laiton cadmié. − 7. Com-
mande des barres de sécurité. − 8. Plaques de réglage. − 9. Ouverture
des canaux. − 10. Canal d'irradiation constitué par des briques mobiles
de graphite. − 11. Canal d'irradiation (blocs de béton mobiles). −
12. Chambres d'ionisation pour les mesures de puissance. − Trois chambres
d'ionisation.

Photo 2.20 Description of the pile Zoé (France, 1948) by a partially sectioned model.
1—Heavy water, 2—Uranium oxide fuel rods, 3—Graphite reflector, 4—Radiation protection in concrete, 5—Neutron diffusing column, 6—Protection of the column made
of a cadium-platted brass door, 7—Safety absorbent rods mechanism, 8—Adjustment
plates, 9—Opening of the channels, 10—Irradiation channel made of mobile graphite
bricks, 11—Irradiation channel (made of concrete mobile blocks, 12—Ionization chamber for power measurements (3 chambers)

illicit (and irresponsible!) behavior in the case of Zoe that fortunately did not
lead to an over-critical situation. If Man is not perfect, constraining procedures and well thought-out devices must force him to excellence, because any
loophole could be borrowed. Let the one who has never crossed the street
outside the limits cast the first stone! Moreover, it should be noted that hierarchical punishment is absurd, insofar as it would lead to hiding one's mistakes, making up one's behavior, looking the other way when a problem is
detected…

Photo 2.21   The Zoé pile inside its hall. The vessel is trapped in the concrete block seen in the photo. Nothing to do with the lack of biological protection of the Vinča assembly (photo CEA)

Santa Susana, a Partial Blockage of the Flow in the Core (California, 1959)

At the end of the 1950s, the US effort in nuclear technology became considerable. Numerous types of experimental reactors were developed. Some improbable concepts were tried. This is the case of the thermalized graphite reactor and cooled by liquid sodium ! Today, sodium is rarely considered except for cooling fast neutron reactors, so the Santa Susana Field reactor, also called *Sodium Reactor Experiment* (SRE), presents rather the disadvantages of the two reactor types, fast and thermalized, than their respective advantages. In any case, in these times of greed for knowledge, this reactor was implemented at the *Santa Susana Field Laboratory*, a complex of industrial research and development facilities located about 11 km northwest of Canoga Park (California, USA) and 48 km northwest of Los Angeles (Photo 2.27).

Photo 2.22 The team of designers and operators of Zoé, the first French atomic pile. Seated from left to right: A. Ertaud (head of the pile physics department), B. Goldschmidt (head of the industrial chemistry department), M. Surdin (head of the electrical construction department), L. Kowarski (technical director), F. Joliot (High Commissioner for Atomic Energy), E. Le Meur (head of the mechanical construction department), J. Guéron (head of the department of general chemistry), S. Stohr (director of the Châtillon center), R. Echard (attaché to the cabinet of the high commissioner). Standing the technicians and engineers: MM Foglia, de Laboulaye, Martin, Beaugé, Pottier, Weill, Berthelot, Rogozinsky, Valladas (photo CEA)

The reactor diverged on April 25, 1957, and produced a thermal power of 20 MWth for an electrical power of 5.8 MWe. The reactor vessel is a cylinder 180 cm in diameter by 180 cm high. The graphite is placed in the vessel in hexagonal claddings coated with a thin layer of zirconium. The sodium circulates at ambient pressure through an external main loop and is circulated by a main electromagnetic pump. This loop can evacuate a power of 20,000 kW. It is therefore a loop reactor concept, in contrast to the *Superphénix* type pool reactors. The primary sodium exchanges its heat with a secondary sodium circuit which heats water through steam generators. This secondary circuit acts as a barrier to the radioactivity of primary sodium (Fig. 2.19).

Because of all these intermediaries, the efficiency of the installation is therefore low (about 29%). A second, so-called auxiliary loop, redundant to the main loop but less powerful (1000 kW), allows the residual power to be evacuated to a small, separate secondary circuit that transfers its heat to a

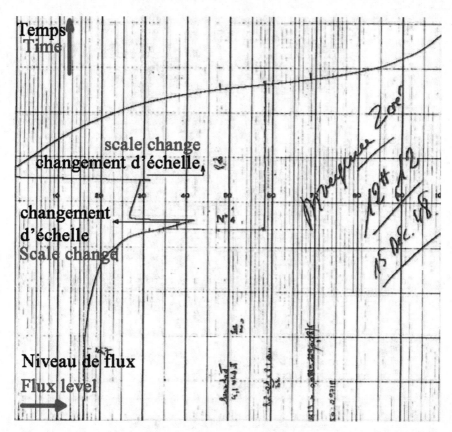

Photo 2.23 The listing of the historical divergence of Zoé on December 15, 1948, 12 h12, signed by the operator Jacky Weil. As the neutron steam flux increases strongly during the divergence, the scale of the counter has been changed so that the signal remains on the graph paper. The scale changes cause a sudden apparent decrease of the signal while the measured neutron flux level increases continuously from the bottom of the image to the top. The comments about the calibration and scale changes were added by me

forced-air heat exchanger. This circuit is only used when the reactor is shutdown. An inert nitrogen atmosphere overlaps the sodium in the vessel to avoid any risk of ignition on contact with air. The inlet temperature of the primary circuit is 260 °C, and the outlet temperature is 516 °C. In the context of the time, the reactor does not have a thick concrete Reactor Building, but a conventional building (Photo 2.28).

The fuel assemblies are suspended from cables inserted from the vessel cover. The reactor core contains 43 assemblies, consisting of 7 fuel rods (Fig. 2.20). These are cladded in stainless steel, measuring about 180 cm in height and containing low-enriched uranium (2.77%). A seal of NaK, an

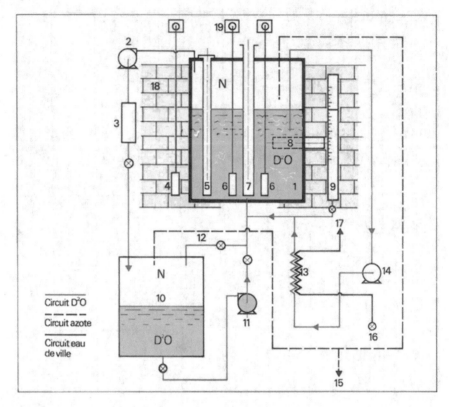

PILE ZOE à 150 kW - CIRCUITS

1 - Cuve Pile en aluminium (H = 2,35 ⌀ = 1,81 m).
2 - Pompe du circuit recombinaison.
3 - Circuit recombinaison.
4 - 2 barres de réglage tangentes à la cuve.
5 - 66 barreaux d'U. naturel.
6 - 2 barres de sécurité.
7 - Canal axial d'expérimentation
    (flux max 8.10¹¹ à 100 kW).
8 - Canaux d'irradiation : 2 tangentiels, 6 radiaux
    (2,5.10¹¹ à 100 kW).
9 - Niveau d'eau lourde (D²0) dans la cuve.
10 - Cuve réserve D²0.
11 - Pompe de remplissage en D²0 cuve pile.

12 - Vanne vidange.
13 - Echangeur de refroidissement de D²0.
14 - 2 Pompes D²0 pour circuit de refroidissement pile.
15 - Vers rampe azote et circuit régulation d'admission
    d'azote dans le circuit pile.
16 - Arrivée eau de ville.
17 - Evacuation à l'égout de l'eau de ville.
18 - Réflecteur graphite autour de la cuva,
    épaisseur : 90 cm avec passages canaux.
19 - Mécanismes des barres de sécurité et de réglage,
    sur le toit pile.

Nota : pression circuit général azote : 3 à 4 gr.

Fig. 2.18 Schematic of the heavy water systems of Zoé. As the power of the pile is much more consequent of that of Vinča (150 kW instead of 50 W), we will note the presence of an exchanger and a real circulation of the heavy water in the vessel via pump 14. The pump 11 is a booster pump for filling the vessel. 1—Reactor vessel in aluminum, 2—Recombination circuit pump, 3—Recombination circuit, 4—2 control rods tangential to the vessel, 5—66 fuel rods in natural uranium, 6—2 safety rods, 7—Axial channel for experimentation (neutron flux 8 10¹¹ n/cm²/s at 100 kW), 8—Irradiation channels (2 tangential, 6 radial, neutron flux 2.5 10¹¹ n/cm2/s at 100 kW), 9—Heavy water level in the vessel, 10—Heavy water tank, 11—Main feeding pump for vessel filling, 12—Discharge valve, 13—Heavy water cooling heat exchanger, 14—2 heavy water pumps of the cooling circuit, 15—to the nitrogen circuit, 16—Light water for cooling inlet, 17—Cooling light water discharge, 18—Graphite reflector surrounding the vessel 90 cm thick with crossing channels, 19—Safety and control rods mechanism on the roof of the pile

Photo 2.24 A photo from the early 1950s (before 1956) of the Zoe hall. Additional layers of protective bricks were added after the power was increased from 5 kW to 150 kW. Steel rods that support the fuels hang from a gantry in the foreground. One of the fans is also visible, which cools the reflector when the pile is in operation. The device for loading and unloading the samples, which are loaded into the dedicated irradiation channels, moves on a rail (photo CEA)

Photo 2.25 The Hall of Zoé has now become a museum. A plate, updated regularly and screwed on the external wall of the pile, indicates the contact dose, which has become extremely low and without danger for the visitor (photo CEA)

Photo 2.26 An amusing souvenir from Zoé: a portion of heavy water caught in a block of Plexiglas. It remains to be seen whether this heavy water has really been subjected to neutron radiation, in which case beware of tritium! One can doubt it

alloy of sodium and potassium that is liquid at room temperature[22] whose 22%(Na)-78%(K) eutectic only vaporizes at 785 °C, provides the thermal bond between the fuel and its steel cladding, an innovation that takes into account the thermal creep of the metal fuel and its significant thermal expansion while maintaining a good heat exchange.

On July 13, 1959, a blockage of some sodium channels led to the partial melting of 13 fuel assemblies. The cooling channels were blocked by products of decomposition at high temperature of the oil, tetralin[23], used to cool the seal of the primary circuit pump (Fig. 2.21). In fact, the vertical axis of the pump rotates inside of a bearing. This bearing is isolated by a technological trick. A frozen sodium film seals the pump body. This is done by cooling the bearing from the outside with liquid tetralin, a special oil that does not react with sodium, to ensure that the sodium in the film solidifies approximately in the middle of the vertical bearing. This oil seeped through the seal of the primary pumps into the primary circuit. It decomposed at about 426 °C into hydrogen, naphthalene, and carbon which, by aggregating, clogged some very narrow cooling channels in the core. When the temperature rose due to the lack of cooling, the uranium and iron in the cladding steel produced a low melting point eutectic (725 °C), which facilitated the degradation of the core.

Curiously enough, and probably because of a lack of instrumentation, the operators did not realize that the fuel had melted until the end of the test cycle on July 26, during dismantling. Eyewitnesses reported a certain amateurism

---

[22] This eutectic has a melting point of –12 °C. It has a density and viscosity close to water, but its heat capacity is lower than water and its thermal conductivity is higher. It should also be noted that this eutectic is corrosive with cadmium, antimony, lead, tin, magnesium and even silicone. The only metals with which it is satisfied are chromium, nickel, or steels...

[23] Tetralin (tetra-hydro-naphthalin $C_{10}H_{12}$) is a hydrocarbon obtained by catalytic hydrogenation of naphthalene. It is an excellent heat transfer agent that has little affinity with sodium (absence of oxygen).

Photo 2.27 The site of Santa Susana in the 60s

in the fuel management of the reactor (several attempts to restart the damaged core during cycle 14 after the accident, despite the strong temperature variations, sealing radioactive gas leaks with adhesive tape!) We can imagine today that their radioactivity detection system was ineffective or in any case, largely insufficient, insofar as the released radioactivity is estimated at about fifty curies. After 14 months of repair, the reactor restarted in September 1960 and operated without problem until 1964. The tetralin was replaced by kerosene, water being of course prohibited because of the risk of sodium-water interaction. The reactor was finally dismantled between 1976 and 1981.

Such an extraordinary reactor would certainly have deserved abundant and reliable instrumentation. This accident perfectly illustrates the risks of loss of cooling caused by a closed channel plugging, a situation that can be encountered in fast neutron reactors whose technology is similar. Instantaneous total blockage (BTI) of a hexagonal tube of a fast neutron reactor is a design accident that must be checked to ensure that it does not lead to a propagation of the melting to the six neighboring tubes.[24] The hexagonal tube is a casing that

---

[24] Detecting a BTI is difficult, especially if not all channels are instrumented. It is necessary to be able to guarantee the shutdown control rod drop if the meltdown spreads to the neighboring channels. In fast neutron reactors, the assemblies are closed (hexagonal tube) to be able to regulate their flow, and thus their power, which makes it possible to have zones of the core with variable flow rates, and thus to regulate the power shape.

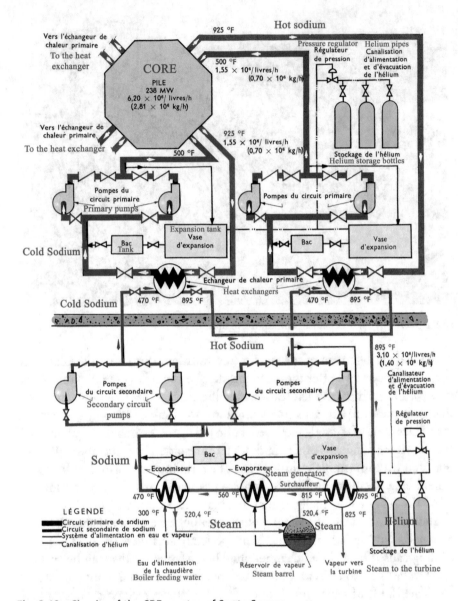

Fig. 2.19 Circuits of the SRE reactor of Santa Susana

makes the assembly sodium-tight with respect to its neighbors (no lateral flow of sodium). An accident of this type happened on October 5, 1966, in the fast neutron reactor Enrico-Fermi-1. A migrating body, namely a zircaloy plate, partially blocked two sodium cooling channels, causing the partial melting of

Photo 2.28 Operators handle the fuel loading machine for the assemblies above the Santa Susana reactor vessel cover. The plugged housings of the assemblies can be seen. The small size of the radial dimension of the core can be seen in relation to the height of a man. It can also be seen that the reactor building is only a conventional building made of superimposed cast concrete slabs

the two assemblies. This plate came from a set of six triangular plates welded in the shape of a lemon press at the entrance to the lower plenum and intended to separate the corium in the event of a core meltdown and its relocation in the cold manifold. The purpose of such a partitioning of the corium was to limit the risk of recriticality of the corium at the vessel bottom. This accident led to improvement on the design of the assembly feeder vents, which must consider the risk of clogging. Following this accident, the reactor was shut down for 4 years until 1970, only to be restarted for two more years of operation. The risk of clogging is much less acute in pressurized water reactors where the geometry of the assemblies remains open (there is no casing surrounding the fuel rods).

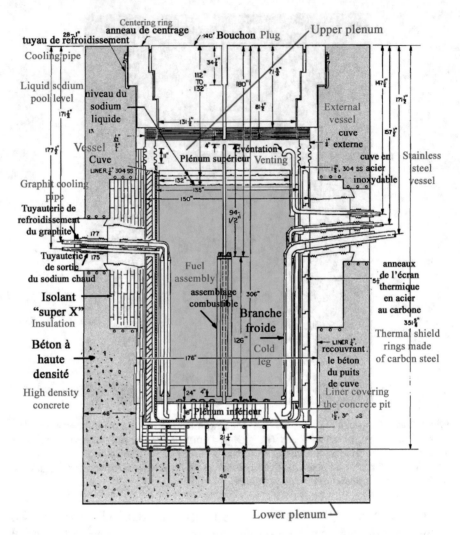

Fig. 2.20 Geometry in vertical section of the Santa Susana reactor (California, USA)

For the record around the Santa Susana case, five members of the cast of the hit family TV series "*Little House on the Prairie*" (205 episodes from 1974 to 1983) unfortunately developed cancer, four of whom, including star actor Michael Landon (1936-1991) who plays the role of the benevolent family father and main actor of the series, died of the consequences of the disease (pancreatic cancer for Landon). For a long time, the origin of these illnesses was attributed to the set, which would have been contaminated (without any proof by measurement) by radioactive fallout from the Santa Susana reactor

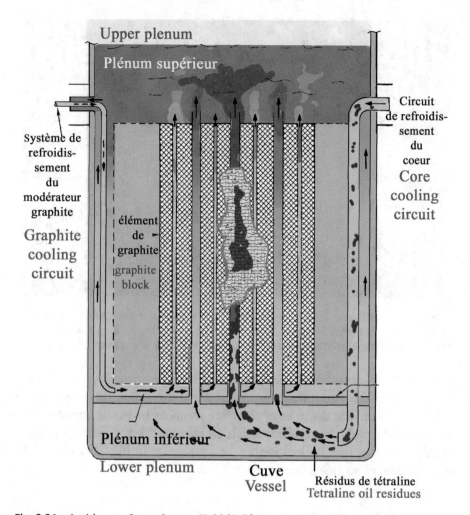

Fig. 2.21 Accident at Santa-Susana Field (California, USA, July 13, 1959). The clogging of a cooling channel due to coagulated residues of oil used for cooling and isolation of the pumps and seeping into the primary circuit, caused the melting of about 30% of the reactor core (13 fuel elements out of 43). Curiously enough, the accident was not discovered until the end of the test cycle on July 26, 1959, despite a significant release of radioactive fission gas. The radioactive releases were estimated to be about 300 times the dose released during the TMI-2 accident

meltdown in 1959. The interior sets were located at Paramount Studios in Los Angeles, but the exteriors were shot at the *Big Sky Movie Ranch*, northwest of Los Angeles. The filming location was just north of Simi Valley, while the reactor is just south. However, despite the obvious proximity of the sites (Photo 2.29), these facts can easily be explained by the risk of cancer deaths in the United States (215 per 100,000 inhabitants in 1991). It should be

Photo 2.29 Location of the filming sites of "Little House on the Prairie" and the Santa Susana reactor (adapted from a Google map). This proximity, although real, does not stand up to a factual statistical analysis of cancer risk in the population of the region

noted that no increased risk has appeared in the population of Simi Valley, which is located between the two sites, and that the radioactive releases have been very small. The controversy therefore seems to be hardly supported by scientific facts, and it is very likely that it is a sad coincidence. Michael Landon readily admitted that he had been a heavy smoker in his life and that he enjoyed alcohol outside of the play set.

Idaho Falls, a Control Rod Ejection (USA, 1961)

The accident of the *Stationary Low-Power Plant n° 1* test reactor, (SL-1) in the Idaho Falls site (Figs. 2.22 and 2.23, Photo 2.30), on January 3, 1961, was the deadliest nuclear accident on American soil. The SL-1 is an experimental boiling water reactor built by Argonne National Laboratory on order of the American army, with the objective of providing energy and heat for a possible arctic installation. The reactor is a direct cycle reactor, without secondary circuit to save space, producing steam by natural circulation, with a net power of 3 MWthermal, for an electrical production of 200 kWe.

Construction of the reactor began in 1957. The site is integrated into the National Reactor Testing Station in a desert part of Idaho, where in 1954, a

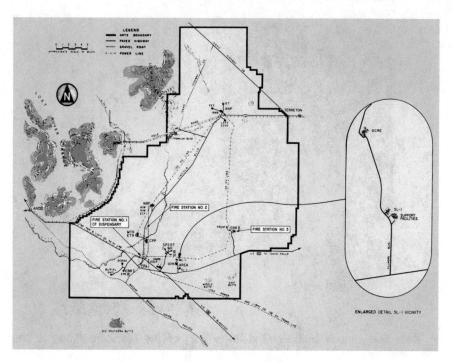

Fig. 2.22 Location of the SL-1 reactor (Idaho Falls), adapted from (Tardiff 1962 (*[Tardiff, 1962]*: A.N. Tardiff: *Some aspects of the WTR and SL-1 accidents*, Reactor safety and hazards evaluation techniques, proceedings of the symposium, Vienna, 14-18 May 1962, IAEA STI/PUB/57, pp. 43–88 (1962)))

power excursion was deliberately induced on the BORAX reactor for experimental purposes. From February 1959 on, the reactor will carry out its task of training military personnel and providing feedback on operational experience. The reactor was built on support "posts" to simulate its planned construction in a permafrost region. The lower part of the reactor is filled with gravel, also readily available in these latitudes, which serves as biological protection around the reactor vessel. The core (Figs. 2.24 and 2.25), very compact, is approximately 90 cm square, containing 40 fuel assemblies (Fig. 2.26) with 5 cruciform control blades (Fig. 2.27). The blades are made of cadmium. The assembly contains 14 kg of highly enriched uranium. The core is designed to last without reloading for at least 4 years (refueling in the arctic zone being inherently difficult). For nearly two years, the reactor operated without any particular problem. On December 23, 1960, the core was shut down for routine maintenance.

The maintenance of the neutron flux detectors began during the night of January 3, 1961. This operation requires unhooking the control rod clusters

Fig. 2.23 SL-1 reactor building: the lower part of the reactor building is filled with gravel

that are in the way of access to the detector housings. The three operators[25] on watch are preparing to lower the water level to its normal level, to put the plugs back in and to reconnect the control rod clusters mechanisms (Fig. 2.28).

At 9 h01 p.m., the procedure indicates to manually raise a few centimeters the control rod cluster to hang it up on its gripper, which was undoubtedly carried out by one of the operators (Richard Legg). It is thought that this raising was too sudden, causing a power excursion. In four milliseconds, the power of the reactor reached 20 GW (Fig. 2.29) and the violent steam explosion that followed expelled the control rods. The reactor vessel itself "jumped" in its housing by a vertical movement, dragging gravel! (Photo 2.31). The first

---

[25] John Byrnes (25 years old), Richard McKinley (22 years old) and Richard Legg (25 years old) were very young Army or Navy personnel in training on SL-1. As soon as the emergency services arrived, the level of radioactivity was such that they could not immediately enter the building. It was not until 10:30 a.m. that the rescue team discovered two mutilated bodies, one dead, the other still alive but particularly contaminated, and which was to die during its transport to the hospital. A macabre detail, it took several days to extract the third man, the shift supervisor Richard Legg, who was literally crucified like a butterfly on the ceiling of the reactor hall, directly above the reactor, by an ejected control rod (Fig. 2.23). His recovery was extremely delicate, with the help of a protective net, in part because of the fear that his fall into the gutted reactor could cause material displacement and a criticality feedback. The record of McKinley, who was buried at Arlington Military Cemetery in a lead casket and placed in concrete containment, states that his body is contaminated with long-lived isotopes and that his body cannot be moved without the explicit approval of the Atomic Energy Commission.

Photo 2.30 The SL-1 building

phase of the accident analysis was to determine whether the reactor had experienced a neutron excursion. In fact, the Hurst gold foil dosimeter located at the entrance to the control room level measured a thermal-neutron fluence of about 2×10^8 neutron/cm^2 (reaction $^{197}_{79}Au + ^1_0n \rightarrow ^{198}_{79}Au$). The analysis of the brass lighter of one of the men indicated a neutron fluence of 9.3×10^9 neutron/cm^2. This analysis was confirmed by the measurement of activity after dissolution of the gold ring of the shift supervisor. The degradation of the core was confirmed by the presence on the crew's clothing of uranium and fission products, confirmation evident by the photos taken under difficult conditions in an extremely dosing environment (authorized time of 30 s!) during the initial phase of the search for the missing third man. The blast, because of the lateral biological protections, was channeled upwards above the reactor, just where the operators were. Radioactivity was measured between 5 and 10 Gray/hour[26] near the top of the reactor.

---

[26] 1 Gray = 1 Joule/kg = 100 Rad. The gray is the official unit of energy deposit since 1986.

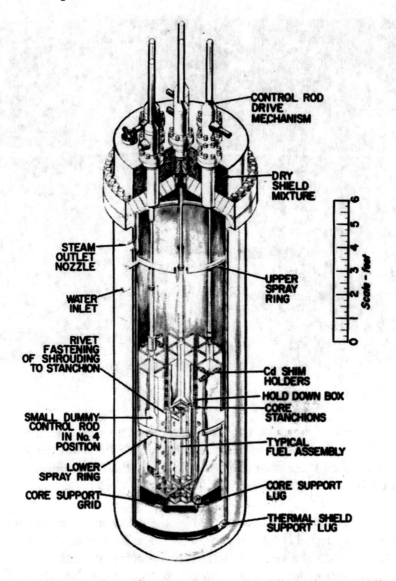

Fig. 2.24 The vessel and core of SL-1, according to (*[Tardiff, 1962]*: A.N. Tardiff: *Some aspects of the WTR and SL-1 accidents*, Reactor safety and hazards evaluation techniques, proceedings of the symposium, Vienna, 14-18 May 1962, IAEA STI/PUB/57, pp. 43–88 (1962))

The inspection of the reactor will be done by a shielded camera which showed that 4 of the control rods remained in place, and that only the central rod was violently ejected. The progressive dismantling of the reactor, first by extruding the vessel with its core still inside, showed that the central part of

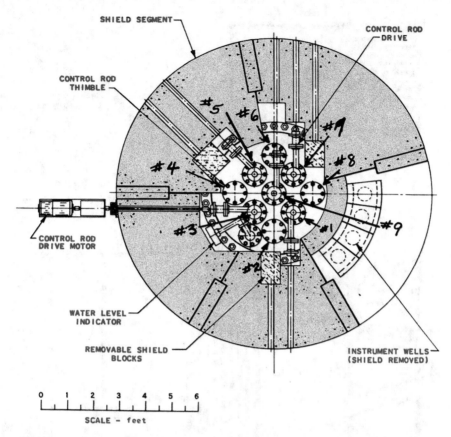

Fig. 2.25 Top view of the SL-1 core: you can see the horizontal views to the cluster control motors and the 5 control rod clusters

the reactor had melted and that 20% of the core was totally destroyed (Photos 2.32 and 2.33).

What were the causes of the accident? The desire to have a small core for easy transport led to a reduction in the number of assemblies and control rods. As a result, the plant's control rod cross was found to carry a very high anti-reactivity weight (Fig. 2.30). Moreover, a careful analysis showed that it was sometimes necessary to help the introduction of the rods mechanically, friction preventing a rod from going to the bottom thrust. In fact, personnel were accustomed to random difficulties due to friction blocking the free movement of rods. The usual procedure was to lift the rod only 4 inches to reconnect it, but there was no thrust to actually limit this lift. Based on the last critical rod position measurement, there should have been 12 inches of margin to criticality, but visual evidence showing scratches tends to prove that

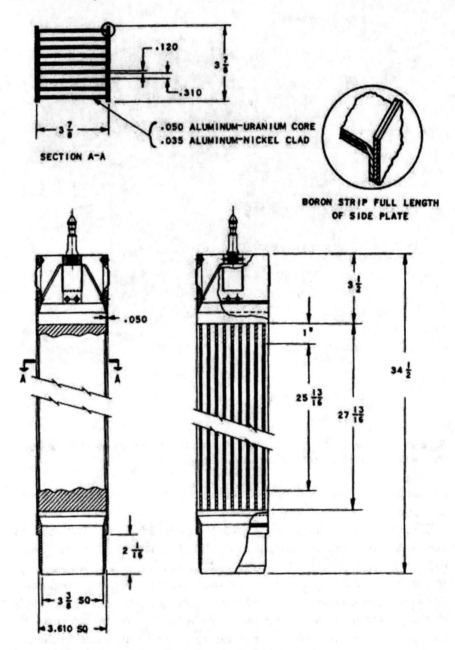

Fig. 2.26 SL-1 fuel assembly using plate-shape fuel

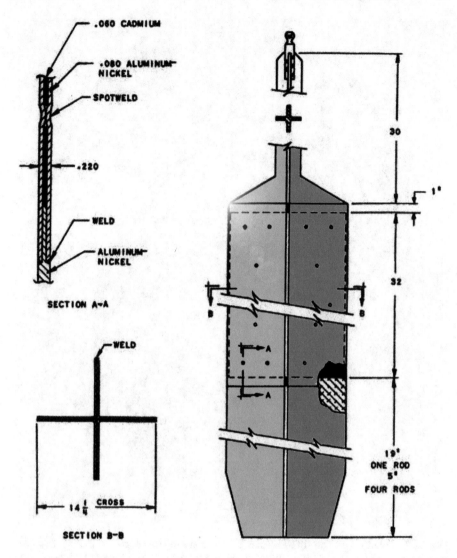

Fig. 2.27 SL-1 Control cross

the operator raised the rod at least 16 inches. It is conceivable that the operator forced the rod out of its socket and, unaware of the danger, pulled it back
too far, carried away by his own inertia? In fact, the ejected rod did pierce his
stomach, as if he had bent over for the vertical pull, as a classical position to
pull up a heavy weight.

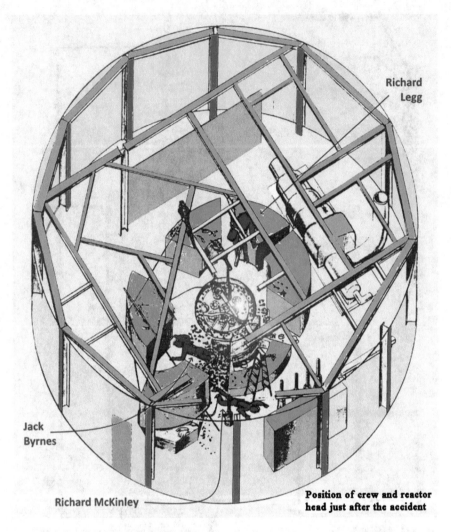

Fig. 2.28 Location of the three bodies of the crew in the reactor building. Only one was still living but died in the ambulance while transport to the hospital. The body was so radioactive that he was left in the ambulance waiting for a leaded coffin. Richard Legg was pinned under the roof of the building where he was found several hours after emergency, causing a false rumor that he was at the origin of the accident

In any case, when the explosion occurred, the reactor was destroyed by a pressure wave estimated to peak at 10,000 psi[27] with great uncertainty, i.e., about 700 bars, and a massive water hammer propelling the water at about 50 m/s, which jammed the central rod cross with a shrinkage of about 20

---

[27] 1 pound per square inch = 6894 Pascals.

puissance en GW

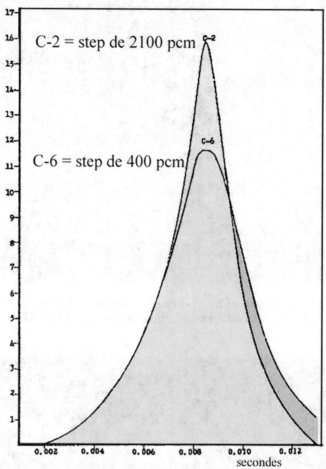

Fig. 2.29 Simulation of SL-1 response to a reactivity step reconstructing the accident (power in GigaWatt)

inches and completely twisted the substantial bolts of the vessel cover (Photo 2.34).

The steam explosion was powerful enough to lift the vessel more than 10 feet. Curiously, the vessel then fell back into its housing in roughly its original position, but pieces of the thermal shield littered the floor. The nuclear energy released by the power excursion caused by a reactivity of 2400 pcm is estimated at between 80 and 270 MJ over less than 10 ms, making it completely impossible for the operators to react. It is the dispersion of materials (loss of critical geometry) and the effect of neutron feedbacks by the

Photo 2.31 Upper part of the SL-1 reactor: one recognizes the gravel of biological protection, initially around the vessel, strewn on the ground (INEL)

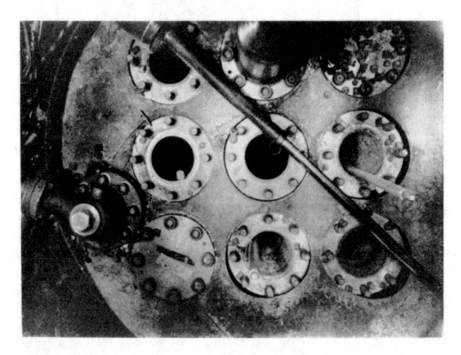

Photo 2.32 Top view of the vessel cover after the accident

Photo 2.33 View of the core during disassembly (1962, INEL)

appearance of void fraction (the average void coefficient was -0.1 pcm/cm³) which definitely stopped the accident.

What lessons can be drawn from this dramatic accident? The fact that the partial withdrawal of a single rod (see Fig. 2.31 for the mechanism) can inject reactivity greater than the fraction of delayed neutrons in the core ($\beta_{eff} \approx 700$ *pcm*) is a most serious design error, which goes against the Single Failure Criterion (SFC)), an absolute dogma of modern safety. The current evaluation criterion of the Shutdown Margin where all rods dropped except the most anti-reactive one, also called the *single rod criterion*, follows directly from the SFC because of the blocked rod penalty. On the other hand, the possibility of using a boric acid injection system in the core water could be engaged manually at the operator's discretion. If a more realistic assessment of the *Reactor Shutdown Margin* had been established at that time, it could have been increased by this easy and safe means of poisoning the core. It should be noted that the operational procedures were not very formalized by documents and that a large part of the initiative was left to the operators. On the other

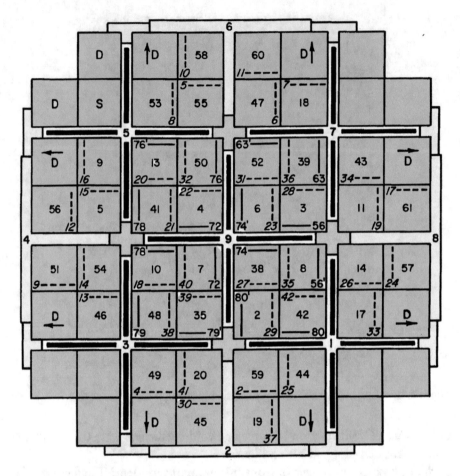

Fig. 2.30 Fuel loading pattern of the SL-1 core. It is easy to see that each control blades will weigh by its size, compared to the size of the core, an important anti-reactivity weight

hand, the weakness in the number of staff in the shift team (a shift supervisor, an experienced operator, and a junior trainee) was based on the need of the military to evaluate what was the critical size of a maintenance team, always in this idea of arctic operations.

Manual dummy rod extraction reconstructions were conducted a posteriori to assess whether a human could extract a rod quickly. They clearly showed that it was possible to extract 23 or 24 inches, i.e., the whole height of the rod, in a time short enough for the reactor period to reach 5.3 ms and to engage the reactor in an exponential power progression by an excess of 1800 pcm of reactivity. If these tests strongly confirm the hypothesis of an unfortunate displacement of the rod, the mystery remains as to the cause of this

Photo 2.34 An impressive photo of the twisted SL-1 vessel-cover bolts after the water hammer (INEL)

withdrawal: error of judgement? ... One even evoked a suicide attempt, not very credible in the context.[28] The hypothesis of an exaggerated movement to counteract friction finally remains the most credible of the hypotheses. The concept of the SL-1 reactor was finally abandoned by the American army, which had other preoccupations as US army was sinking in the Vietnam war. The only positive point is that this steam explosion showed that a reactor, by losing its geometry, cannot behave like an atomic bomb, where everything is done to contain the explosion at its very last point. The vessel and its highly radioactive core were extracted from the reactor pit with a mobile crane for repository (Photo 2.35). Nothing much remains on the site of this accident except diffuse radioactivity disseminated through the building openings (Photo 2.36).

---

[28] When asked afterwards by the scientists about their knowledge of the fact that the removal of the plant rod could cause a prompt-critical accident, they were told somewhat cheekily: *"Of course! We had discussed what we would do if Russians showed up at our radar station... We would have blown it up!"*, Anecdote reported in Susan Stacy's Proving the principle (Stacy 2000).

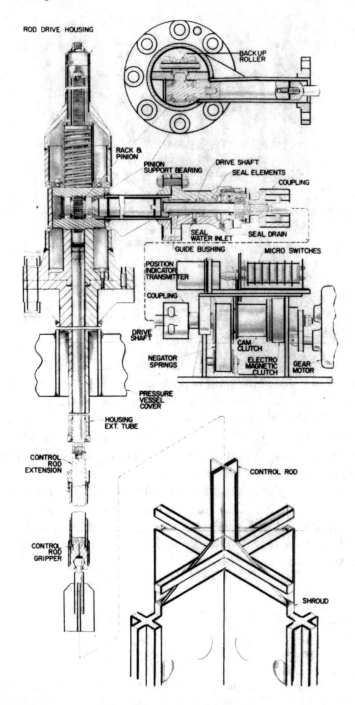

Fig. 2.31 SL-1 reactor control rod cluster mechanism

Photo 2.35   Extraction of the vessel through the reactor dome with a mobile crane. A trailer truck carries a transport cask for the vessel

Photo 2.36 The fence of the SL-1 site and a warning stone with clear depiction

Barentz Sea, the Submarine K-19 Suffers a Loss of Primary Coolant Accident (USSR, 1961)

On July 3, 1961, incredibly the same day of the SL-1 accident described just before, the Soviet submarine K-19 (Photo 2.37) was diving during the Polyarni Krug exercise in the Barents Sea when a leak appeared in the primary circuit (LOCA or Lost Of Coolant Accident) of the starboard reactor. The K-19, 114 m long, has two VM-A pressurized water reactors of 70 MW powering two turbines that propel it at 26 knots in diving. The submarine was launched on 8 April 1959. The accident happened on the first day of its service at sea. The K-19 is a nuclear-powered ballistic missile submarine of the "*Hotel*" class in the NATO breviary. It carries R-13 ballistic missiles, which can only be fired from the surface, unlike the American submarines of the time, which could fire under the water, thus in a much stealthier way.

The incident immediately raised fears of a reactor meltdown because of the residual power. The relative pressure of water in the reactor fell to zero and caused a shutdown of the primary circuit (cavitation?). A separate accident deactivated the long-range radio system, so the submarine could not warn Moscow of the damage. Although the control rods were lowered automatically by scram, the temperature of the reactor continued to rise uncontrollably, reaching 800 °C. No emergency water supply system having been foreseen

Photo 2.37 The K-19 on the surface (left) and a modern model of the submarine. The submarine was nicknamed the "*widow maker*" or "*Hiroshima*" by the sailors. Numerous accidents occurred during the construction of the ship, causing a dozen deaths even before she was commissioned (fire, asphyxiation, crushing..)

at the time of the design of the reactor (exclusion of a large LOCA break at the design!), the commander Nikolaï Zateïev ordered his sailors to fabricate a new cooling system by diverting some of the fresh water stored on board through the ventilation system, thus cooling the reactor. A team of welders took turns in the partially submerged and heavily contaminated boiler compartment to line a new water supply train while being exposed to high radiation. The primary circuit failure resulted in a large release of contaminated and highly irradiating effluent, which spread throughout the building through the ventilation system. Thanks to the courage of the crewmen, the improvised system allowed the reactor to be cooled. A conventional diesel submarine, the Soviet S-270, managed to pick up a distress signal and reached the K-19 to help.

It was said that the cause of the rupture was due to a pressure test of the primary circuit at the reception of the primary circuit. During this test, the pressure was increased to 400 bars (i.e., twice the permissible design pressure of the primary circuit) because of the omission of an operational pressure measurement system. The incident was hidden or glossed over so as not to hinder the progress of the project or for fear of possible sanctions. In any case, no measurements, even non-destructive ones (X-rays), were taken to verify the conformity of the primary circuit and the real effect of this overpressure. The Russian government will later declare to have found evidence of a defective welding (?), which is difficult to doubt from the survivors' account. The real question is to know if this failure was structural at the origin (what were the radiography control procedures at that time in the USSR?), or if this failure was induced by the overpressure of the primary circuit test. The accident of July third caused at least eight deaths by severe irradiation in the following two weeks and about 15 in the two years following. The submarine, however

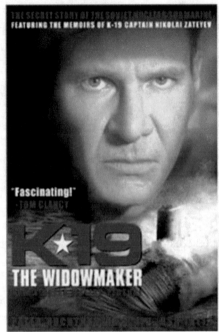

Photo 2.38 The poster of the film from the events and the real Nikolai Vladimirovich Zateyev (Николай_Владимирович_Затеев). Harrison Ford has the good part in the film. The real origin of this case remains darker

nicknamed "Hiroshima," was later rehabilitated, and the reactor compartment was cut out and replaced by a new one during operations which lasted two years. The irradiated compartment was simply drowned in the Kara Sea. Dose reconstructions give figures that are chilling: 54 Sieverts for Lieutenant Boris Kochilov, commander of the group of welders and in the front line (he will die "only" on July 10, 1961, despite this appalling dose that should have killed him before), and doses higher than 10 Sv, the lethal dose, for many sailors who died also in July 1961 despite bone marrow transplants whose technique was initiated by Professor Mathé on the irradiated scientists of Vinča in 1958. The K-19 was deleted from the soviet naval fleet lists on April 19, 1990. The American film K-19 by Kathryn Bigelow (2002) with Harrison Ford and Liam Neeson (Photo 2.38) relates these dramatic events by giving the good role to the commander of the ship.

Fermi-1, Fuel Melting in a Sodium Cooled Reactor (1966, Michigan, USA)

The Fermi 1 reactor, located in Michigan, underwent a partial core meltdown on 1966, October 5. This reactor was a prototype breeder reactor, launched in the 1960s while France was developing its own fast neutron reactor known as *Rapsodie*. Fermi-1 was the world's first commercial fast neutron reactor, followed two other experimental reactors of the same type built in USA, EBR-I and EBR-II (Photos 2.39 and 2.40).

The site also houses a 1170 MWe Fermi-2 boiling water reactor (Photo 2.41). In 2016, NRC renewed the operating license of Fermi-2 for an additional 20 years through March 2045. In July 2019, NRC ordered an inspection assessing the potential of degraded paint inside a portion of the reactor possibly to impede safety systems. The inspection aimed to assess if the degraded paint inside a portion of the reactor containment at the plant could

Photo 2.39 Fermi-1 plant in the sixties

Photo 2.40   Fermi-1 pant in the seventies. New buildings appear

affect certain safety systems in accident conditions. The move followed the US NRC's recent engineering inspection, which reported a deprivation in the paint inside the torus, a donut-shaped component of the reactor containment located below the reactor vessel (Fig. 2.32). The torus, which is filled with water, is designed to absorb energy from the reactor or supply water to safety systems during an accident. According to the regulator, the torus' loose paint chips could potentially impede the water flow to safety-related equipment at the time of an accident (adapted from Kondapuram Rani from NS Energy).

Unlike thermal neutron reactors, which must slow down the neutrons by collision on a moderator atom in order to favor fissions (water in the case of PWRs, graphite in the case of French UNGGs or Soviet RBMKs), fast neutron reactors do not use a moderator. Indeed, while a thermal reactor uses the fissile property of Uranium 235, fast neutron reactor will rather target the fertile property of Uranium 238. This isotope being non-fissile to thermal neutron, the aim of fast breeder reactors is to keep the neutron spectrum as fast as possible to benefit from the fission of "even" isotopes such as ^{238}U or ^{240}Pu, called fertile isotopes. Thus, in a reactor containing fertile as well as fissile material, the ratio between the consumed and the fertile nuclei converted

Photo 2.41 The Fermi-2 air coolers. The Fermi 2 reactor is a 1170 MWe boiling water reactor (BWR), commissioned in 1985, initially for 40 years (2025) and built by General Electric which is owned by DTE Energy and operated by its subsidiary Detroit Edison

into fissile material is called the conversion factor. For example, if for every ten U-235 nuclei, eight ^{238}U nuclei are converted into ^{239}Pu, the conversion factor is 0.8. In a thermal reactor, by definition, the conversion factor is less than 1 as thermal reactor mainly consume initial fissile nuclei. In a CANDU (thermal) reactor, the conversion factor is about 0.8. In all types of reactor, a neutron capture by ^{238}U induces rapidly a production of ^{239}Pu through the equation:

$$^{238}_{92}U + {}^{1}_{0}n \rightarrow {}^{239}_{92}U \rightarrow {}^{239}_{93}Np \rightarrow {}^{239}_{94}Pu$$

The half-life of ^{239}U is 23 min as the half-life of ^{239}Np is 2.3 days. Since ^{239}Pu is a fissile isotope under thermal and fast neutron, ^{238}U is therefore a large source for future fission of ^{239}Pu in the fuel. It is also possible to design a reactor with a conversion factor greater than 1. This is called a breeder reactor, i.e., a reactor that produces more fissile material than it consumes, because the harden neutron spectrum favors conversion. The extra-plutonium produced

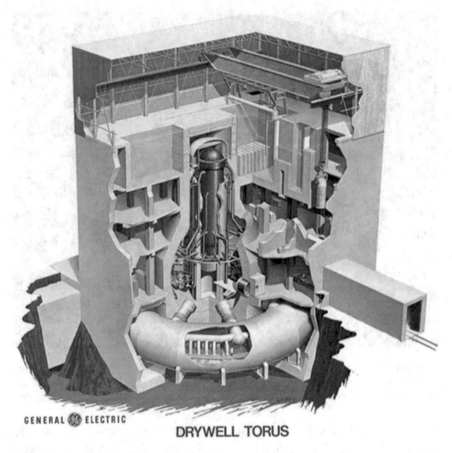

GENERAL ⚙ ELECTRIC

DRYWELL TORUS

Fig. 2.32 The Fermi-2 Containment building showing the drywell torus on which the defective paint was found

in breeders can efficiently fuel thermal reactors or even other breeders. It is necessary to understand that breeders should not moderate the neutrons like PWRs to avoid thermalization of neutrons, thus it is necessary to avoid water as coolant. Hence the use of sodium or lead-bismuth as coolant. The liquids are bad moderators as their constitutive isotopes are much heavier than hydrogen.

Historically, the first nuclear reactor to produce electricity was a fast neutron reactor. On December 20, 1951, in Idaho, the National Reactor Testing Station (NRTS) commissioned the Experimental Breeder Reactor-1 (EBR-1), which produced enough electricity to power four 25 W bulb-shape light. The first real power reactors based on the principle of breeder principle were the British Dounreay Fast Reactor at Caithness in the north of Scotland, the EBR-2 in Idaho and the "*Detroit Edison Fermi 1*" reactor near Detroit in Michigan, named after Enrico Fermi. One of the fundamental difficulties of

FBRs is that 400 times more neutrons are needed than a thermal neutron reactor to produce fission. A greater density of neutrons is then required. As we said previously, it is essential that these neutrons should be slowed down as little as possible. The core of an FBR must therefore contain no moderator and a minimum of other structural materials to avoid parasitic captures.

Nevertheless, the use of sodium has its drawbacks. As every chemist knows, sodium reacts strongly with water. Therefore, even if sodium is not under pressure, the open surfaces in a Liquid Metal Fast Breeder Reactor (LMFBR) are covered by an inert gas such as argon. Unlike gases or water, sodium is opaque, which makes remote inspection of the reactor particularly difficult. Of course, sodium must not be brought below its melting point, i.e., 97.5 °C once in the circuit, otherwise, it will solidify. Moreover, even if sodium does not easily absorb fast neutrons, when this capture takes place, sodium 23 self-activates in sodium 24, which is an intense gamma radiation emitter. Its half-life is only 15 h, but it has a high activity. Therefore, the sodium primary circuit must be completely surrounded by the biological barrier of the core. Practically, this requires a second sodium circuit with a heat exchanger inside the biological barrier. This secondary sodium circuit, protected against neutrons, carries the heat from the primary circuit through the barrier to the second heat exchanger, where steam is generated. The steam generators, in which the sodium and water are only separated by thin tube walls must be manufactured according to very strict standards. The steam generators are among the most troublesome features of LMFBRs.

Prototype breeder reactors, such as Fermi-1, use oxides as fuel because of the high melting point. The fuel is not only made of uranium oxide, but a mixture of uranium and plutonium oxides. The uranium is not enriched. The low thermal conductivity of the mixture of oxides requires the fabrication of individual small-diameter stainless steel cladding (less than 6 mm in diameter). The core is enclosed in an open tank of liquid sodium (hot plenum), itself inside a larger tank of molten sodium (cold plenum). The sodium passes through the fuel elements and then flows through an intermediate heat exchanger, where it transfers its heat to the secondary circuit, containing non-radioactive sodium. Three intricate circuits are required. The first sodium circuit is completely inside the vessel and cools the active core. The secondary sodium circuit transports the heat outside the biological barrier, where a third water circuit produces steam via steam generators in order to run the turbine. The second circuit avoids sending water directly in the vessel in order to eliminate any risk of massive sodium fire. No pipes or other penetrations enter the primary circuit below the sodium level, thus avoiding reasonably the risk of loss of primary coolant.

The Fermi 1 reactor (Photos 2.42 and 2.43) was the first commercial LMFB (liquid metal cooled breeder) reactor, the only one and the last one built in the

Photo 2.42 A rather threatening image from the Fermi-1 presentation brochure from 1970, where we see that public communication is in its infancy

Photo 2.43 Aerial plant view (from the brochure)

United States. Fermi-1 followed the experimental breeder reactors EBR-I and EBR-II. Note that EBR-1 experienced a severe core meltdown in November 1955 (see Chap. 1). The Fermi I reactor was a prototype fast reactor designed with a power of 430 MWth or 94 MWe, but the first core A was limited to 200 MWth or 69 MWe. The nuclear plant was located on the western shore of Lake Erie, at Laguna Beach in Monroe County (State of Michigan), halfway between the city of Detroit and the city of Toledo (Ohio). Construction of Fermi-1 began on January 8, 1956, achieved criticality in 1963, produced its first MWe in December 1965. Fermi-1 was connected to the grid on May 8, 1966. Its design was the work of two subsidiaries of the consortium Atomic Power Development Associates (APDA), Dow Chemical and Detroit Edison. The latter was the operator. Fermi-1 was built by Power Reactor Developments Company (PRDC).

FBRs include two different designs. For both designs, two heat exchangers are used in order to isolate the primary sodium from the rest of the installation. In the pool design, an intermediate sodium–sodium heat exchanger is drowned under sodium inside the vessel, so that the primary sodium never leaves the core vessel (Fig. 2.33). On the other hand, in the loop-design reactor, the primary hot sodium flows out of the vessel to feed the intermediate heat exchanger (IHX) and the IHX is out of the vessel (Fig. 2.34). Pool-design is universally considered safer than loop-design as no active sodium leaves the vessel, but pool-design requires immerged electromagnetic pumps to push the primary sodium in the active core.

Fermi-1 (Fig. 2.35) was of the loop-design, i.e., the intermediate heat exchanger is located outside the primary circuit. The core is located inside the reactor building as well as the Intermediate Heat Exchanger for evident radioprotection purpose (Figs. 2.36 and 2.37). The reactor building has an easy-recognizable hemispherical dome. The vessel is hidden by a top roof that can be lifted with the polar crane, allowing the transfer cask car to approach for refueling. The car moves on rail to reach the rotating plug closing the vessel and protecting operators from sodium vapor. Steam generators stand right in the auxiliaries building next to the reactor building and provide steam to the turbine located in the turbine hall. The technology of electromagnetic pumps was not mature at that time, hence a loop-design. The primary circuit was filled with sodium in December 1960. The reactor reached criticality for the first time in August 1963. The reactor operated at very low power during the first years of operation. Once the authorization to operate at high power was received, the power tests began immediately in December 1965.

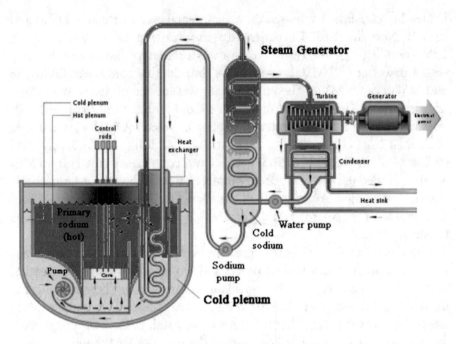

Fig. 2.33 Pool-type FBR. The primary sodium is highly radioactive but stays in the vessel, transferring its heat to the secondary non-radioactive sodium through an immerged intermediate heat exchanger. Doing that way requires that electromagnetic primary pumps lay inside the vessel. The secondary sodium heats water in the steam generator. The steam runs the turbine

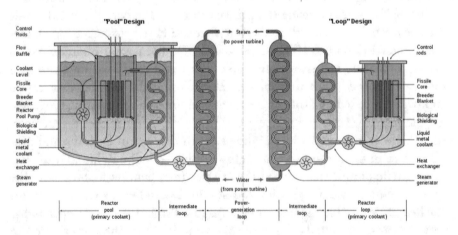

Fig. 2.34 Pool-design FBR (left) compared to loop-design FBR (right). The Intermediate Heat Exchanger (IHX) is located inside the vessel in the pool-design and outside the vessel in the loop-design. Primary pumps of the loop-design are classical volumetric pumps

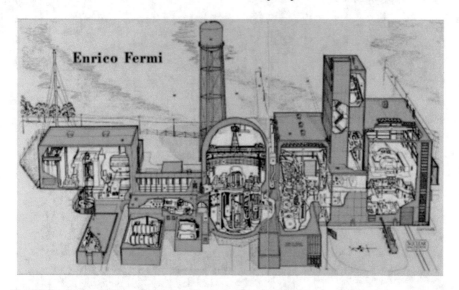

Fig. 2.35 General plan of Fermi-1 (from the brochure)

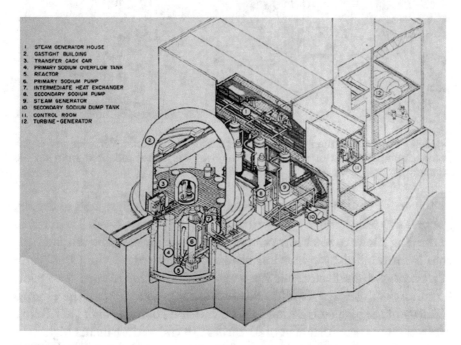

1. STEAM GENERATOR HOUSE
2. GASTIGHT BUILDING
3. TRANSFER CASK CAR
4. PRIMARY SODIUM OVERFLOW TANK
5. REACTOR
6. PRIMARY SODIUM PUMP
7. INTERMEDIATE HEAT EXCHANGER
8. SECONDARY SODIUM PUMP
9. STEAM GENERATOR
10. SECONDARY SODIUM DUMP TANK
11. CONTROL ROOM
12. TURBINE-GENERATOR

Fig. 2.36 Period plan of the reactor building and the nuclear auxiliaries

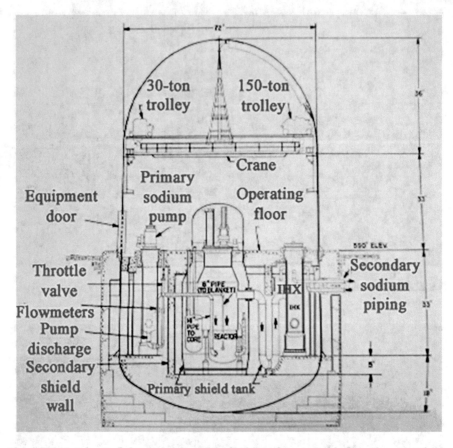

Fig. 2.37 Axial cut of the Fermi-1 reactor building from an old plan. The size of the primary pumps is impressive, almost as high as the vessel (adapted from an old blueprint)

The active core is placed in a stainless-steel vessel (Figs. 2.38 and 2.39, Photo 2.44) sealed atop by a rotating shield plug (Photo 2.45). The rotation of the plug allows to reach any position in the core for refueling purpose. The core is surrounded by cylindrical blankets of depleted uranium (99.7% in ^{238}U) acting as a reflector, but the main interest of this blanket is to allow breeding as all neutron captures in the blanket produce ^{239}Pu. The external diameter of the blanket is 80 inches (203 cm) and 70 inches high (177.8 cm). The active core has a diameter of 31 inches (78 cm) and 31 inches high.

The sodium enters the "*cold plenum*," crosses the core from bottom (288 °C) to top (427 °C) under 8.27 bars before ending up in the "*hot plenum*." The sodium then exits the vessel to feed the intermediate heat exchanger. The average temperature of the coolant inside the core was about 310 °C. Before the

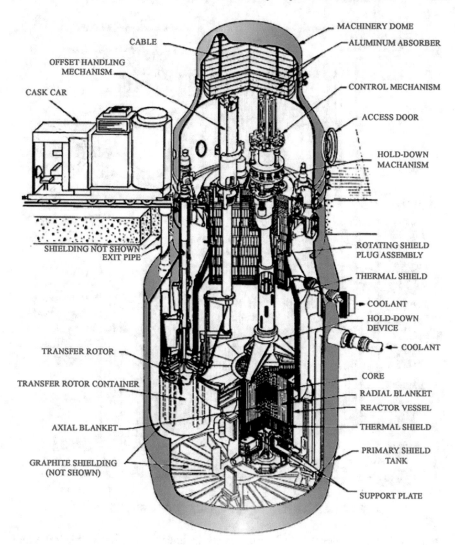

Fig. 2.38 Cut view of the Fermi-1 vessel

incident, the core used an uranium metal fuel surrounded by a zirconium cladding. The core contained 105 assemblies in total. One can also see the different protections surrounding the core, in particular the thermal barrier and the different fertile blankets in depleted uranium (Fig. 2.40). The fuel elements were 4 mm in external diameter, 79 cm high, and were arranged in a square lattice of 2.646 inches (Fig. 2.41).

Fermi-1 was subject to a partial meltdown of the core during its power up on October 5, 1966. A few weeks before the incident, abnormally high

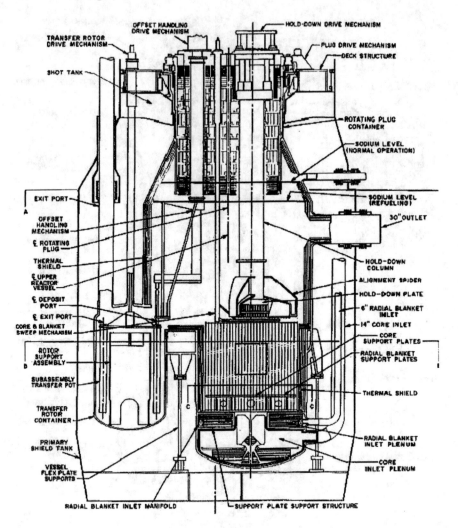

Fig. 2.39 Complete view of the Fermi-1 vessel, including the subassembly intermediate storage barrel on the left of the core. This ingenious device allows a direct transfer from the barrel to the core without lifting the core rotating plug. This technic was also used on the *Superphenix* French FBR (1200 MWe) but could not be longer used as the barrel rapidly appeared to leak. The French barrel was made of carbon steel instead of stainless steel due to economic reasons

temperatures at the level of assemblies were observed by the thermocouples located at the outlet of the coolant. In June, these temperatures were 20 to 25% above normal, then in August from 40 to 47% above normal. The operations were then carried out at low power. In addition, another thermocouple placed above one of the assemblies indicated an abnormally low temperature

Photo 2.44 Entrance of the vessel in the Fermi-1 containment through the hatch. Surrounding buildings are still to raise

of the coolant compared to normal conditions. The reading of the thermocouples seemed suspicious. To verify the validity of the measure, the reactor was shut down, and the assemblies indicating abnormally high temperatures were reinstalled under different thermocouples, using the intermediate storage barrel. This was done to determine if the anomaly was due to thermocouples or the fuel assemblies themselves. It was then observed that the location of the abnormally high-temperature data varied at each start-up but was not correlated with the movements of the fuel assemblies. The reactor operated without incident at a power of 100 MWth. Then on October 5, 1966, the power was lowered at 67 MWth, then again at 20 MWth at 3 a.m. The operator then observed a control signal indicating an erratic neutron population. The problem had already occurred sometime earlier and was thought to be an electrical fluctuation in the control system. The reactor was placed under manual control, and when the fluctuations disappeared, the control system was returned to automatic control. At 3:05 a.m., the power was restored to 27 MWth, the error signals were again observed. It was noticed later that the control rods were pulled out further than the normal location. Two of the assemblies showed temperatures of 370 °C. This was much higher

Photo 2.45 Top view of the vessel cover and the rotating plug

than the ordinary range of coolant temperature around 315 °C. At 3:09 a.m., the alarms in the upper part of the building began to sound indicating a damage of the fuel and a release of radioactive fission products. These alarms were triggered by ionizing radiation. The building was immediately isolated. A radiation emergency plan was declared. The power increase of the reactor was stopped at 31 MWth, followed by a power reduction. At 3:20 a.m., the reactor power was reduced to 26 MWth and it was manually shutdown. Over a one-year period, many assemblies were moved in order to perform examinations. The cause of the incident was considered *"relatively trivial."* The

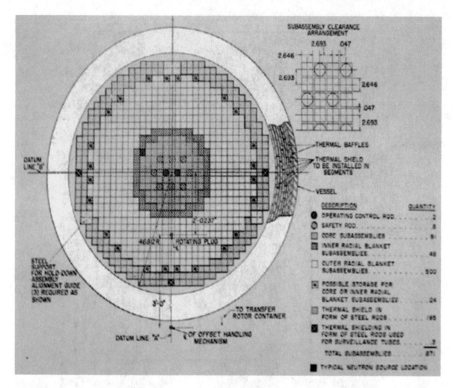

Fig. 2.40   Map of the core

examinations revealed that the partial meltdown occurred in two adjacent fuel assemblies. A third (possibly fourth) assembly was deformed but without internal damage. On September 11, 1967, a piece of debris defined as a "*foreign body*" was found stuck to the inlet plenum. The investigation showed later that this debris was a zirconium plate of the "*melt down section liner*" located originally in the vessel bottom of the reactor. In fact, at the bottom of the core, six zircaloy plates were welded to the inlet of the plenum to solve the problem of re-criticality in the event of core meltdown. This recommendation was made by the Advisory Committee on Reactor Safeguard, the safety authority, in 1959, in order to "divide" the mass of the falling molten core by spreading on a conical corium-flow divider, and to ensure the subcriticality of the resulting corium fragments. Two of its six plates broke off, became loose and one caused a blockage of the coolant flow at the inlet of the assemblies. The zircaloy plate was carried by the coolant and moved between different positions until it partially or completely obstructed the inlet channels of the various assemblies during the shutdown and restart phases of the reactor. The coolant flow would have been limited, through the assemblies concerned, to

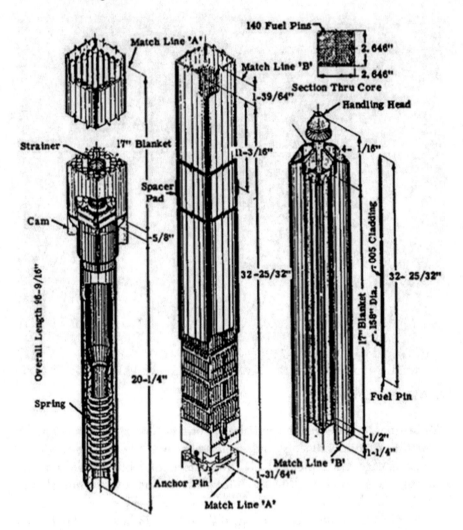

Fig. 2.41 Fermi-1 fuel bundle and control blades

between 3% to 30%. Figures 2.42 and 2.43 show in detail the zone between the vessel bottom and the fuel core. Figure 2.44 shows where were found the loose plates (top view).

On January 30 and 31, 1968, the Joint Committee on Atomic Energy led a congress to shed light on the partial meltdown of the core. On February 2, PRDC formally notified that the foreign body that had blocked the neutron flux was from one of the six triangular-shaped pieces of metal installed at the bottom of the vessel, in this case the zirconium plates. To repair the damage inside the core, it was not necessary to use a tool specially designed to operate

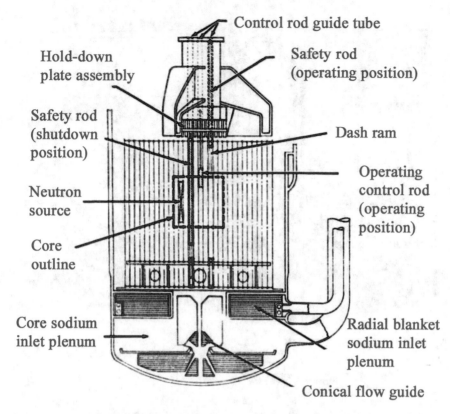

Fig. 2.42 Simplified sketch of the vessel internals

in highly irradiating environment. The metal fuel core was then removed and replaced by a uranium oxide core. On December 16, 1968, the last of the six zirconium plates was removed from the entrance of the plenum. On February 10, 1970, PRDC was authorized to restart the reactor. The reactor then restarted on July 18, 1970, four years after the incident. Its restart, initially planned for May, was delayed until July, due to a sodium fire. Its operation ceased in 1972. There were no injuries, and no radioactivity was released into the environment. The safety system revealed an activity of 10,000 Curies due to fission products released inside the coolant. Fortunately, the damage did not spread to adjacent assemblies, and the accident did not reach the worst-case scenarios. The incident put in light the problems associated with the coolant blockage. The zircaloy plates that were found were not intended to be included in the original design of the reactor, and plates were unfortunately chosen for budgetary reasons.

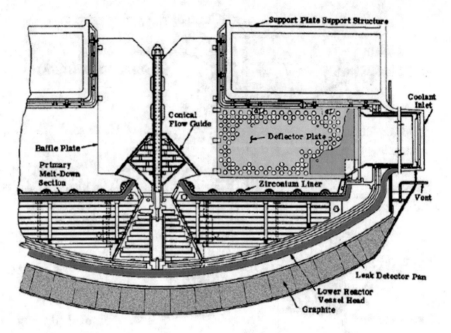

Fig. 2.43 Vessel lower internals

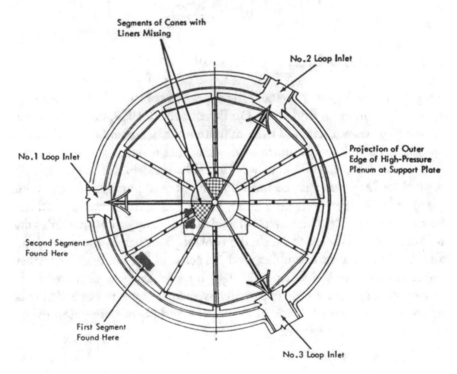

Fig. 2.44 Location of the loose fragments of the corium-divider cone liners spread in the lower plenum

Since the Fermi-1 incident, the fuel assembly inlet nozzles, at the level of the tubes, include multiple by-pass passages for the coolant that make impossible the total blockage by external debris. Research and testing of internal and external blockages have been undertaken to quantify and understand the damage caused by such mechanisms. The scenario of internal or external blockage of an assembly has been taken into account in reactor design. Their design must follow different recommendations at the level of:

1. The design of the assemblies: provide several coolant orifices inside the assemblies.
2. Inlet plenum design: ensure coolant flow distribution and the feeding of the assemblies.
3. Instrumentation design: detection by multiple thermocouples, delayed neutron detectors, gas "beacon" detector.
4. The design of the fuel handling equipment (technology of spent fuel concrete casks).

In addition, other lessons were learned, notably concerning the parts of the reactors likely to be damaged by vibrations. They must be carefully designed and monitored to prevent possible release of debris. In some countries, the scenario of fuel assembly jamming was adopted as a "design basis accident" for fast neutron reactors.

This partial meltdown of the core, even if it led to the shutdown of the Fermi-1 reactor for nearly 4 years, did not prevent the reactor from returning to service in 1972. Fermi-1 was decommissioned in 1975, once plans for a new facility were proposed. However, this project was abandoned in the 1980s by the American authorities, who preferred to develop the treatment of spent fuel. This incident nevertheless allowed the reactor type to benefit from important feedback concerning blockage issue. Thus, many improvements of the reactor design have been proposed, in order to avoid a single piece of debris blocking the flow of coolant. LMFBRs technology is still considered in the GEN-IV program for future reactors.

The obvious conclusion is that the technology of laterally closed fuel tubes, which is widely used in fast neutron reactors, is a design weakness that has led to numerous accidents of partial or total plugging of one or more assemblies. This problem is also recurrent in the case of concepts with separate cooling channels like CANDUs or RBMKs. In the case of FBRs, these housings impose the desired rigidity of closely spaced rod bundles and allow the liquid metal flow in each channel to be adjusted for better flattening of the power sheet. The idea of introducing cooling feedthroughs into the hexagonal tubes of the power FBRs is being considered, but this will likely result in a loss of stiffness and even the risk of vibration induced by fluid jets. In nuclear technology, one must be wary of jumping to conclusions and making risky arrangements. The message given in the American Nuclear Society about Fermi-1 ("*New age for Nuclear Power*," Photo 2.48) was rather obscured by the partial meltdown of the reactor core.

The episode at Fermi 1 in Frenchtown Township was the subject of the 1975 anti-nuclear book (Photos 2.46 and 2.47), "*We Almost Lost Detroit*," written by John Fuller, and the inspiration for a song of the same name by the late Gil Scott Heron. The song was more recently covered by the Detroit indie band JR JR that still regularly plays the tune before audiences around the world (source https://eu.freep.com/story/news/ local/michigan/) (Photo 2.48).

Photo 2.46 The provocative "*non-novel*" book "*We almost lost Detroit*" written by John Fuller, and the provocative answer from Detroit Edison "*We did not almost lose Detroit*". Believe it or not? Judge by yourself. At least, left cover is much more commercial!

Photo 2.47 Alan Lenhoff and Jan Prezzato talk about *"death toll"* in the Ann Arbor Sun, June 17, 1976. The text, also introducing the accident of SL-1 in Idaho, is a strong support to Fuller's book. The photomontage shows a radioactive death cloud spreading over the city of Detroit. Real facts are rather short

Chapelcross, a Carbon Dioxide Flow Blockage and Magnesium Cladding Melting (1967, Great Britain)

Great Britain chose very early on to develop a national reactor type based on the choice of the first plutonium reactor at Windscale for the military program. This reactor type was named Magnox (for Magnesium Non-OXidizing). This type of reactor uses natural uranium metal, moderated by graphite and cooled with CO_2 carbon dioxide. Construction was spread out from 1953 to 1971, the first being the Calder Hall reactor, which was inaugurated in 1956 by Queen Elizabeth II (Photo 2.49), and the last one was the Wylfa plant. Since the end of 2015, they are all out of service because they have been replaced by an evolution of the concept: the AGR (Advanced Gas-cooled Reactor). They used this famous *"stainless magnesium,"* which was a magnesium-aluminum alloy used to clad the fuel in this type of reactor. The uranium must not be in direct contact with the carbon dioxide that cools it to limit contamination by radioactive fission products.

Chapelcross is a site near Annan in the province of Dumfries and Galloway in southwest Scotland (Fig. 2.45), with 4 Magnoxes of 48 MWe each (182 MWth, thus a modest efficiency of 23%), which began construction in 1955

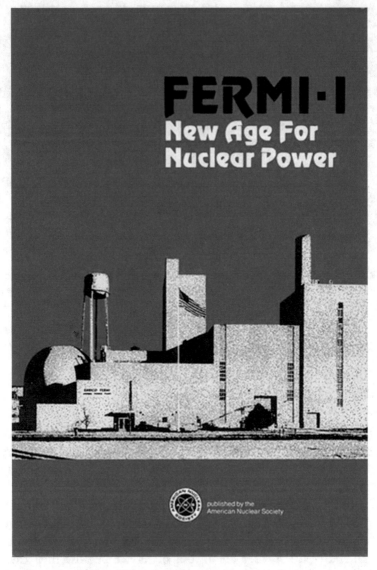

Photo 2.48 A much more scientific text about Fermi-1 from the American Nuclear Society, unfortunately, obscured by the partial meltdown event of 1966

and was coupled to the electric grid in 1959. Chapelcross is the sister reactor of Calder Hall (Photo 2.50). Note that the fuel at Chapelcross is slightly different from that at Calder Hall (more on this later). Another difference from Calder Hall is that Chapelcross has its own cooling pool for spent fuel, which is necessitated by its distance from Windscale. The core consists of 1696 channels in a 203 mm pitch lattice for loading fuel rods. The active core has a diameter of 9.45 m and a height of 6.4 m. 112 channels allow the insertion of

Photo 2.49 Queen Elizabeth II inaugurated the Calder Hall reactor on October 17, 1956. This view shows the fuel loading machine (in white in the background), which runs on rails. A sealed part of the machine is connected to the primary circuit by means of conduits. The circular fuel caps and the asperities in counter-relief that allow unscrewing, are clearly visible on the ground

Photo 2.50 The four reactors of Chapelcross are particularly standardized except for a color inversion of the SGs siding (photo NDA)

Magnox stations worldwide

13 réacteurs Magnox ont été construits principallement en Grande-Bretagne, mais aussi en Italie et au Japon. Le combustible a été fabriqué en Angleterre à Springfields

Fig. 2.45 Implementation of the Magnox reactor type in the world. Chapel Cross, the site of the 1967 accident, is located in Scotland (green dot on the map)

control rods. The core is cooled by a carbon dioxide CO_2 at a pressure of 7 bars that enters at 140 °C and exits heated to 336 °C with a flow rate of 891 kg/s. The fuel is in the form of natural uranium metal in the form of six 1016 mm cast rods, each clad with Magnox-C, a magnesium alloy. The six rods are embedded in an assembly. The total mass of metallic uranium is 120 tons. The vessel containing the fuel and the graphite moderator blocks is made of steel (Low-term A-kill mild steel) with a cylindrical shape closed by two domes (internal diameter 11.28 m, external 21.3 m, thickness of the dome plates 51 mm). The vessel itself is contained in a concrete compartment (called a caisson or leak tight housing) (Fig. 2.46). The caisson is contained in a conventional building with windows (Fig. 2.47). Four centrifugal fans (total 5.4 MWe) move the coolant CO_2 (Fig. 2.48) through 4 primary loops (Fig. 2.49). Four steam generators act as heat exchangers CO_2/H_2O and produce 180 tons/hour of steam at 310 °C under 14 bars (Fig. 2.50). This steam turns a turbine and two alternators of 23 MWe each at 3,000 rpm.

The reactor was loaded via a fuel loading machine that runs on rails placed in a lattice on the loading face (Fig. 2.51). The machine moves and positions itself in front of a pressure tube. The mast containing the fuel assembly to be loaded is connected to the "tulip-shape socket" which closes the pressure tube and unscrews the cylindrical socket-shaped plug, which seals the primary circuit. In doing so, carbon dioxide from the primary circuit enters the sealed mast, and a new fuel can be lowered without any gas leak, or an existing fuel can be removed. The aim of this machine is to avoid the spread of radioactive gases on the service desk. The operations are automated as much as possible to avoid human presence and radiation protection risks.

The fuel rods of an assembly are in the form of a uranium metal "rod" entirely cladded with a magnesium alloy. The cladding alloy has a complex shape with cooling blades to improve heat exchange with the carbon dioxide that flows past the outer face of the cladding (Fig. 2.52).

On May 11, 1967, a fuel element in a channel of Reactor 2, which was loaded with fuel elements being evaluated for the future AGR commercial reactor program, suffered a partial carbon dioxide flow blockage, attributed to the presence of graphite debris. Due to the burn-up and high temperature, the graphite that makes up the core of the reactor was deformed until a portion broke off. As it fell, the graphite debris became blocked in a loading channel, partially obstructing it and greatly disturbing the flow of CO_2. With the cooling impaired, the fuel elements in the channel rapidly rose in temperature, until the cladding failed, and fission products escaped, contaminating the core. In practice, the fuel overheated and the Magnox cladding, made of a magnesium-aluminum alloy, failed, leading to a deposit of contamination

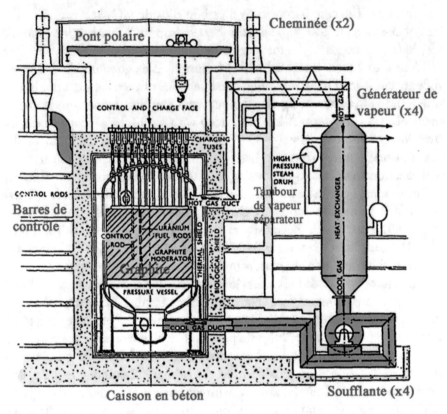

Fig. 2.46 Axial section of the reactor (Calder Hall and Chapelcross). We notice that the steam generators (imposing!) are located outside the building. They are surrounded by a siding in grating, which allows a visual inspection but has no real function of external protection

in part of the core. It should be noted that the use of magnesium cladding limits the temperature of the carbon dioxide to 360 °C because the melting points of magnesium (650 °C) and aluminum (660 °C) are relatively low. The triggering of a radioactivity alarm in the carbon dioxide caused the shutdown of the reactor. After depressurizing the reactor's primary circuit (initially to 7 bars), cameras were inserted into the offending fuel channel, which revealed a blockage caused by the melting of a Magnox cladding. As the gas could not circulate in the channel, the magnesium ignited, starting a fire in the reactor. Due to the deformation of the fuel, the personnel could not adopt the usual method of clearing the channel with the top unloading machine, which

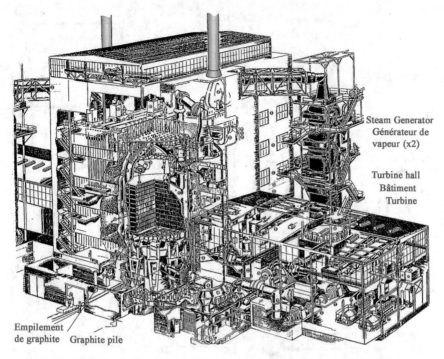

Steam Generator
Générateur de
vapeur (x2)

Turbine hall
Bâtiment
Turbine

Empilement
de graphite   Graphite pile

Fig. 2.47   General plan of the Calder Hall and Chapelcross reactors

required the development of a special technique. After the accident was stopped, volunteers had to enter the concrete caisson and steel vessel to install a containment tray under the failed fuel elements to ensure that no graphite and/or cladding fragments fell further down into the reactor. Senior staff members, Dr. J.H. Martin, Director of Health, Physics and Safety, Mr. David MacDougall, the Assistant Superintendent and Mr. L Clark volunteered to enter the reactor to perform the operation (Photo 2.51). This maneuver had been carefully rehearsed and closely timed on a realistic model before being attempted. A special isolation hatch was built around a duct leading to a cooling air access, where Dr. Martin and Mr. MacDougall entered dressed in PVC suits. Their only link to the outside was an air hose, a radio communication cord, and a nylon lifeline attached to their waists. There were no windows in the caisson, and it was like a black pitch with no lighting, an extremely stressful situation. Upon exiting, the men commented: *"The suits were cumbersome,*

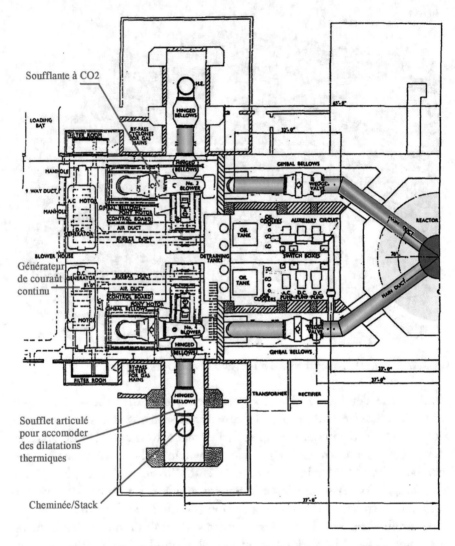

Fig. 2.48 Diagram of a half-circuit for the circulation of carbon dioxide. The same half-circuit is found symmetrically on the other side of the reactor

and we were sweating a lot... we never thought about the danger... we felt a bit like men on the Moon". Mr. Clark was the sentry at the entrance to the duct, while Dr. Martin went in to take final radioactivity measurements, and Mr. MacDougall made sure the recovery bin was in place. (adapted from Sarah Harper's article https://www.coldwarscotland.co.uk/chapelcross-almost-chernobyl-chapelcross-fire-1967).

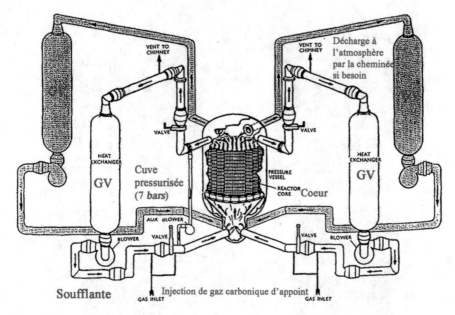

Fig. 2.49 Diagram of a 4-loop Chapelcross primary circuit

The deformation of graphite under temperature and irradiation is a known phenomenon, as well as the release of energy (Wigner effect) when the graphite is irradiated at too low temperature (see the importance of the Wigner effect in the Windscale accident described above). After the release of Wigner energy in the Windscale No. 1 pile in 1952 and after the severe Windscale fire in October 1957, the UKAEA Energy Commission recommended that graphite temperatures in the C.E.G.B. power reactors be increased to a level that would prevent the accumulation of stored Wigner energy over the lifespan of the reactors. This recommendation resulted in the inclusion of removable graphite sleeves in the fuel channels of the Chapelcross No. 2, 3, and 4 reactors. The inclusion of sleeves creates a thermal gradient that increases the temperature of the graphite in the core of the moderator blocks, particularly in the lower part of the core. The graphite sleeve completely surrounded the rod and its cladding without being in contact with the cladding to allow cooling carbon dioxide to pass through. The temperature increase, due to the insulation of the fuel channels from the cooling gas, comes from the gamma and neutron heat up. In this way, the graphite moderator is intrinsically self-healing (permanent thermal annealing) with respect to the Wigner Effect. However, it should be noted that the graphite and gas temperatures are higher in this situation, compared to a more open geometry. The choice of magnesium as cladding material may seem curious when one knows the risks of

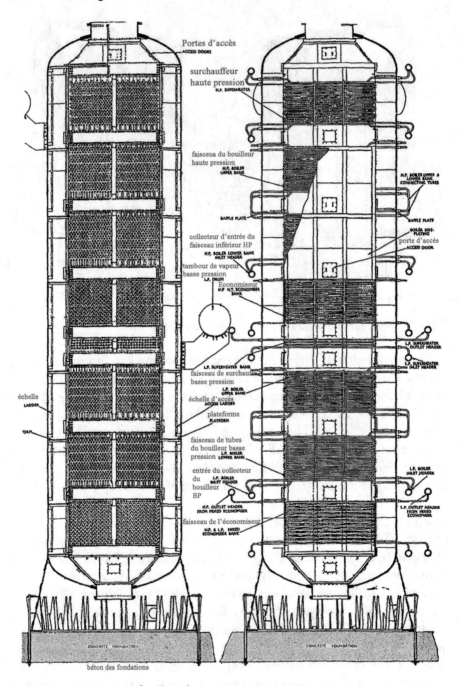

Fig. 2.50 Axial section of a Chapelcross steam generator

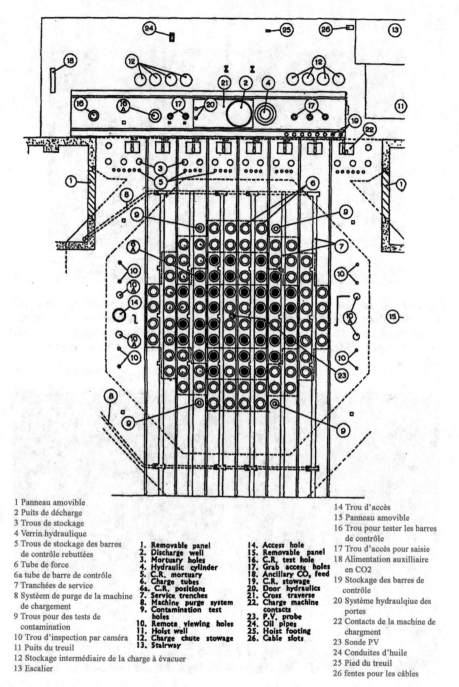

1 Panneau amovible
2 Puits de décharge
3 Trous de stockage
4 Verrin hydraulique
5 Trous de stockage des barres
  de contrôle rebuttées
6 Tube de force
6a tube de barre de contrôle
7 Tranchées de service
8 Systèem de purge de la machine
  de chargement
9 Trous pour des tests de
  contamination
10 Trou d'inspection par caméra
11 Puits du treuil
12 Stockage intermédiaire de la charge à évacuer
13 Escalier

1. Removable panel
2. Discharge well
3. Mortuary holes
4. Hydraulic cylinder
5. C.R. mortuary
6. Charge tubes
6a. C.R. positions
7. Service trenches
8. Machine purge system
9. Contamination test
   holes
10. Remote viewing holes
11. Hoist well
12. Charge chute stowage
13. Stairway

14. Access hole
15. Removable panel
16. C.R. test hole
17. Grab access holes
18. Ancillary CO₂ feed
19. C.R. stowage
20. Door hydraulics
21. Cross traverse
22. Charge machine
    contacts
23. P.V. probe
24. Oil pipes
25. Hoist footing
26. Cable slots

14 Trou d'accès
15 Panneau amovible
16 Trou pour tester les barres
   de contrôle
17 Trou d'accès pour saisie
18 Alimentation auxilliaire
   en CO2
19 Stockage des barres de
   contrôle
20 Système hydraulqiue des
   portes
22 Contacts de la machine de
   chargment
23 Sonde PV
24 Conduites d'huile
25 Pied du treuil
26 fentes pour les câbles

Fig. 2.51 View of the loading face. The loading face is located above the reactor. On this face moves the fuel loading machine that can load and unload the reactor while it is running. This so-called "continuous" loading makes this type of reactor particularly interesting to produce weapons-grade plutonium, insofar as it is possible to unload spent fuel with an isotopic percentage of plutonium 239 greater than 95%. The more the fuel is irradiated, the more the plutonium will contain even isotopes of plutonium 240 and 242) that are detrimental to the optimal functioning of a nuclear weapon

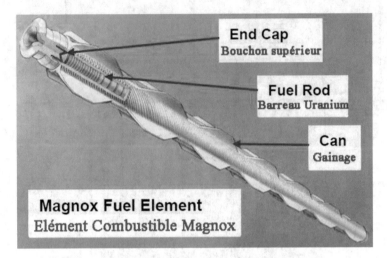

Fig. 2.52 A Magnox fuel element. These fuels have evolved constantly and significantly over the course of Magnox reactors, and it is safe to say that no two reactors in this reactor type will have the same fuels. The external blades are designed to increase the exchange surface, as well as the surface threading, in order to improve heat exchange

Photo 2.51 Three volunteer managers: From left to right: J.H. Martin, David MacDougall, and L. Clark are about to enter the carbon dioxide depressurized caisson at Chapelcross-2. They are equipped with an externally ventilated (non-self-contained) "*Mururoa*" (Mururoa is a French island in the pacific where atomic bombs were tested underground in real conditions. Hence the name of the special suits used by the personnel) type suit (photo *The Annadale Observer*)

oxidation in the presence of water steam and even in pure carbon dioxide, but its neutron capture cross-section for thermalized neutrons (or slow neutrons) is very low. Magnesium, which is lighter than aluminum, is easily spun into tube form, has a good weldability, and does not produce low melting point eutectics on contact with uranium metal. It is also abundant and rather inexpensive. However, the significant coarsening of the magnesium grain on heating is detrimental to the mechanical strength of the fuel elements. It is therefore necessary to consider the addition of other metals to refine the grain, such as zirconium, zinc or, as in this case, aluminum. Its low melting point (650 °C) does not allow the heating of the CO_2 at more than 400 °C in nominal conditions (leaving some margin for incidental situations). But the particularly closed geometry of the fuel element in its graphite channel reinforces the risk of obstruction of the channel in case of channel degradation and debris formation. Unfortunately, magnesium oxidizes in the presence of water steam (see the Lucens accident below), and oxidizes even in the presence of pure CO_2 according to the chemical reactions, driven by the Gibbs free energies ΔG:

$$CO_2 + Mg => MgO + CO \quad \Delta G = -74\,000 \text{ calories}$$

$$CO + Mg => MgO + C \quad \Delta G = -86\,000 \text{ calories}$$

$$CO_2 + 2Mg => 2MgO + C \quad \Delta G = -160\,000 \text{ calories}$$

$$CO + Mg0 => CO_3Mg \quad \Delta G = -800 \text{ calories}$$

And considering traces of air (oxygen + nitrogen) in the CO_2:

$$O_2 + 2Mg => 2MgO \quad \Delta G = -254\,000 \text{ calories}$$

$$N_2 + 3Mg => Mg_3N_2 \quad \Delta G = -77\,700 \text{ calories}$$

It appears that all these reactions are possible under atmospheric pressure from 400 °C to 500 °C except the production of carbonate CO_3Mg, which dissociates upon 420 °C. These reactions are even favored by the pressurization (7 bars) of carbon dioxide. Let us note also that the radiolysis carbon dioxide produces free oxygen, which is very corrosive. In air, magnesium is very flammable. A pure magnesium ribbon will catch fire with a simple match. In powder form, it becomes explosive by increasing the contact surface. The flame produced is strong white and very incandescent. Hence its use in the flashes of the early days of photography, causing many accidents. For these same properties, it is used today in the manufacture of some pyrotechnic materials. Magnesium produces, during what can be called a "fire," a significant amount of heat which is communicated to the surrounding structures and self-sustaining the fire. Magnesium, like aluminum, has a strong reducer character and therefore oxidizes easily, releasing a lot of heat. It is a good fuel in the air. Magnesium is such a good reducer that it can burn in the CO_2, which is usually not a good oxidizer. This reaction produces white powdery magnesia MgO and black solid carbon. Thus, the accident scenario is refined. The loss of geometry of the channel, partially or totally blocked by carbon debris of the graphite channel deformed in temperature, led to a strong temperature increase of the carbon dioxide. This temperature excursion led to an accelerated oxidation of the magnesium-aluminum cladding, which "burned" and melted, releasing radioactive fission products into the primary circuit.

The reactor was restarted in 1969 after successful two-year cleanup operations, and it was the last reactor of its type to cease operation in February 2004.

Siloé, Melting of Fuel Plates (Grenoble, France, 1967)

Siloé was a French nuclear research reactor, of the pile pool light water type with an open core (but covered with water) and a thermal power of 15 MWth at start-up. Built from August 1961 by Indatom on the scientific polygon of Grenoble on the site of the CEA near the city (Photo 2.52), The reactor diverged on March 18, 1963, at 11:15 p.m., one year after the first Grenoble reactor, Mélusine, and eight years before the high-flux reactor (RHF) at the nearby *Institut Laue-Langevin*. A model of Siloé was tested in April 1962 in the Mélusine reactor. The core is composed of plates made of an alloy of uranium metal and aluminum highly enriched to more than 90% in ^{235}U. The core is reflected by beryllium plates. The building housing the reactor-pool consists of a vertical cylindrical concrete body 25 m high and 27 m in diameter. The two floors of this hall are equipped with experimental areas. A hot cell for treating the fuel completes the equipment. The primary function of Siloé was the doping of silicon crystals and the production of medical radioisotopes by neutron irradiation. In 1968, the power was increased to 30 MWth, then to 35 MWth in 1974, and even to 40 MWth in order to carry out tests on materials requiring large neutron fluxes. The core of Siloé is submerged by a large quantity of water which acts as a biological protection against radiation, so that one can operate freely on the reactor service desk during operation (Photos 2.53 and 2.54).

Photo 2.52 Siloé building inside the CEA site. The reactor is located in the white cylindrical building

The civil engineering of the SILOE reactor pool had two compartments: a compartment called the "*main pool*," with a volume of 213 m³, containing the reactor core at the bottom, and a compartment called the "*working pool*," with a larger volume of 322 m³, arranged in a horseshoe shape around the main pool (Fig. 2.53). This pool was used for storage of experimental devices and safe interventions (out of neutron flux) on them. The faces of the pools in contact with the water are tiled in the manner of a real pool, except that the tiles are joined with Araldite glue, which is more waterproof than conventional joints. Nevertheless, this tiling was to pose recurrent sealing problems from 1965 to 1970, until a leak at the foot of the "*stool*" supporting the core was visually detected by air bubbles (Fig. 2.55). A stainless steel plate joined by a synthetic foam was then affixed. The degradation of this foam under irradiation necessitated replacement with a rubber gasket held in place by lead. The problem was only permanently solved in 1972, and the leakage was estimated at 1500 m³ of tritiated water, which must have polluted the groundwater table. To finish with the leaks, a hole of 5 mm in diameter was detected in 1986 in a corner at the bottom of the pool, which had to be repaired. The press echoed these leaks, which caused a stir in the population. Beginning in

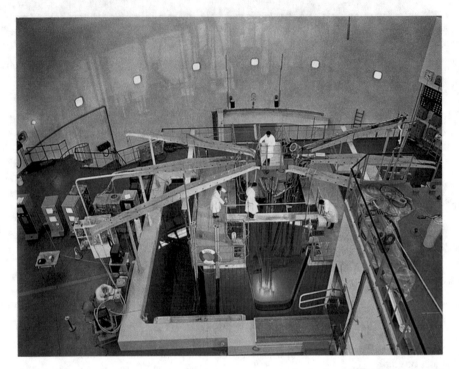

Photo 2.53 The Siloé core during operation. One can see the intense bluish Cerenkov-Mallet radiation that characterizes the core in operation. The operators can handle the core on the bridge (photo *Association des Retraités de l'Institut Laue-Langevin*)

1987, the CEA undertook important work to bring the pile up to standard: a stainless steel casing 3 mm thick was installed on the walls of the main pool, a vessel known as the "BORAX vessel" 7 mm thick, the vessel itself is placed on a stainless steel plate 20 mm thick mounted on shock-absorbing paraseismic pads. The aim of these modifications is to guarantee the strength and tightness of the reactor pool in the event of an explosive accident or earthquake.

A neutron equipment was added later to perform experiments using neutrons (neutron diffraction in powders and crystals, polarized neutron diffraction...). Siloé was thus equipped with neutron exit channels that do not look directly at the core, but at the beryllium reflector which adjoined one of the four sides of the core. These channels are like trenches that promote the leakage of neutrons to the detectors. At the beginning, there were only two radial channels and two devices (DN1 and DN3). After the closure of Mélusine (1988), a tangential channel was added that looked at the beryllium wall through the plant.

Photo 2.54   The core of Siloé (photo CEA)

On November 7, 1967, during a power increase to 42.3 MWth carried out as part of authorized tests in preparation for an increase in the nominal power of the reactor to 30 MWth, a partial fusion of six fuel plates belonging to a fuel element occurred. This test aimed to characterize the phenomenon known as "flow redistribution." When the power of 42.3 MWth was reached, a sudden decrease in power without any pilot action of about 7 MWth in one second was observed, followed by a slower decrease until it stabilized, 20 s later, at 20 MWth. The reactor was manually shutdown 26 s later, by dropping the two-reactor safety elements. A rapid increase in radiation dose rates was then observed (detected by a submerged dose rate measurement chamber, up to 1000 rad/h (10 Gray/h), and on another measurement chamber, located above the pool water, up to a value of 220 rad/h (2.2 Gray/h). This detection led to the evacuation of the reactor building and annex buildings, and the use of iodine traps in the emergency ventilation system. These high values indicate a loss of fuel tightness and the release of radioactive fission products.

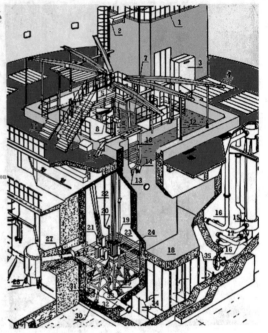

Ecorché de la pile Siloé

1: Toit de la cellule chaude
2: Poutre roulante
3: Porte de la cellule chaude
4: Araignée support de boucle
5: Plongeoir de manoeuvre
6: Gaine de reprise du dessus piscine
7: Grand batardeau
8: Mécanisme des barres
9: Mécanisme chambre de fission
10: Petit batardeau
11: Passerelle de dessus piscine
12: Piscine de travail auxiliaire
13: Chatière pour passage de boucle active
14: Passage de boucle active
15: Echangeur
16: Tuyauterie primaire et secondaire
17: Pompe primaire
18: Plancher piscine de travail et toit du bac de désactivation
19: Coeur: éléments standards et de contrôle, béryllium,
    boites à eau et bouchons
20: Décrochement de boucle expérimentale
21: Bloc de plomb des chambres de mesure
22: Perches des chambres de mesure
23: Chambre à fission
24: Stockage combustible irradié
25: Grille de coeur et bouchons
26: Chaussette amovible
27: Canal pour sortie de faisceau
28: Protection du monochromateur
29: Tabouret du bloc réacteur
30: Tabouret des chambres
31: Clapet de convection naturelle
32: Tuyauterie d'aspiration coeur- bac de désactivation
33: Diffuseur retour circuit primaire
34: Chicanes du bac de désactivation
35: Porte de visite du bac de désactivation

Fig. 2.53 Sketch of the Siloé pile. 1- Roof of the hot cell, 2- Rolling beam, 3- Hot cell door, 4- Loop support spider, 5- Maneuvering plunger, 6- Pool top recovery duct, 7- Large cofferdam, 8- Control rod mechanism, 9- Fission chamber mechanism, 10- Small cofferdam, 11- Pool top walkway, 12- Auxiliary work pool, 13- Cat flap for active loop passage, 14- Active loop passage, 15- Heat exchanger, 16- primary and secondary duct, 17- Primary pump, 18- Working floor and roof of the deactivation tank, 19- Core: standard fuel and control, beryllium, water boxes and plug, 20- Disconnection of the experimental loop, 21- Lead block of the measuring chambers, 22- Measuring chamber poles, 23- Fission chamber, 24- Spent fuel storage, 25- Core grid and plug, 26- Movable sock, 27- Channel for beam output, 28- Monochromator protection, 29- Reactor block stool, Chambers tools, 31- Natural convection flap, 32- Suction pipe to core /deactivation tank, 33- Primary circuit return diffuser, 34- Deactivation tank baffle, 35- Access door to the deactivation tank

During dismantling, once the atmosphere in the hall had returned to an acceptable ambient dose, 187 g of uranium-aluminum alloy (enriched to 93% in uranium 235) melted, corresponding to a mass of 36.8 g of uranium 235, 18 g of which were released into the primary circuit (Photo 2.55). The complement was found in the form of corium relocated at the foot of the control element. Fortunately, the fuel element concerned had a low fission rate (FIMA (Fission per Invested Metal Atom) burn-up of 4%). Nevertheless, 2000 curies of rare gas activity (74 10^{12} Bq) would have been released. The activity of noble gases decreases rapidly with time. Fuel entrained by the

Photo 2.55 Degradation of the fuel element of Siloé in 1967 (view from the bottom). The uranium-aluminum fuel plates are simply encased in a casing (left). The plates are partially perforated by large tears after extraction from the casing (right)

cooling water was subsequently found in the deactivation tanks until 1971 (adapted and commented on from the book *Retour d'expérience des réacteurs de recherche français*, IRSN, available on the web site https://www.irsn.fr/FR/ Larecherche/publications-documentation/collection-ouvrages-IRSN/ Documents/RR-ReacteursRecherche_web-NB-Chapitre-10.pdf). The real cause of the accident is unclear. One immediately thinks of a local overpower, but it was not the hottest part of the core that melted. The boiling temperature of the water at the bottom of the pool (1.5 bars of pressure) is 128 °C, whereas the hot point (in water) did not exceed 115 °C at a nominal flow rate. However, in order to melt, the fuel element had to rise to at least 660 °C, the melting temperature of aluminum (the melting temperature of uranium metal is still higher than 1132 °C). In visual terms, the fuel plates do not seem particularly oxidized (Photo 2.55), which indicates a rapid degradation by drying (local exceeding of the critical heat flux). Given the thinness of the metal plates (more easily cooled than an oxide plate due to higher thermal conductivity), a significant flow loss was required in the incriminated channel. The margin should have been even greater for the incriminated element, which was not at the hot spot. The appearance of corium clearly indicates the drying of the fuel wall. As paint flakes from the structures overhanging the reactor were found several times in the pool water, a postulated scenario was imagined that a (partial?) blockage of a water channel had occurred, which would have reduced the flow. The corium would then have spread into the other adjacent channels. The principle of assembly of the fuel plates means that each cooling channel is isolated from the others. It would have been judicious to provide openings in the design to allow fluid to communicate between the channels, which would have made the assembly more resistant to instantaneous total blockage, a phenomenon much feared in fast neutron reactors, but which probably happened in Siloé.

Corrective measures were taken following the accident. Painted sheet metal was replaced with unpainted stainless-steel structures (no more risk of chipping/flaking). The facility's emergency exhaust system was redundant, and air and water sampling systems were installed for use from outside the building. In addition, a control system for activating the purification circuit was installed in the control room (thus without having to travel near the main pool). The Siloé accident, although perfectly documented by the IRSN, remains a largely unknown accident in France.

Gradually, the CEA is going to shut down all the nuclear activities of the CEA in Grenoble, the site being considered too close to the Grenoble suburb. Siloé was shut down on December 23, 1997. Decree no. 2005-78 of January 26, 2005, authorized the CEA *"to proceed with the shutdown and dismantling of the basic nuclear facility no. 20, called the Siloé reactor, in the municipality of Grenoble."* As of September 2012, the Siloé reactor was dismantled in its entirety because repeated leaks of radioactive water made it problematic to maintain the building in its current state (tritium level). Cardem and Eurovia-Vinci were involved in dismantling the reactor (Photo 2.56). The work began with the cleaning of the elements before the asbestos removal, then the demolition of the four different buildings consisting of a technical wing, office buildings, a "crown" building, and the reactor could proceed. On January 22, 2013, an excavator equipped with a large arm (Photo 2.56) began the demolition of the 27-meter-high reactor, whose raft had previously been made safe by installing a watertight protection system and a 3.80-meter backfill above it. This raft was then demolished under containment in April 2013 to level the ground. The final decommissioning of the facility was pronounced on January 8, 2015, by the ASN.

Photo 2.56 Dismantling of the Siloé reactor building (CEA-Grenoble) (photo Cardem and Eurovia-Vinci)

Lucens, Partial Fusion of a Fuel Rod (Switzerland, 1969)

In the early 1960s, Switzerland wanted to create a 100% Swiss nuclear reactor type, like its French neighbor with its Natural Uranium-Graphite-Gas (UNGG) reactor type, using heavy water as a moderator and carbon dioxide gas as a coolant. This technological choice allows the use of natural uranium that the Swiss hope to find in large quantities in the Alps, a hope that will prove to be disappointed. The heavy water could be produced using electricity from its many hydraulic dams. The basic idea being to use unenriched natural uranium (0.711% in uranium 235), the neutron balance is then very tight, and one can only use graphite or heavy water as a moderator. The parasitic capture of neutrons by light water is too great to hope to operate with natural uranium. The Canadians have developed a national reactor type (CANDU) where the moderator and the coolant are made of heavy water. If the moderator is stored in a calandria without flow in which the fuels are bathed, the coolant circulates in a primary circuit to exchange its heat with a secondary circuit of light water that will produce steam to turn a turbine. This circulation inevitably leads to fluid losses that are very costly when it comes to heavy water. The Canadians will realize this by replacing the heavy water in the coolant with light water. The Swiss retained the idea of a heavy water calandria but, like the French and the British, decided to cool the reactor with carbon dioxide.[29] CO_2 in pressure. This design will lead to the realization of the

---

[29] Carbon dioxide is an inorganic compound with the chemical formula CO_2. Its form is gaseous above -78.48 °C. Carbon dioxide is produced by human respiration, but also by the combustion of carbonaceous materials (graphite, wood...) in the air. The air outside nowadays contains about 0.04% CO_2. From a given concentration in the air, this gas is dangerous or even deadly for humans because of the risk of asphyxiation or acidosis, although CO_2 is not chemically toxic strictly speaking. Unlike vegetable plants, mammals cannot dissociate the CO_2 molecule to use oxygen. The exposure limit is 3% over a period of 15 minutes. Beyond that, the health effects are all the more serious as the CO_2 content increases. Thus, at 2% CO_2 in the air, the respiratory amplitude increases. At 4% (i.e., 100 times the concentration in the atmosphere), the respiratory frequency accelerates. At 10%, visual disturbances, tremors and sweating may occur. At 15%, there is a sudden loss of consciousness, and at 25%, respiratory shutdown leads to death. Regardless of the risks to humans, carbon dioxide has significant industrial benefits. Thanks to its low impact on the environment compared to other refrigerants currently used (up to 3800 times less impact on the environment than the Hydrofluorocarbons HFCs initially used in the refrigeration industry), carbon dioxide is used in the industry because it has no impact on the ozone (it has an ODP (Ozone Depletion Potential) index of 0) knowing for example that the R404A fluid has a GWP of 3800, and little direct impact on the greenhouse effect (GWP index (Global Warming Potential) of 1) knowing also that the R12 fluid has a GWP of 10,900. It is non-flammable (used as a gas in fire extinguishers), non-corrosive, compatible with all materials and non-chemically toxic. However, it forms acids when mixed with water, which suggests extensive dehydration of the circuits before commissioning. To become a good heat transfer medium (modest thermal capacity at 20 °C of 840 J/kg/K against 5193 J/kg/K for helium, even air has a better thermal capacity of 1004 J/kg/K but air feeds fires), it is necessary to increase its pressure inducing well protected circuits.

Fig. 2.54 The Diorit reactor at the Paul Sherrer Institute in Würelingen. Moderated and cooled with heavy water and with a power of 30 MWth. It was started in 1960 and operated until 1977. It is the first reactor of Swiss design and construction. Diorit is the direct ancestor of the Lucens reactor

underground reactor of Lucens (pronounced *Lussan*) started in 1968, with an electrical power of 6 MWe.

Lucens is not the first reactor installed in Switzerland. In August 1960, researchers put into operation the "*Diorit*" reactor » (Fig. 2.54, Photos 2.57, 2.58 and 2.59) on the site of the Federal Institute for Reactor Research (IFR) in Würelingen. This facility was used to test various reactor concepts and to produce radioactive isotopes for medicine, research and industry. In 1957, the

Photo 2.57 View of the top of the Diorit reactor in its vessel pit at the level of the pressure tubes-(1959) (photo IFR)

Würelingen site saw the commissioning of another experimental reactor, the *"Saphir"*. The "Saphir" reactor was not a development of Swiss industry, but had been acquired from the USA. The reactor used light water as a moderator. Rudolf W. Meier, a renowned Swiss physicist and president of the Swiss Federal Commission for Energy Research (CORE) from 1986 to 1991, summarized the importance of the Diorit project for the development of nuclear know-how in Switzerland as follows: *"The construction of Diorit took place at a time when there was a strong desire to develop the use of nuclear energy in Switzerland from our own industrial power plant. Entrepreneur Walter Boveri and ETH Professor Paul Scherrer were very determined to support this concept, and their credibility in business and scientific circles provided the necessary additional weight in political circles."* Werner Zünti and other pioneers (Fritz Alder, Walter Hälg, Paul Schmid) launched the P34 project for an experimental-skill acquisition reactor. The preliminary project was completed in 1955. The foundation of the Reactor AG with Rudolf Sontheim as director ensured the financing and construction of a completely new institute in Würelingen within five years, until the first commissioning of Diorit (1960–1977) (source https://www.nuklearforum.ch/fr/actualites/e-bulletin/il-y-40-ans-diorit-etait-mis-en-service). This nuclear reactor was operated by the EIR from 1960 to 1977. The moderator was heavy water (D_2O). In addition, heavy water was used as coolant. The initial reactor, commissioned in 1960, had a thermal power of 20 MWth without producing electricity. The fuel used in the research reactor was initially natural uranium, then enriched uranium. The 2-meter-long, aluminum-clad, nickel-clad fuel elements were manufactured by the Canadian company AMF Atomics Canada Ltd.

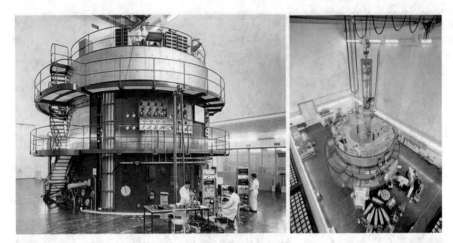

Photo 2.58 View of the service desk of Diorit. This part is very similar to what will be Lucens. Reaktor AG was founded in 1955 on the initiative of the two large Swiss companies Sulzer Winterthur and BBC Brown Boveri Baden. Among the shareholders were more than 100 Swiss companies. According to its articles of association, the company's purpose is *"the construction and operation of experimental reactors for the creation of scientific and technical bases for the construction and operation of industrially usable reactors ..."*. Reaktor AG received financial support from the Swiss federal government from the beginning. The company had a heavy water reactor built, which went into operation in 1960 and was called Diorit. The reactor was to serve as a precursor to a Swiss power reactor, i.e., a Swiss nuclear power plant technology was to be developed, which was not only to be used for domestic power generation, but also for export. Relatively quickly, however, it became clear that the financial outlay for private industry was becoming too high. The facilities of Reaktor AG—including the experimental Diorit reactor—were handed over to the federal government in 1960, which established the Swiss Federal Institute for Reactor Research (EIR) as an adjunct institution of ETH Zurich. The issue of what happened to the plutonium produced by Diorit in its early days was the subject of controversy in Switzerland in 2016, as was the transfer of 20 kg of unpurified plutonium powder (less than 92% plutonium) to the United States for safe storage. The very existence of this plutonium shows an initial desire by Switzerland to produce an atomic weapon

Switzerland's interest in nuclear power was then expressed in projects launched by three industrial groups. Many people supported the idea that the district heating plant at the ETH Zurich should be replaced by an atomic reactor to produce thermal and electrical energy. The *"Consortium,"* a group of companies, took on the task of implementing this project. The model followed was that of the ÅGESTA plant (natural uranium oxide moderated with heavy water and cooled with light water), which went into operation in 1954 near Stockholm. It was then decided to build the reactor in a cavern 42 m underground, near the main building of the ETH. Cooling was by means of an overhead cooling tower drawing water from the Limmat River. At the same time, electricity producers were working on the construction of a nuclear power plant. In 1957, the project company "Suisatom A" was founded. This

Photo 2.59 The vessel (calandria) of Diorit (Photo PSI-Würelingen)

plant was also to be housed in a cavern near Villigen, but in contrast to the Zurich project it was to be used exclusively for electricity generation. The third project was carried out by the industrial group "Enusa." This project aimed to build an experimental nuclear power plant for the Expo 1964. The equipment was to be installed in a cavern dug in the rock, near Lucens. The structure of the sandstone of the local geological layers was homogeneous and facilitated the excavation of the cavern. The reactor should have been built according to American plans.

But in September 1959, the Federal Council asked Enusa, Suisatom and the Consortium to merge their projects to develop a Swiss experimental reactor. Thus, in July 1961, the National Society for the Promotion of Industrial Atomic Technology (SNA) was created. The experimental plant was to be the intermediate step for the later development of a large nuclear power plant for commercial use "made in Switzerland" with export ambitions.

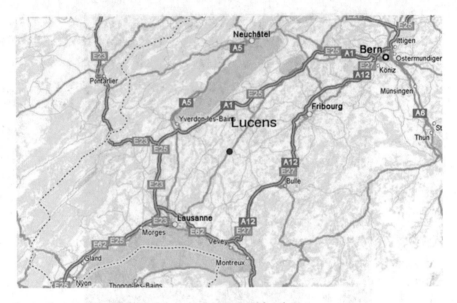

Fig. 2.55   Location of the Lucens plant in Switzerland

In May 1962, the SNA, as project owner, decided to build the Lucens Experimental Nuclear Power Plant (CNEL), with a thermal power of 30 MWth, for a gross electrical power of 8.5 MWe and a net power (after removal of the electricity used on the plant itself) of 6 MWe. For five years, the plant was built two kilometers southwest of Lucens (Fig. 2.55, Photo 2.60, Fig. 2.56), 25 km north-east of Lausanne and 60 km from Bern, on the left bank of the Broye, the river that was to supply the secondary cooling circuits of the reactor.

An access gallery 100 m long (Photos 2.61 and 2.62) led to three caverns respectively for the reactor (Fig. 2.57), the turbine and the fuel element storage pool (Photo 2.63, Fig. 2.58). A ventilation chimney dug in the mountain allows the ventilation of the underground parts. A chimney on the surface evacuates the stale air at altitude. The supplier of the reactor was Ther-Atom. Ther-Atom was also part, with three engineering offices of the *Groupe 8 de Travail de Lucens* (GTL). Their mission was to supervise the studies, the construction management, and the tests of the CNEL. From a technical point of view, the CNEL reactor was a development of the Diorit reactor already mentioned. The Federal Commission for the Safety of Nuclear Installations, founded in 1960, was the first nuclear supervisory authority of the Swiss Confederation (CSA). It accompanied the licensing process of the CNEL from the beginning. However, the CNEL was a real challenge, because the

Photo 2.60 The surface buildings of Lucens in 1969

CSA could only rely on very limited experience in reactor core design and containment layout.

The Lucens reactor (Fig. 2.59) used slightly enriched uranium (0.93%) as fuel, 99.75%-pure heavy water as moderator contained in a calandria, and carbon dioxide (CO_2) as coolant. A light enrichment is still necessary because the small size of the core induces significant neutron leakage (13,000 pcm leakage). By increasing the size of the core to reduce these leaks to 5000 pcm, one could have used only natural uranium. The fuel assemblies were made in the same way as those used by the British and French graphite-gas reactor

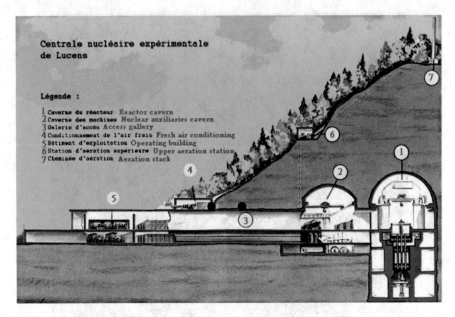

Fig. 2.56 Section of the underground plant of Lucens (Switzerland)

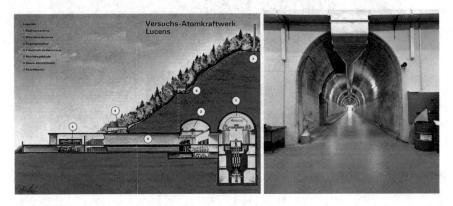

Photo 2.61 The access tunnel to the cavern reactor

types, that is, the uranium metal rods were housed in magnesium alloy cladding. Each fuel element was itself housed in its own pressure tube where the coolant circulated. This design made it possible to obtain a particularly compact reactor, requiring a containment of restricted dimensions. This containment consisted of a wall of concrete, asphalt, and aluminum about 60 cm thick that lined the artificial reactor cavern. The reactor core contains 73 fuel elements that are bathed in the heavy water calandria made of aluminum, 3.10 m in diameter and 3.16 m high (Photo 2.64, Fig. 2.60). The core is

Photo 2.62   View of the operating building on the surface of the Lucens plant (photo 24 Heures)

divided into two concentric regions of different assembly pitch to flatten the radial power sheet by playing on the moderation ratio. The pitch of the outer region (29 cm) is wider than the pitch of the inner region (24 cm).

The heavy water in the calandria must not exceed 80 °C and must not boil. The heavy water does not have a heat removal function in nominal operation, it is the role of the carbon dioxide to ensure this function. Above the calandria there is a metal caisson of light water, whose function is the biological protection of the service desk area where the operators are located. Water is a very good "shielding" against neutrons.[30] The calandria and the biological protection caisson are penetrated vertically by 73 aluminum tubes (channels) 14.5 cm in diameter welded to the bottom and top of the calandria. It is in each of these tubes that a complex system will be inserted assembling a pressure tube that channels the flow of carbon dioxide under pressure (60 atmospheres), the 7 axial nuclear fuel rods per element, a column of graphite support which rigidifies the fuel rods and also serves as a moderator, a system of bayonet coupling in the upper part and a double connection of the inlet

---

[30] Protections against gamma rays are usually made of heavy materials such as lead. On the other hand, hydrogenated materials protect against neutrons: water is a cheap and very effective representative.

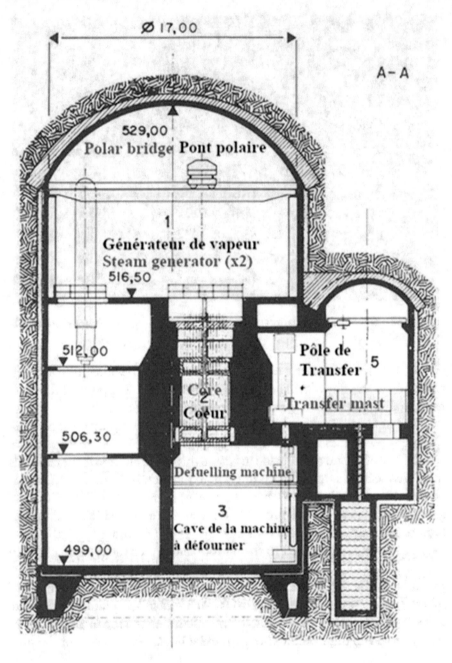

Fig. 2.57 Axial section of the Lucens reactor cavern

Photo 2.63 Model of the reactor cavern and the turbine cavern. One recognizes well above the reactor the two vertical steam generators of great height

and outlet pipes of the gas coolant. The double connections are connected to cold gas manifolds (headers) for the inlet and hot gas manifolds for the outlet. Customized valves per assembly allow for individual dosing of the carbon dioxide flow to each fuel element (Photo 2.65).

The diameter of the pressure tube is of course smaller than that of the channel into which it is inserted, and the gap between the two tubes is filled with carbon dioxide, which acts as a thermal insulation against the heavy water in the calandria. The cold coolant (220 °C) descends into the assembly, licking the inner face of the pressure tube, and is then forced upwards to contact the fuel rods to cool them. The gas is heated up to 385 °C even though the primary circuit is sized up to 520 °C. The pressure tubes are therefore closed at the bottom. The graphite support is inserted into the pressure tube and locked at the bottom by a bayonet device. The support is pierced by 7 channels in which the 7 fuel rods are inserted (Fig. 2.62). The graphite support serves as a guide for the coolant and to support the rods. Each fuel rod is an assembly of 4 segments screwed one on the other with a height of 2.765 m. The rods are

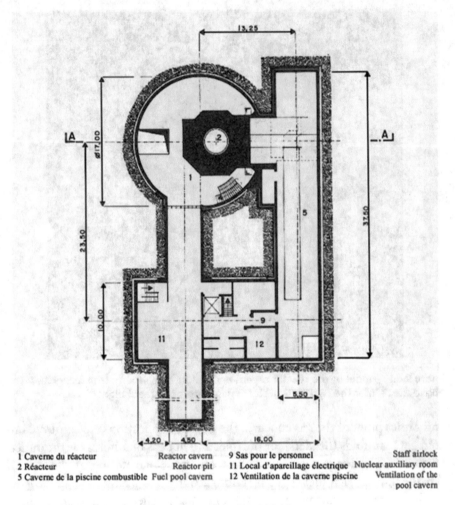

1 Caverne du réacteur   Reactor cavern   9 Sas pour le personnel   Staff airlock
2 Réacteur   Reactor pit   11 Local d'apareillage électrique Nuclear auxiliary room
5 Caverne de la piscine combustible Fuel pool cavern   12 Ventilation de la caverne piscine Ventilation of the
                                                                                                pool cavern

Fig. 2.58 Map of the Lucens cavern at elevation 508.30

integral with the graphite support at the bottom and free to expand at the top. The segments are made of low-alloy uranium metal (1% Molybdenum) 17 mm in diameter and 650 mm high, isolated from the carbon dioxide by a 1.75 mm magnesium cladding. Once heated by the rods, the CO_2 transfers its heat to a secondary circuit through two Steam Generators (SGs) that work as a heat exchanger to vaporize light water (Photo 2.66, Fig. 2.60).

The superheated steam from the secondary circuit then feeds a turbine, which is itself coupled to an alternator to produce electricity. The vessel is radially surrounded by cylindrical steel shields (gamma protection) placed inside a 2.8 m thick concrete caisson (biological shield against neutrons

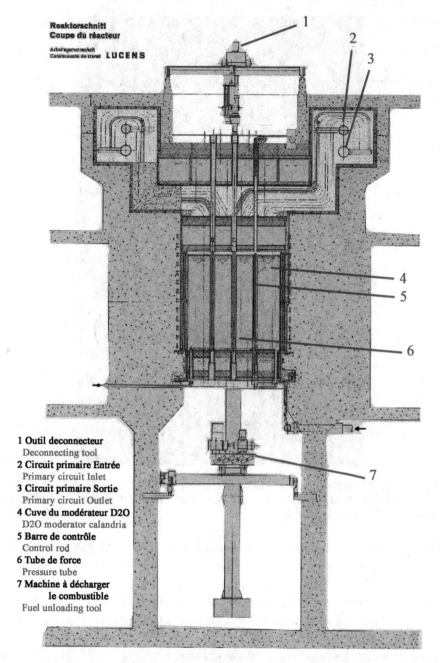

Reaktorschnitt
Coupe du réacteur

Arbeitsgemeinschaft
Communauté de travail **LUCENS**

1 **Outil deconnecteur**
  Deconnecting tool
2 **Circuit primaire Entrée**
  Primary circuit Inlet
3 **Circuit primaire Sortie**
  Primary circuit Outlet
4 **Cuve du modérateur D2O**
  D2O moderator calandria
5 **Barre de contrôle**
  Control rod
6 **Tube de force**
  Pressure tube
7 **Machine à décharger**
  le combustible
  Fuel unloading tool

Fig. 2.59 Section of the Lucens reactor (Switzerland). The nuclear fuel is "bathed" in a calandria of heavy water that slows down the neutrons. The cooling is ensured by pumps blowing carbonic gas which circulates in pressure tubes. A carbon dioxide channel can be isolated to insert fuel (adapted from *Bulletin Technique de la Suisse Romande* n°13 of 30 June 1962)

Photo 2.64 Placing a fresh fuel assembly in a storage rack. The core contains 73 such assemblies. The spent fuel assemblies are evacuated without contact by the unloading machine to the deactivation pool

because concrete contains about 6% water) (Fig. 2.61). The gap between the steel shields, the concrete and the pipe room, is in a slightly over-pressurized stagnant CO_2 atmosphere to prevent air ingress. Towards the bottom, a lower caisson protects from radiation the unloading machine room, which allows the extraction of a spent fuel element. This area is inaccessible during nominal operation, as is the vessel pit of a Pressurized Water Reactor. The pressure tubes and their fuel element are deflected together by the unloading machine after a burn-up of about 3000 MegaWatt-days/ton and after disconnection of the pressure tube from the top by the disconnection machine. For this purpose, the gas pressure in the shutdown reactor is lowered, and the residual

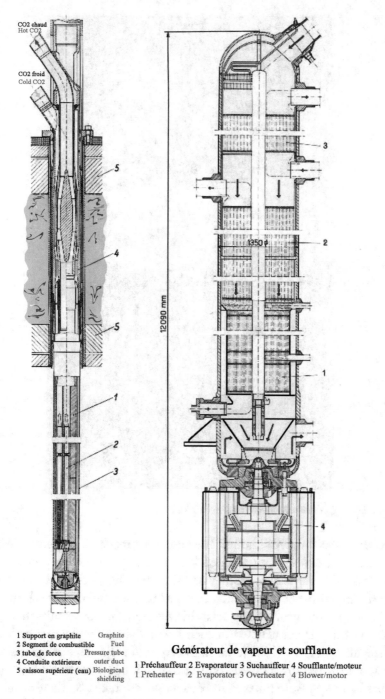

CO2 chaud
Hot CO2

CO2 froid
Cold CO2

1 Support en graphite — Graphite
2 Segment de combustible — Fuel
3 tube de force — Pressure tube
4 Conduite extérieure — outer duct
5 caisson supérieur (eau) — Biological shielding

Générateur de vapeur et soufflante

1 Préchauffeur 2 Evaporateur 3 Suchauffeur 4 Soufflante/moteur
1 Preheater   2 Evaporator 3 Overheater 4 Blower/motor

Fig. 2.60 Fuel assembly (left) and steam generator (right) at Lucens

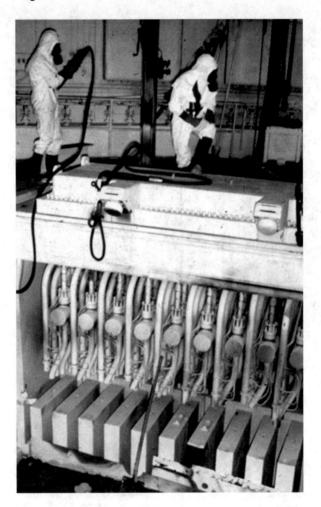

Photo 2.65 CO$_2$ control valves for assemblies

power decreases for several hours. The unloading machine is positioned below the orifice of the fuel to be unloaded. The tube is then sealed to prevent radioactive leakage in the case of leaking cladding, and the machine then takes it to a transfer pit with a hood in which the pressure tube enters, towards the deactivation pool. The heavy water calandria rests on this lower caisson, and the whole assembly is constructed in such a way that the vessel can be lowered into the unloading machine room in the event of major repairs. At the service desk above the reactor is the disconnect tool that operates the bayonet fasteners that fix the pressure tubes to the dual connection heads. There are also the winches for operating the 12 control rods located in the intermediate ring.

Photo 2.66 The floor at SGs level after the accident

Four shutdown rods by gravity drop are located more in the center of the reactor (Photo 2.67). The rods consist of two concentric tubes, the central part containing a cadmium-silver alloy, two materials that are highly neutron absorbers. The rods are cooled by a flow of CO_2.

The hot gas (temperature 385 °C mainly limited by the nature of the fuel and its cladding) is sent to two steam generators to vaporize light water. The SGs have centrifugal steam dryers. The light water circulates from bottom to top in helical tubes. CO_2 flows in counter-current. Steam is produced at 370 °C under 21.5 atmospheres. The gas blowers are located just below the SGs. The flow rate of the throttle valves at the outlet of the blowers can be adjusted. The blowers are connected by flanges to the SGs, which allows room for differential expansion between SG and blower. The rest of the primary circuit is fully welded. The blowers are driven by asynchronous motors at 6 kV and 3000 rpm. The turbine produces 8.55 MWe net at 3000 rpm. After letdown in the turbine, the steam passes through a condenser. After reheating at low pressure, the water is sent to a feeding tank which serves as a third reheater, then the water is injected by two feeding pumps which send it back to the two SGs at 147 °C. A tertiary circuit cools the condenser by taking

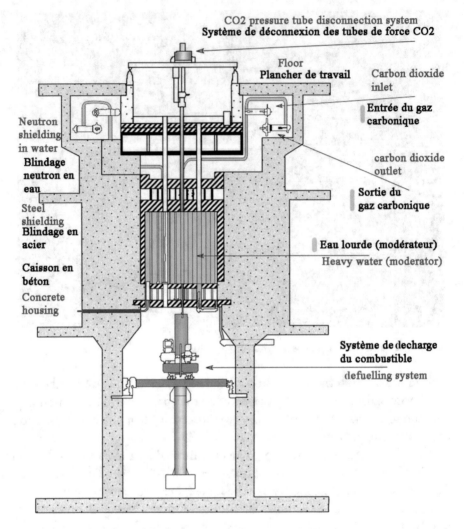

Fig. 2.61 Axial section of the Lucens reactor

water from the Broye (100 liters/s). A tank of 500 m^3 can continue to supply the tertiary circuit in case of loss for a short time allowing the scram (Fig. 2.62).

The air for the ventilation of the cave is sucked, filtered, and air-conditioned in a room located above the entrance of the access gallery. The necessary flow of 12 m^3/h is provided by a fan with a second one as backup (Photo 2.68). Some of the air is sent to the access gallery, the equipment room, and the decontamination room. The turbine building, the deactivation pool room and the electrical equipment room are each equipped with a closed-circuit ventilation system. These closed circuits evacuate the heat up of the electrical

Photo 2.67 View of the upper side of the Lucens reactor vessel (photo IFSN). The control rods and shutdown control rods, wrapped in transparent plastic, can be seen emerging from the vessel cover

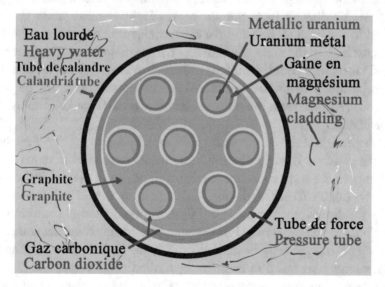

Fig. 2.62 Pressure tube and fuel element principle (dimensions are not respected)

Photo 2.68 View of the aeration station of the Lucens reactor. The reactor is located in depth (photo IFSN)

equipment by cooling with the tertiary circuit. The reactor hall is supplied with fresh air directly from the air intake. Tightly sealed safety valves are installed in the fresh air line and the exhaust cladding to hermetically isolate the reactor hall. In the event of a shutdown, the contaminated air is treated by a closed emergency circuit with fans and particle/aerosol filters located in the turbine building, before being released to the outside stack for dispersion (Fig. 2.63) (adapted from the *Bulletin Technique de la Suisse Romande* n°13 du 30 juin 1962).

For the design of the reactor vault, SSN took into account the "*hypothetical worst-case accident*" (by hypothetical, it is meant reasonable) in terms of pressure, temperature and radioactive releases, and not, as in the rest of the world, the "*worst-case accident imaginable*," i.e., a massive loss of primary coolant. During the licensing procedure, the CSA imposed various conditions and obligations on the developer. For example, the commission demanded that pressure and leakage tests be carried out on the reactor cavern. When the measurements later failed to confirm the desired almost complete tightness, an emergency effluent venting system with activated charcoal filters was installed. These filters are placed to filter out the most dangerous aerosols, such as iodine and cesium. The aerosol-bearing gases are sent to a chimney

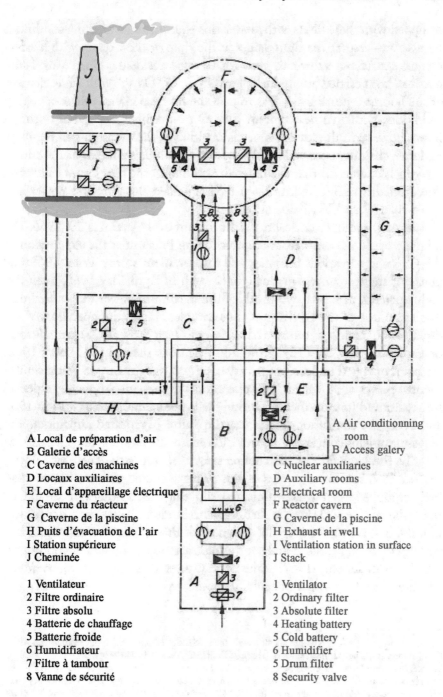

A Local de préparation d'air
B Galerie d'accès
C Caverne des machines
D Locaux auxiliaires
E Local d'appareillage électrique
F Caverne du réacteur
G Caverne de la piscine
H Puits d'évacuation de l'air
I Station supérieure
J Cheminée

1 Ventilateur
2 Filtre ordinaire
3 Filtre absolu
4 Batterie de chauffage
5 Batterie froide
6 Humidifiateur
7 Filtre à tambour
8 Vanne de sécurité

A Air conditionning room
B Access galery
C Nuclear auxiliaries
D Auxiliary rooms
E Electrical room
F Reactor cavern
G Caverne de la piscine
H Exhaust air well
I Ventilation station in surface
J Stack

1 Ventilator
2 Ordinary filter
3 Absolute filter
4 Heating battery
5 Cold battery
6 Humidifier
5 Drum filter
8 Security valve

Fig. 2.63 Ventilation system of the underground plant in Lucens. This system will play its role well by isolating the contaminated caverns from the beginning of the accident

equipped with these filters at the base. This equipment created the conditions necessary to respect the limit values at the time of the releases with a safety margin considered sufficient, even for extreme accidents. Since May 1966, tests had been carried out in the helium circuit of "Diorit" with a fuel element of the Lucens type. The aim was to gain the first experience under operating conditions with this new type of fuel. On November 16, 1966, during a power increase in the reactor, a partial melting of the uranium and the magnesium[31] cladding occurred. CSA required a thorough analysis of the anomaly, which concluded that the incident was due to the rapid power increase. Based on this result, Therm-Atom recommended slow start-up power and speed changes for CNEL operation.

The reactor reached criticality for the first time on December 29, 1966. The following year was devoted to commissioning tests under the supervision of the CSA and to various finishing and improvement works. From 1968, the power of the reactor was gradually increased. In April/May 1968, a ten-day endurance test at nearly two-thirds of the maximum power was carried out. The test phase of the CNEL was thus completed, and the operation of the experimental plant was transferred to *Energie Ouest Suisse* (EOS) for industrial operation on May 10, 1968. From mid-August to the end of October 1968, the plant operated under a temporary continuous regime up to its maximum thermal power of 30 MWth. The operation was then interrupted by a period for repairs and improvements. During the experimental reactor tests in 1967 and 1968, the circulation system was the source of repeated difficulties. The biggest problems concerned the two carbon dioxide blowers, the term used to describe the large fans of the cooling circuit, specially designed for Lucens. Indeed, to ensure the tightness of this primary circuit, the two blowers had been equipped with water-lubricated slip rings as bearings. It should be remembered that CO_2 is a gas that is fatal to humans, hence the strict control of leaks. The seals were specially designed for this application and tested in a test rig for a long time. However, during a test under operating conditions, which began in May 1967, some of the water that seals the rotary joints migrated into the primary circuit. It should be remembered that water and

---

[31] Magnesium (symbol Mg) is a light alkaline earth metal (density 1.738) with atomic number 12, white-gray in color. It has been known since the dawn of time (Magnesia is the name of a region of Thessaly in Greece), but recognized as a chemical element by Joseph Black in 1755 and isolated in its pure metallic form by Sir Humphry Davy by electrolysis in 1808 from a mixture of magnesia MgO and mercury oxide HgO. The melting points of magnesium (650 °C) and aluminum (660 °C) are relatively low, but above all magnesium ignites easily (it was used historically for the flashes of the first cameras). While its low weight is a definite advantage in the cladding of nuclear fuel assemblies, its low melting point and pyrophoric properties have caused it to be phased out in modern nuclear fuel designs.

carbon dioxide produce carbonic acid according to the following reaction for which K_h is the chemical production constant:

$$CO_{2(aq)} + H_2O_{(liq)} \rightarrow H_2CO_3 (aq),$$

$$\text{with } K_h = [H_2CO_3]/[CO_2] \approx 1,70 \times 10^{-3} \text{ à } 25°C.$$

In October 1968, the reactor had to be shut down again after an extended endurance test following a new water intrusion in the primary circuit. On October 24, 1968, the blowers were modified and improved during work that lasted several months. On December 23, 1968, *Energie Ouest Suisse* received the final operating permit, based on an expert opinion from the CSA, among others. In its report, the safety authority considered the fuel elements used in the reactor to be rather unreliable (risk of magnesium fire), but finally agreed with Therm-Atom's proposals to operate the reactor with the least possible brutal thermal cycles. The safety authority was finally to give the final operating license, but not without ensuring that the safety measures to protect the population had been taken. The reactor was to operate until the end of 1969. The date of January 21, 1969, corresponds to the definitive start-up of the experimental installation. At about 4:23 a.m., the reactor reached criticality, then the power was gradually increased. The primary circuit had undergone hot drying the previous two days because of excessive humidity. At about 6.15 a.m., the operators in the control room noticed a small defect in the cyclic monitoring of the carbon dioxide temperatures in the core and a strong background noise on some channels due to the supposed detection of cladding rupture. All this was corrected at the end of the morning without any disturbance to the test phase. The power was then increased from 9 to 12 MW and at 5:14 p.m. the 12 MWth was reached. At 5.20 p.m., the pressure in the primary circuit suddenly dropped and the carbon dioxide, which acts as a coolant, escaped into the cavern. At the same time, a large loss of heavy water showed that the aluminum vessel of the moderator could be damaged. The instrumentation recorded a significant increase in radioactivity in the containment. The emergency shutdown was triggered to the surprise of the operators in the control room (Photo 2.69), and the control rods dropped into the core. The ventilation check valves are closed to isolate the cavern. In the control room, the shift operators applied the appropriate emergency procedure and called back the team they had just replaced. This shutdown was associated with the tight closure of the ventilation pipes in the reactor cavern due to the detection of radioactivity. At 5:40 p.m., Jean-Paul Buclin, technical director of the plant, was notified by telephone while he was in Würenlingen for a

Photo 2.69   Control room of the Lucens plant. Clearly, the operator must be standing

meeting of the nuclear safety commission. For half an hour, he went over the events and emergency procedures in detail with the control room. He decided to carry out a rapid draining of the heavy water and "*to save 20 million Swiss francs.*" He arrived in Lucens at 9:30 p.m. and in the meantime, the personnel had put on masks and protective clothing following an increase in radioactivity in the access gallery. At 9:45 p.m., the cooling of the reactor was well underway, and the radioactivity in the access corridor was decreasing. From 0:00 to 3:00 a.m., experts checked and copied all records, while others measured the radioactivity outside the site. At 6:30 a.m., a press release was issued. Afterwards, the accident was brought under control and Jean-Paul Buclin declared: "*This was a perfectly controlled incident.*" He will also say: "*We would never have acquired such a complete, fast, and basically inexpensive experience without Lucens. There is no reason to be embarrassed by this adventure,*" an a posteriori justification that will not prevent the development program of the heavy water reactor type from being literally and figuratively buried.

Let us return to the accident phase itself, to focus on the state of the core. At the same time as the pressure drop in the primary circuit, the latter let out a gaseous mixture with a large proportion of highly radioactive contaminated CO_2 into the cavern, empty of any human presence, immediately suggesting a degradation of the fuel cladding. This outgassing process lasted nearly 15 min until the primary circuit, operating at 60 atmospheres, had released enough coolant into the reactor cavern to reach the common equilibrium pressure of 1.2 bar. It seems that the cause of the Lucens accident dates back to October 24, 1968. Indeed, it was on this date that maintenance work had taken place on the rotating joints of the coolant recirculation blowers. It was at this time that several liters of back pressure water from the rotating joints,[32] would have escaped into the primary circuit. After the water infiltrated the primary circuit, carbonic acid was produced. Carbonic acid is a weak acid found in carbonated beverages and produces the "pungent" effect on the tongue. Carbonic acid is also the cause of the acidification of the oceans due to the production of CO_2 by man.[33] This acidic water attacked the magnesium cladding of several fuel elements. The equation for the oxidation of the cladding by water vapor is given by:

$$Mg_{(metal)} + H_2O_{(steam)} \rightarrow MgO_{(solid)} + H_{2(gas)}$$

This oxidation produces hydrogen gas release and flammable magnesium oxide powder. Magnesium oxide is known to be unstable. Moreover, its instability tends to increase with high temperatures. The corrosion products formed then fell into the heat transfer gas circulation channels. It is postulated that the debris fell to the bottom of the pressure tubes and partially clogged them, reducing the flow in some channels. Fuel element 59 was insufficiently cooled due to the reduced CO_2 flow rate inherent in the pressure tube plugging. Several of the seven fuel rods in fuel element 59 (Photo 2.70) thus underwent an overheating that went unnoticed at first because not all the fuel

---

[32] The seal back pressure technique isolates the downstream portion of the seals from the upstream portion that contains a potentially radioactive liquid or gas. A higher pressure is applied downstream of the seal labyrinth to contain the radioactivity by imposing a flow from downstream to upstream of lower pressure, in the direction of decreasing pressure. This is what is successfully applied on the primary pumps of Pressurized Water Reactors, but in this situation, one has to inject non-active water into a circuit of possibly active water. In this case, the backpressure water must not penetrate the primary circuit of carbon dioxide, at the risk of acid formation, hence the complex technology of the implementation of the backpressure.

[33] The more than 40% increase in atmospheric CO_2 concentration from 280 ppm (parts per million) in 1750 to 400 ppm in 2015 and 403.3 ppm in 2016 increased the dissolved CO_2 in the form of carbonic acid in the ocean, increasing its acidity by 26%, as measured by its pH, which decreased by about 0.1 from 8.2 to 8.1 (source https://reseauactionclimat.org/acidification-rechauffement-ocean-dangers-demultiplies/)

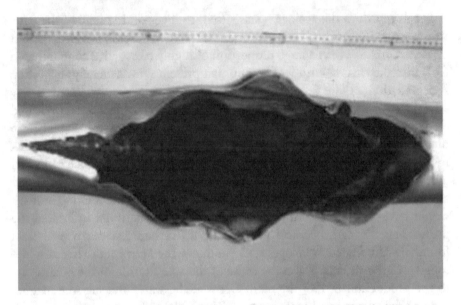

Photo 2.70 Impressive view of the cladding of fuel element 59. The cladding is completely torn

elements were equipped with a temperature probe in the uranium. When the temperature reached 600 °C, the magnesium cladding of the central fuel rod melted, followed shortly after by the uranium metal (melting at 1135 °C) that it protected. Thus, a column of molten metal was formed (with the heavy uranium at the bottom of the column and the magnesium above). This melting process then spread from one to the next to the neighboring fuel rods (Photo 2.71). And the metal eventually ignited in the CO_2, causing a massive release of radioactive fission products into the coolant and the Automatic Reactor Shutdown (ARS). The ARS shut down the nuclear chain reaction, but not the fire in fuel element 59. The graphite column bent, met the nearby pressure tube, overheated it and caused it to burst under the effect of the 50-bar pressure that prevailed there when the temperature reached between 700 °C and 800 °C. This explosion initially ruptured one of the five rupture discs responsible for limiting the pressure of the heavy water tank (the calandria). Through this opening, 1100 kg of heavy water, a molten mixture of magnesium and uranium, and contaminated coolant were projected into the reactor cavern. About a second later, a thermal reaction between the heavy water and the molten metal triggered a second explosion. This is called a corium-water interaction or steam explosion. The shock wave caused the control rods, which had already been lowered during the scram, to jam in their guide tubes, but without touching the particularly well-protected (by reinforced tubes) safety

Photo 2.71 Double connection of channel 59 (left) and channel deformation (right)

Photo 2.72 Channel 59 in top view

control rods. The overpressure led to the rupture of the four other rupture discs of the moderator calandria tank, with new projections of radioactive material into the biological shield made of water. This process continued over the next few minutes until the decompression of the primary circuit in the reactor vault was completed. (adapted and commented from https://www. ensi.ch/fr/2012/05/31/serie-de-lucens-analyse-profonde-de-laccident). Post-mortem examinations showed that the cladding of fuel 59 had completely burst. The dismantling from above showed the damage caused by the explosion on the double connection of channel 59 (Photo 2.72).

The investigation report, published in 1979, revealed to the public that there had been a partial meltdown of the core. The cause was said to be moisture in the cavern (!!!) and leaking seals that caused water to accumulate. The part between the humidity and the water intrusion in the previous test was not specified. The accident would have been classified at level 4 of the INES scale nowadays, the scale having been created only in 1990 (accident not involving significant risks outside the site). Clearly, the cavern has saved from a more severe classification. During the accident, it is true that all safety devices worked as intended. Neither the operating team quickly equipped with self-contained breathing apparatus, nor the environment, were exposed to unacceptable radiation doses. The ASPEA (*Association Suisse Pour l'Energie Atomique*) declared that this "*unintentional breakdown was rich in lessons learned.*" The word breakdown is a mild euphemism in this case. The official investigation report mainly details the technical causes of the accident and not much about its radiological consequences. However, the measurements taken that night and in the following days show that the level of radioactivity did not increase significantly. Moreover, the contamination of the access corridor immediately after the accident was due to two isotopes with a half-life not exceeding three hours. Radioactive gases did escape into the cave but were contained in the rock. The Federal Office of Public Health, responsible for monitoring radioactivity in the Swiss environment, has been monitoring radioactivity around the Lucens plant since the accident. Drainage water samples from the former plant are collected every two weeks and analyzed at the Institute of Radio Physics in Lausanne. The radionuclides monitored are tritium (present in the irradiated water, half-life 12.32 years), strontium 90 and gamma emitters. The strontium 90 contents (half-life 29 years) measured over the last ten years are below the detection limit, which is about five milliBecquerels per liter (tolerance value for drinking water: 1000 millibecquerels per liter). Tritium activity is detectable and averages around 10 Becquerels per liter (tolerance value for drinking water: 10,000 Becquerels per liter). Gamma emitters such as cesium 137 (half-life 30.1 years) and cobalt 60 (half-life 5.27 years), for example, have not been detected. Some studies, which show an increase in intestinal cancers in the Broye district between 1970 and 1990, are, however, used to contradict the official version. Professor Matthias Bopp, co-author of one of these studies, said: "*In men, the general excess mortality in the Broye has the same components as in neighboring regions, i.e., diseases related to alcohol consumption, accidents and lung cancer. In women, heart disease was the cause of the additional deaths. It is therefore impossible to deduce a link with the nuclear accident of 1969, especially since intestinal cancer is not among the cancers suspected to be caused by irradiation.*"

The post-decommissioning monitoring program consisted of collecting two samples of water from the drainage system every 15 days, one from the pond collecting drainage from the nine main drains in the cavern (Fig. 2.64). The second is in the control chamber, which is located just before the release into the Broye. Until the beginning of 2010, the measured tritium contents were between 10 and 20 Bq/l (average value of approx. 15 Bq/l), whereas surface water usually does not exceed 3 Bq/l. At the end of 2011, a notable increase in tritium activity was noted (up to 230 Bq/l, Fig. 2.65) relayed by

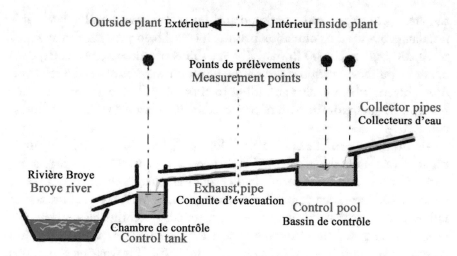

Outside plant Extérieur ←⋯→ Intérieur Inside plant

Points de prélèvements
Measurement points

Collector pipes
Collecteurs d'eau

Rivière Broye
Broye river

Exhaust pipe
Conduite d'évacuation

Control pool
Bassin de contrôle

Chambre de contrôle
Control tank

Fig. 2.64 Activity measurement points (every 15 days) of the collected water from the Lucens plant

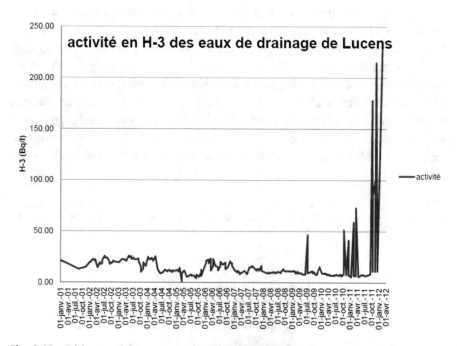

Fig. 2.65 Tritium activity measured at the release in the Broye. A strong increase is noted from October 2011 onwards, which has caused a controversy in Switzerland

the Press (newspaper 24 heures) and anti-nuclear associations. However, it is not dangerous for the population because it is well below the regulatory limits of 10,000 Bq/l or 20,000 Bq/day. The variability of a drainage water depends of course on the rain regime and on the intrusion water towards the contaminated cavern which evolve according to time. It will take about 100 years from the accident for the tritium produced by the reactor to disappear almost completely.

The dismantling of the Lucens plant, led by the Director Jean-Paul Buclin,[34] began a year after the accident. It was necessary to wait for the regulatory authorizations, and some people had crazy ideas, such as drowning the cave, which would have been counterproductive given the very likely exfiltration of radioactive water that this would have generated. Patiently, the workers protected by suits (of the "*Mururoa*" type with ventilation) dismantled and cleaned the plant (Photos 2.73, 2.74, 2.75 and 2.76). The working conditions were Dantean, with workers losing up to 4 liters of sweat per hour of work. The dismantling of the plant lasted until the end of 1972.

The fuel assemblies were sent to the Eurochemic plant in Mol, Belgium. Most of the radioactive waste was transferred to the Paul Scherrer Institute, except for various large parts. However, it was not until September 2003 that the last low-level radioactive elements left Lucens for the temporary storage center for nuclear waste at Würenlingen in the canton of Aargau. The work for the decommissioning consisted essentially of the installation of a drainage system around the underground structures (caverns), the installation and commissioning of a specially protected pipe for the direct discharge of the water collected by the drainage system into the Broye, the filling of some of the caverns with concrete in 1992, the installation of a fence delimiting the

---

[34] This recognized expert was later contacted by the Soviet Union during the Chernobyl accident to delimit the dangerous zones. He is nowadays considered as an expert in the field of dismantling and remains one of the great craftsmen of the feat that was the accidental plant of Lucens.

Jean-Paul Buclin interviewed on Swiss television RTS.

Photo 2.73   Verification of breathing apparatus for cleaners

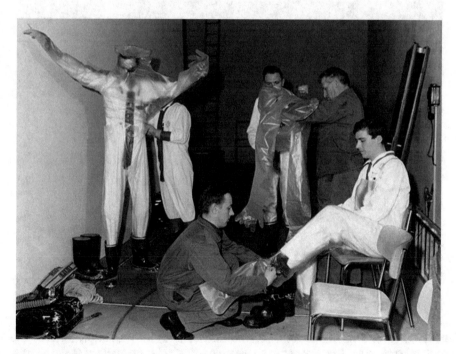

Photo 2.74   Dressing of the cleaners in ventilated protective suits

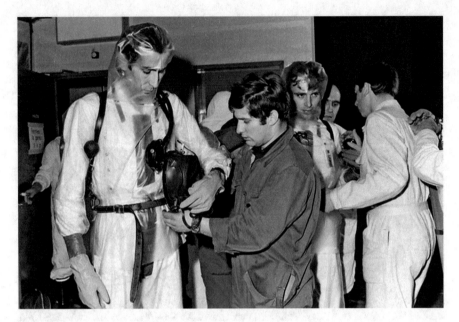

Photo 2.75 Preparation of the cleaners. Checking the ventilation systems

Photo 2.76 Two operators from Lucens are waiting in their uniforms to enter the contaminated area. They appear to be wearing a dosimeter on their chest and a radiation detector in their hand

plot intended for the casks and controlling access to it, with the construction of a shield wall intended to complete the radiological protection against radiation from these casks. Most of the plant was decommissioned in April 1995 and the Federal Council decided on December 3, 2004, that the former experimental nuclear power plant in Lucens was completely decommissioned and no longer constituted a nuclear installation within the meaning of the former Atomic Energy Act. The cost of the final decommissioning was 16 million Swiss francs. Since October 1997, the premises have been used as a storage facility for various museums and cultural institutions in the canton of Vaud. The site is reconverted into storage for stuffed animals! (Photo 2.77).

The Lucens accident, although little known by the public, is considered as one of the most serious accidents in the field of civil nuclear power in the world. The personnel and the local population were not irradiated, or only slightly. The radioactivity measurements carried out after the accident did not reveal any significant contamination of the environment. However, the cavern was largely contaminated. During the following years, the reactor was dismantled, and the cavern was decontaminated. In 1992, the cavern was partially filled with concrete.

Lucens sounds the death knell for the hopes of a Swiss national reactor type. Following the Fukushima accident, the Swiss Federal Council

Photo 2.77   The present of Lucens. A decontaminated part is used to store stuffed animals (DR)

confirmed, on May 25, 2011, the gradual phase-out of nuclear power by deciding not to renew the nuclear power plants in operation and opting for their definitive shutdown once they have reached 50 years, i.e., between 2019 and 2034. On September 28, 2011, the Council of States confirmed the shutdown of the construction of new nuclear power plants while demanding the continuation of research in the nuclear sector, a wishful thinking which does not really make sense insofar as the competences disappear very quickly due to the lack of job opportunity.

What lessons can be learned from the Lucens accident? The main cause of the accident was the use of a magnesium cladding, which is highly oxidizable by water, has a relatively low operating temperature and is likely to "burn" in an unfavorable atmosphere. Magnesium was quickly abandoned by fuel designers throughout the world in favor of zirconium. A highly aggravating phenomenon is the fact that the geometry of the rod cluster is not open, unlike in PWRs. Any total or partial blockage cannot be compensated by coolant from another nearby channel. This is a defect that is also found in the Canadian CANDU or Russian RBMK reactor design. Let's recall that operators use this property to flatten the power sheet by playing on the opening of the carbon dioxide valve in each assembly. This tactic is also used in fast neutron reactors cooled with sodium or lead-bismuth, where the assemblies are isolated from each other by a hexagonal tube. Detection of these blockages requires instrumentation of each assembly with thermocouples capable of rapidly detecting an abnormal increase in temperature, which was not the case for the Lucens reactor (economy?). The penalty was immediate and definitive for the operator. The cavernous situation in which the reactor is housed, nevertheless, allowed the avoidance of significant radioactive releases and spared the population from certain contamination. In France, only the Chooz-A reactor, shut down in 1991, had this core cavern configuration.

Saint Laurent des Eaux A1 (France, 1969), Saint Laurent des Eaux A2 (France, 1980)

The most serious accident that took place on French territory remains without question the partial fuel meltdown accident in the A1 reactor of the Saint-Laurent des Eaux plant on October 17, 1969. The Saint-Laurent-A1 plant, located on the banks of the Loire (Photos 2.78 and 2.79), is part of the last wave of Natural Uranium-Graphite-Gas (UNGG) construction in France, coupled to the grid in 1969 and shutdown in 1990, for a capacity of 1662 MWth (480 MWe net).

Photo 2.78 Saint-Laurent-des-eaux site. The two Natural Uranium-Graphite-Gas (UNGG) reactors are located at the far left, recognizable by their cubic shape. Two air-cooling towers in operation are located at the end of the site on the right. The two PWRs that they cool in addition to the Loire River are located just to the left of the towers, recognizable by the hemispherical shape of the reactor buildings and the next large parallelepipedal turbine buildings

Photo 2.79 The two reactor blocks of Saint-Laurent-A1 and A2. The welded structures that surround them are characteristic of the French Natural Uranium-Graphite-Gas (UNGG) of the late 1960s

The reactor (Fig. 2.66) contains 446 tons of natural uranium in the form of metal clad with a magnesium-zirconium alloy (43,865 fuel elements called cartridges in the core, 600 mm long, hollow rod 23 mm internal, 43 mm external, surrounded by a graphite fuel jacket 112 mm internal, 137 mm external). The moderator, designed to slow down the neutrons that become more efficient, is made of a stack of 2572 tons of graphite 9 m high (Photo 2.80).

Fig. 2.66 Artist's cut view of the Saint Laurent-A1 plant. The interior of the concrete caisson shows the core (1) and the CO_2/water exchangers (3). Note that the direction of the carbon dioxide flow is from top to bottom, which may seem surprising at first, because it is the opposite of the chimney effect of hot gases. But the idea here is to cool the inside of the concrete biological protection caisson. The blowers that circulate the gas are shown in 9, the condenser in 7 and the turbine in 5

Photo 2.80 Building the graphite pile: channels through the graphite sleaves are clearly visible

The flow of CO_2 gas cooling the core is 8747 kg/s at a pressure of 26.5 bars. The inlet temperature of the gas is 217 °C for an outlet temperature of 400 °C. The circuit contains 185 tons of carbon dioxide. The plant produces 2178 tons/h of steam at 390 °C and 33.6 bars at turbine inlet. The reactor is controlled by 138 absorber rods, including 3 safety rods, 24 neutron flux-shape control rods, 12 pilot rods, 81 short-term reactivity compensation rods and 18 long-term reactivity compensation rods. The upper desk slab has 109 loading holes that can be opened by the fuel loading machine under pressure. The concrete caisson is cooled by water flowing through tubes between the sealing plate and the concrete.

On October 17, 1969, at 7:08 a.m., seven months after the coupling of the plant, during the handling of fuel elements in channel n°21 of the pit in naval battle position F9 M15, the reactor was at 80% of nominal power. The loading/unloading device (Main Handling Device DPM controlled by a punched card displacement system in the computer context of the time) mistakenly places a flow control device, normally intended for other uses, above ten fuel elements already loaded, on top of which five graphite logs have been placed. The flow control device consists of a graphite rod with a 20 mm hole, which causes a pressure drop at the passage of the carbon dioxide, a pressure drop 40 times higher than that of a normal fuel. The classic use of these carbon dioxide flow reduction devices allows to play on the flow between different channels to homogenize the temperatures radially in the pile; The DPM is located above the loading platform and moves on a guide rail to be placed above the pit to be unloaded (Photos 2.81 and 2.82). The machine connects to the channel pit to be treated, unscrews the sealing plug and then, thanks to a telescopic arm, removes the cartridges and replaces them with new fuel placed on standby in a barrel of the DPM.

The partial obstruction of the channel, caused by this loading error, was sufficient to cause a rapid increase in temperature (less than 10 s) such that it led to the melting (and ignition in the CO_2 coolant) of the magnesium-aluminum fuel cladding, then the melting of the uranium fuel (Photos 2.83 and 2.84). The contamination of the channel with fission products was immediate, leading to an increase in activity measurements at 7h08min00s by the activity measurement system of the fuel loading machine still in position, then by the general cladding rupture detection system at 7h08min10s, the loading being carried out during operation of the reactor. This detection led to an emergency shutdown at 7h08min11s. Tests were underway in a hot test channel in which so-called degaussing logs were tested, i.e., graphite logs (3 different types) without fuel elements. The objective of the test was to find out if these logs did not present any disadvantages in operation. In fact, the core

Photo 2.81   Natural uranium graphite gas reactor (first French reactor type). The Saint Laurent-A loading platform and the Main Handling Device (DPM). The rails of the fuel loading machine and the tulip-shape plugs (with circular top) closing the CO_2 channels are visible in front of the photograph (photo: Bouchacourt-Foissote-Valdenaire-ENSIB)

was not loaded in the usual way. The core and hot channel were loaded and unloaded by the same fuel loading machine that had had many failures in the past. It is probable that a human error took place on the hot channel, consisting in removing a false graphite fuel jacket (empty fuel log). The discovery of a missing element on October 8 led to the fabrication of a punch card for the fuel loading machine program. This card had to contain the location address that was not detected on rereading, and channel 21 was loaded by a fuel jacket

Photo 2.82 "tulip-shape" plug of a channel and load face. The damaged channel is isolated by the access barrier (photo: Bouchacourt-Foissote-Valdenaire-ENSIB)

with a reduced passage section. The continuation of the loading operations alternated manual and automatic phases so that a cell of the barrel of the reloading machine was found empty of logs, whereas it should have contained 5. The operator then thought of a shift in block on 6 cells. Only a weighing system placed on the fuel loading machine could have detected this error, but the operator had not yet been trained to interpret this indication. Unfortunately, probably to save time due to numerous delays, no flow measurement was made once the wrong fuel cartridge was placed, and the heat up occurred. The F9 M15 channel was first gassed with 60 °C CO_2 at the time of loading, which delayed its heat up, then it was increased to 225 °C (at the nominal CO_2 circulation temperature), which precipitated the accident. As soon as the control rods dropped, the reactor caisson was "deflated" from 35 bars to 1 bar of CO_2. Analysis of the measurements of the leak detection system made it possible to locate the incriminated channel and the releases were filtered by iodine filters. The dosimeters of the EDF agents present measured 2.3 mSv for the safety officer on duty, 0.6 mSv and 0.2 mSv for two operators and less than 0.1 mSv for all the others. At the time, the public exposure limit was 5 mSv/year.

The corium fell into the "*debris catcher*," also called the "*garbage can*" placed at the bottom of the channel and spurted out through the holes (orifices) of the catcher, normally intended for the passage of carbon dioxide. The attack on the steel of the corium catcher by the molten corium led to the erosion of the "*garbage can*" (that is the official term) at its weakest point, namely the holes in the core catcher, which are pierced in "lace-like" fashion to allow the carbon dioxide to pass. CO_2 (Photo 2.85). Human error was most likely the cause of the accident. The DPM was programmed by a punch card system that was difficult to verify and the plant did not have in-house

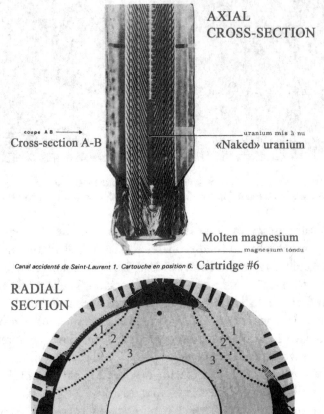

AXIAL
CROSS-SECTION

coupe A B ⟶
Cross-section A-B

uranium mis à nu
«Naked» uranium

Molten magnesium
magnesium fondu

Canal accidenté de Saint-Laurent 1. Cartouche en position 6. Cartridge #6

RADIAL
SECTION

Zone
1- T<645 °C
2- 645°C<T<710°C
3- T>710 °C

Uranium
uranium
Molten magnesium
magnésium fondu (T$_{100}$)

Isothermes de changement de phase
Zone (1) : α + γ   θ < 645°C
Zone (2) : β + γ   645°C < θ < 710°C
Zone (3) : γ   θ > 710°C

gaine
Cladding

Cartouche en position 6.
Coupe AB.
Cartridge #6

Photo 2.83 Degradation of a fuel cartridge (vertical section and section) analyzed by the CEA. On the top picture, we see the melt magnesium flows. On the lower picture, we see the radial extent of the magnesium flows as well as the cooling blades around the cartridge (in black)

Photo 2.84 A fuel jacket debris placed on the support area in front of the heat exchangers and below the core (the exchangers are placed below the active core). Large debris such as this could be removed by remotely operated tongs. The finishing work had to be done manually after re-entering the caisson. The fuel, almost fresh, was (relatively) not very radioactive

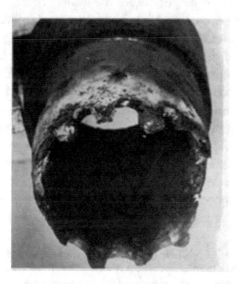

Photo 2.85 Photo of the damaged "trash can." The "recuperator" part has completely disappeared when all the full zones of the circular honeycomb part have melted

computer-trained specialists. It was also claimed that there was a fuel loading pattern error that could never be proven. Five elements in the lower part of the core (out of a column of 10) were destroyed, producing about 50 kilograms of corium. 14 kg were retained by the debris collector located below the active core. The rest was thrown out and spread to the surrounding structures (the support area) and to the lower structures (the heat exchanger tubes) by gravity (Fig. 2.67). At high temperature, the uranium was partially oxidized by contact with the CO_2, and uranium oxide dust was carried by the gas and deposited on the intact fuel elements, causing significant pollution. This pollution resulted in a strong increase in the background noise of the cladding rupture detection device (DRG), which measures the radioactivity of the fission gases leaking from the fuel cladding. The fact that the fuel that had just been loaded into the channel was fresh meant that its fission products content was very low.

After the accident, the pollution of the reactor was estimated at 100 grams of uranium deposited on the surface of the cladding, instead of 6 grams for a new core (there is always some uranium powder on the surface of new fuel because of the manufacturing process). This had the consequence of increasing the count rates by a coefficient of fifty.[35] As a result, the DRG had to be recalibrated to ensure that it was still capable of performing its function.

The rehabilitation of the reactor required the solution of several problems: the cleaning of as much corium and uranium debris as possible (Photos 2.86 and 2.87). EDF used to the maximum the devices controlled remotely from the upper slab such as suction hoses and remote-controlled gripping (pneumatic clamp) of the most voluminous pieces (47 kg of large debris, of which 15 kg were collected thanks to a scraper allowing to make heaps accessible to the clamp). But a human intervention was necessary to remove some debris adhering to the structures. Each of the people who entered the caisson stayed there less than 8 minutes. Because of the limitation to 3 rems per person, 105 people[36] intervened in the caisson maintained under vacuum, after preparation on a scale model, as well as on reactor no. 2 then under construction. They passed through the top of the exchangers after dismantling and cleaning the cell closest to the damaged channel. Dose rate predictions were carried out by the CEA on this occasion, to program the shifts between the teams. The operators entered the caisson through airlocks located at the height of the

---

[35] A. Grauvogel, J.P. Le Noc: *Saint-Laurent 1 – Incident du 17 octobre 1969, Pollution du réacteur et modification de la DRG*, Bulletin d'information de l'association technique pour l'énergie nucléaire n°92 (1971).

[36] Possibly women (secretaries?), according to one of the former participants in the affair, a detail that I have not been able to confirm from other sources. Perhaps it is an urban legend that captures the attention of listeners! I personally find it hard to believe.

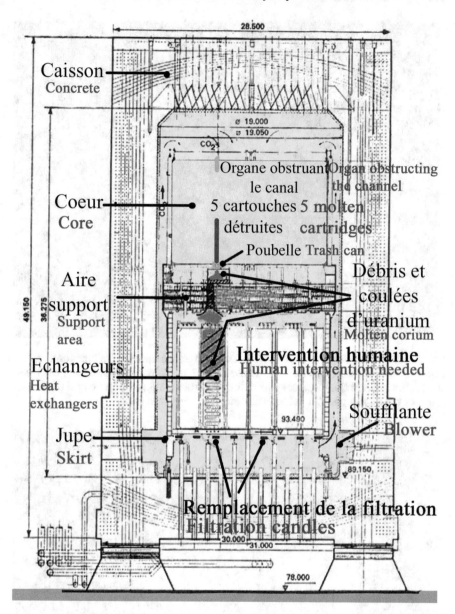

Fig. 2.67 Plan of the core and the damaged channel of Saint-Laurent-A1. The carbon dioxide rises around the core and then passes through the core from top to bottom before passing through the heat exchangers. Note that the core is placed above the heat exchangers, which is antagonistic to natural convection, but which allows the concrete of the caisson to be cooled by cold carbon dioxide, compared to a situation where the core would be placed below the heat exchangers with a cold gas rising and then hot gas descending on contact with the concrete (*jupe* = skirt, *poubelle* = garbage can, *soufflante* = CO₂ blower, *coeur* = core, *échangeurs* = heat exchanger)

Photo 2.86 Installation of a suction system from above around the damaged channel at the level of the reactor slab. A controlled zone surrounds the whole. One can clearly see on the ground the displacement rails of the fuel loading machine and the circular plugs of the access pits. "Tuyauterie d'aspiration"=venting duct

Photo 2.87 Cutting operation on Saint Laurent-A1 during post-accident repair. The operators are wearing ventilated clothing indicating a possible source of contamination

turbofans. Climbing along the heat exchangers in total darkness except for a little artificial lighting, they had access to the lower support area of the reactor, and, after 2 weeks of cutting work in a containment and extremely stressful environment, they were able to access the damaged channel.

The dust around the incriminated channel could then be sucked up. The risk of ignition of the uranium debris in the air following the opening of the caisson (uranium and in particular its hydrides can ignite in air) was also considered before being invalidated. At the end of these cleaning operations, 47 kg were removed from the reactor, which left about 10 kg, including 5 to 8 kg of uranium, trapped in the exchangers and certain cells. Pierre Way, at the time a technician in the uranium store, reported[37] that the Pegasus casks, intended to transport the spent cartridges to La Hague, were poorly adapted to mate with the orifice of the container system (known as *Mecca*): "*With two colleagues with strong nerves, we tied a sailor's knot around the blocks and rushed everything into the cask. Everything happened very quickly. We were not irradiated.*" The fit-up system was adapted later. It was then necessary to create a sufficiently efficient filtration system to prevent the reactor from becoming polluted again over time. This filtration, made of cartridges (known as *glass wool candles*) and metal sieves, had the task of recovering the residues partially oxidized by the CO_2 at 400 °C, which would not fail to be carried away by the heat transfer gas. This filtration proved to be disappointing as only 1.5 kg of material could be recovered. The remaining material was never located more precisely. Are they the cause of the pollution of the Loire? The last filters were removed in 1978. Thermochemical studies have shown that oxidation only in the presence of carbon dioxide (i.e., without oxygen) was finally quite slow. A granulometric analysis of the debris made it possible to predict approximately the rate of de-scaling of the core from the knowledge of the flow of CO_2 (9 tons per second). A metal casing was then introduced under the exchangers with a seal against the skirt (the structure that channels the gas flow). On this frame were fixed baskets containing about ten filtering candles (so-called because of their shape) of 17 cm diameter and 75 cm height, for a total of 1600 candles. Beyond design basis tests were performed to evaluate the efficiency of this device outside the core. As a last resort, the DRG cladding breakage measurement system had to be adapted to take into account the high level of parasitic noise due to the dust, the quantity of which decreased over time. Indeed, the restart of the reactor was conditioned to the possibility of detecting cladding failures later. Tests at low power showed that the DRG could still perform its function effectively, without even renewing the contaminated fuel. An "auxiliary computer" was developed to check the punched

---

[37] As reported in *Génération SPT* n°21, *Journal de la production thermique d'EDF*, July-August 1988.

cards of the fuel loading machine, i.e., a card reading system that shows the fuel loading pattern as actually punched. In addition, the fuel loading machine was equipped with a continuous weighing system and a camera for identifying the elements being loaded.

The time required to rehabilitate the reactor, resulting from effective collaboration between EDF and the CEA, was short enough for the reactor to be reconnected to the grid on 16 October 1970, exactly one year after the accident. The filtration allowed the removal of an additional 1.5 kg of dust, in a very localized area vertically above the damaged channel. By today's standards, this accident would probably have been classified 4 on the INES scale for describing nuclear accidents (see Appendix 1). Indeed, when the caisson was opened and ventilated, contaminated carbon dioxide was released into the atmosphere after filtration, and it is estimated that the presence of Very High-Efficiency filters retained most of the particles with a diameter greater than 0.3μm.

The financial cost of the rehabilitation is estimated at ten million francs in 1969, to which must be added the cost of the loss of operation (electricity sales), which is about the same. Measures to improve safety were taken afterwards: the damaged channel was simply condemned, perforated bells were installed at the head of the channel to ensure permanent cooling by CO_2, a gas turbine was added to secure the electrical source. A last resort panel was settled, and an additional cooling pump was installed. The reactor was then able to operate without any particular problems until its final shutdown in 1990.

A new partial meltdown accident occurred on March 13, 1980 at 5:40 p.m. in the other Natural Uranium-Graphite-Gas (UNGG) reactor Saint Laurent-A2 (Fig. 2.68). Following an increase in radioactivity in the coolant indicating the presence of fission gas, the reactor was shut down. Visual inspections showed that a pressure transducer holding plate of about 0.5 m^2 which had become loose because of corrosion and obstructed a dozen channels in the F05 M19 cell and in neighboring cells, out of 3,000 cooling channels. This obstruction led to a major melting of two fuel elements, about 20 kg of uranium and magnesium.

Corrosion is a major problem in Natural Uranium-Graphite-Gas (UNGG). It is due to radiolysis[38] by carbon dioxide, which produces oxidizing free radicals.

---

[38] Radiolysis consists in the decomposition of a molecule under the effect of radioactive radiation. Thus, liquid water can be transformed into hydrogen and oxygen, and carbon dioxide into carbon and oxygen. It is oxygen that corrodes metal structures. The oxidation is exacerbated by the temperature: the higher the temperature, the faster the oxidation.

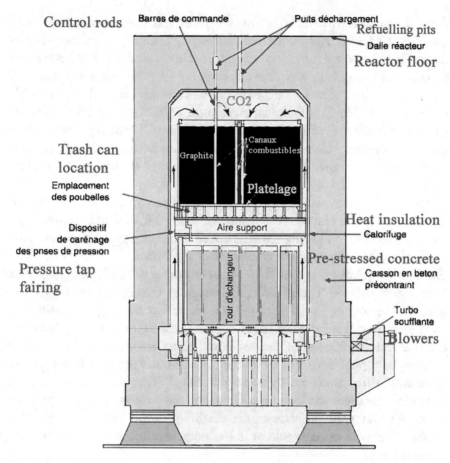

Fig. 2.68 Axial section of the Saint Laurent des Eaux-A2 reactor

$$CO_2 \rightarrow CO + \underset{\text{radical libre}}{0}$$

This accident rendered the plant unavailable for two and a half years and was rated 4 on the INES scale. Although less uranium was melted in this accident than in 1969, the fuel was much more burned-up, resulting in a greater release of radioactivity. Early signs of corrosion had been reported in September 1976 at the twin plant of Vandellos in Spain, but in a report in Spanish that did not attract the attention of the French. Deterioration in January 1980 of the pressure sensors of Saint Laurent-A2 due to unidentified corrosion/detachment of the fairing sheets did not attract attention either.

After the accident in March 1980, the plant was authorized to restart in October 1983, after 500 people had intervened,[39] cleaning of the support area under the core (as in 1969). The cumulative releases remained low because the deflation of the caisson was delayed, knowing that the fuel was spent fuel. Releases were limited to 1.5 mCi of iodine and aerosols (for authorized weekly maximums of 15 mCi and annual maximums of 0.2 Ci) and 775 Ci of noble gases (for maximums of 1200 Ci weekly and 8000 Ci annually). As in 1969, glass wool candles were placed to filter the carbon dioxide, making it possible to recover mainly small pieces of graphite. The final shutdown of the plant took place in 1992.

To complete the history of abnormal situations in the French Natural Uranium-Graphite-Gas (UNGG) plants, a level-2 incident occurred on January 12, 1987, following the freezing of the water supply from the Loire River, the cold source for the condenser of the Natural Uranium-Graphite-Gas (UNGG) turbofans (problem of supercooling of the liquid water to a temperature below 0 °C explaining the very large quantity of ice at the water intake). The cooling function of the condenser was no longer ensured, and there was a leak of steam from the condenser, detected by the fire systems as a fire outbreak, which led to the last liquid water resources in the auxiliary building being emptied by automatic spray. The water supply was re-supplied by demineralized water from the neighboring PWR plant. Afterwards, the army men destroyed the ice dam and restored the water supply. This incident clearly illustrates the problem of the reliability of the cold source on river plants, whether in extreme cold or in low water conditions during a period of severe drought.

It is for technical reasons, but above all for economic reasons, that Natural Uranium-Graphite-Gas (UNGG) will come to an end in France. Indeed, the operating costs per kilowatt and per year were twice those of PWRs (440 French Francs/1988 for 200 French Francs/1988). Moreover, due to a lack of standardization, each plant used a different fuel, which posed specific problems for each plant. The aging of the plants justified their shutdown. Chinon-A1, very recognizable thanks to its steel sphere, became a museum in 1985, accessible to the public.

On May 4, 2015, the encrypted French television channel *Canal* + presented a program entitled " *Nuclear power, the politics of lies?*" This program presents the accidents of October 17, 1969, and March 13, 1980, as a scoop on alleged severe accidents hidden and buried by the EDF and the government. This very "tabloid" presentation of the facts is rather curious since the

---

[39] Because of the radioactivity, the exposure was limited to 20 minutes in order not to exceed 30 mSv.

first French edition of this book dates from 2011 and was largely based on public facts, so it could not decently be called a scoop. The government responds to these accusations with a mission of inquiry requested by Ségolène Royal, the French Minister of Ecology, Sustainable Development, and Energy, to Philippe Guignard, Chief Engineer of the bridge, water, and forests corps, and Serge Catoire, Chief Engineer of mines corps. While the mission agrees that there was no concerted plot, it appears that the decree that governed the releases of alpha emitters did not mention precise limits on March 13, 1980, and on December 13, 1980, new decrees were issued for the start-up of two new PWR reactors at Saint Laurent, vaguely specifying that "*these liquid or gaseous releases must in no case add alpha emitters to the environment*," which is scientifically impossible to respect insofar as small quantities of fuel remained trapped in the caissons. The discovery of traces of plutonium (less than one gram in total, 10–20 milliCuries of activity, and 10 Sv/mg of radiotoxicity) in the sediments of the Loire (in millions of tons) and its average water flow (1000 m^3/h raises the question of the origin of this pollution, although everyone agrees that the risks for the environment and the population are almost nil. Was it the bursting on 21 April 1980 of a cask that had transported a damaged fuel element, or was it the water from the desiccation of the carbon dioxide released into the Loire by the Natural Uranium-Graphite-Gas (UNGG) plants, which could certainly have contained traces of plutonium? It should be noted that the isotopic analysis of the plutonium makes it possible to know whether it is military plutonium from atmospheric fallout dating from the time of atmospheric testing (the plutonium is then practically pure in plutonium 239), or plutonium produced in power reactors (the plutonium then contains isotopes 238 to 242 and americium 241, and the plutonium 239 isotopy is much lower than that of military plutonium). The mission noted the good faith of the operator and of the authorities in the administrative context of the time, while noting an uncertainty about the norms concerning releases of alpha emitters, an uncertainty that has been slow to be resolved.

As a matter of curiosity, a forum was set up during the fiftieth anniversary of the 1969 accident by an association from Orléans of the "1901 law" type, somewhat pompously called the "*Collège d'Histoire de l'Energie Nucléaire et de ses aléas*," proposing to reveal alleged "secrets" about the accidents at Saint Laurent des Eaux. Not having participated personally, I cannot give my opinion on the level of information delivered, but the sensationalism of the poster (Photo 2.88) bodes well for information that is, to say the least, biased. Let's bet that this association had at least read the first 2011 edition of this book!

Photo 2.88 The poster of a forum on the accidents of Saint Laurent. Not so secret as that! The teasing is the following: "*Did you know it? 50 years ago, on October 17, 1969, there was a NUCLEAR FUSION accident involving 50 kg of uranium, as at Three-Mile-Island, which was kept secret for over 40 years! What was the contamination of the living and the environment? Few know*". The words "*NUCLEAR FUSION*" are rather misleading in this context

Bohunice A1: (Czechoslovakia, 1976, 1977)

Czechoslovakia embarked on a nuclear program in the late 1950s. The site chosen was Jaslovské Bohunice, 60 km from Bratislava (Photos 2.89 and 2.90). Besides the A1-reactor, 4 WWER-440 reactors of Russian design (2 models 440/230 and two more recent models 440/213) were built in 1972. The construction of the reactor, a model called "KS-150" of Soviet design but entirely manufactured by the Czech company Skoda, began in 1958 (Photo 2.89). The reactor (Fig. 2.69) used natural uranium (4.5 tons in all) contained in an assembly 12 m high (!), moderated with heavy water contained in a calandria (Photo 2.91) and cooled by carbon dioxide at 65 atmospheres (65.9 bars) which passes through pressure tubes. The reactor went into operation on December 25, 1972. The reactor has a power of 143 MWe gross (93 MWe net, 560 MWth), which heats the primary circuit of carbon dioxide (Fig. 2.70) at 410 °C. The secondary circuit is composed of 6 SGs, the steam of which turns three turbo-alternators of 50 MWe each. The core can be loaded continuously, with the reactor running, just like the French Natural Uranium-Graphite-Gas (UNGG) or English MAGNOX models. A fuel loading machine moves on the loading side to

Photo 2.89 Beginning of the A1 reactor construction (circa 1959)

Photo 2.90 The site of Bohunice. The A1 reactor is recognizable with its very high hall with white roof and by the chimney on its left. The eight air coolers in operation serve the 4 WWER-440 reactors on the right of the picture

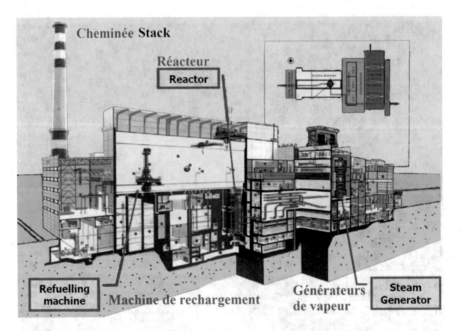

Fig. 2.69 The A1-reactor. The fuel loading machine being very high to contain assemblies of 12 m high, the hall must be very important in size

Photo 2.91 KS-150 reactor calandria inserted into its housing (left) and manual insertion of fresh fuel for the start-up core (right)

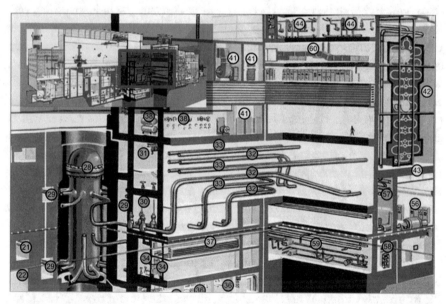

Fig. 2.70 The primary circuit of the A1 reactor: 28- Reactor pressure vessel, 29- Primary pumps of carbon dioxide, 32- Six cold loops of carbon dioxide, 33- Six hot loops of carbon dioxide, 42- one of the 6 steam generators (image adapted from the site of the Slovak company Javys). The reactor vessel has 6 cold inlets at the top of the active core and 6 hot outlets at the bottom of the active core, which means that the primary circuit is complex and winding because the SGs are located far from the core, requiring an extremely large amount of plumbing, and welding work inducing very significant risks of leakage

connect to a pressure tube in a tight way and inserts or extracts an assembly (Photo 2.92 and Fig. 2.71).

A serious incident took place on January 5, 1976 during the reloading of a pressure tube by a new fuel. The shift supervisor, Viliam Pačes, directs the loading maneuver. The electronic control on the loading mast, which tests the correct closing of the pressure tube, shows that the connection is tight. In fact, operator Martin Slezàk lifts the fuel assembly slightly as required by procedure. However, the connection was not tight, and the pressurized carbon dioxide (65.86 bars) ejected the assembly, which was 12 m high. The failure of the closing system of the carbon dioxide channel thus causes the ejection of the new fuel that had just been placed. The force of the flight is such that the ejected fuel will hit the crane of the fuel loading machine. Even some of the steel cubes used to block the assembly start to fly away. The whistling sound of the carbon dioxide depressurization is frightening, far superior to a full-powered alarm siren. After a moment of fright, the shift supervisor rushes into the control room to alert and retrieve gas masks, then returns with a man from the radiation protection team to the reactor hall. Operator Slezàk is injured more seriously and evacuated. For 10 to 15 min, Pačes, assisted by Milan Antolík, will heroically struggle to evacuate the fresh fuel assembly around the fuel loading machine and close the channel despite the flow of gas which cannot be stopped because it must evacuate the residual power, even if the reactor is shut down. The radioactive carbon dioxide then escapes into the reactor hall during this time. As the fuel is fresh, there is little irradiation of personnel outside the reactor hall, but two unmasked operators near the hall airlock who did not evacuate were asphyxiated to death by the carbon dioxide. This accident will remain largely ignored in the West until 1998 and 2006.[40] It would probably have been classified at level 3 on the INES scale. The reactor will be shut down until the end of 1976 for modifications. An investigation of the accident by the Czechoslovak security services was conducted. Antolik explains: « *We all knew why. It was simply impossible for Soviet technology to fail. Even when we were later debriefed by the StB secret service and the criminal police, all questions were directed towards finding a culprit. They pushed the search for a saboteur in order to be able to qualify the accident as a deliberate act of sabotage. At that time, it was inconceivable to say that a Russian reactor had been damaged*». Viliam Pačes spent his entire career on Slovakian reactors. Pačes apparently received a high dose during the accident, as he experienced nausea afterwards, but he is still alive today. He is currently retired. When Fukushima, he was asked about the 1976 accident. He reports some elements on what would have turned to an even bigger tragedy:

---

[40] An excellent synthesis of Jozef Kuruc and Lubomir Màtel: *Thirtieth anniversary of reactor accident in A-1 Nuclear Power Plant Jaslovcske Bohinice*, XXVIII Dny radiačni ochrany, November 20–24, 2006, Luhačovice, Czech Republic, Sbornik rozsirenych abstraktu, pp. 159–162, ISBN 80-01-03575-1, from which we draw most of the illustrations in this paragraph.

Photo 2.92 The reactor hall. The fuel loading machine (in orange at the bottom of the photo) moves on the loading floor and is connected to a pressure tube of which one sees the tight closing system. Contrary to the French system where the fuel loading machine moves on rails, this machine moves thanks to the crane (yellow) which runs on rails placed at about 3 m from the ground and clearly visible along the walls. A lateral movement along the crane allows to reach all the pressure tubes that fill the octagon of the core. The height of the machine (and its weight) is considerable because of the height of the assembly (12 m), whereas the French have chosen a fuel in the form of a cartridge, which is much easier to handle

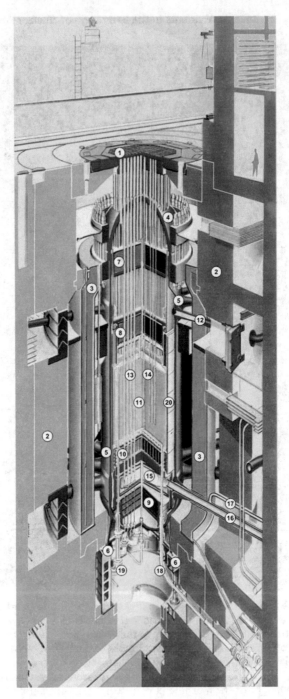

Fig. 2.71 KS-150 reactor: 1- Loading face, 2- Concrete biological protection caisson, 3- Biological protection water tank, 4- Pressure vessel cover, 5- Pressure reactor vessel, 6- Pressure vessel support, 7- Upper biological protection, 8- Middle biological protection, 9- Lower biological protection, 10- Graphite reflector of the core support plate, 11- Heavy water calandria in the active zone of the core (about 12 m high), 12- Cold leg for carbon dioxide entry from the steam generators, 13- Pressure tubes containing the

« At the time the fuel assembly came out of the reactor and the gas escaped, I didn't have an oxygen mask. So, of course, I inhaled some. But I got exhausted pretty quickly and I had to get out of the hall, otherwise I wouldn't have survived. Then the evacuation started in the reactor building. We agreed with the plant management that the only way to avoid a catastrophe was to seal the leaking pipe. So, I took an oxygen breathing apparatus and went back quickly, because the gas was leaking at a tremendous rate under high pressure. The dosimetrist went with me, but there was not much he could do on the spot. The radiation in the reactor hall was so enormous that it exceeded the capabilities of the measuring instrument he was using. Carbon dioxide was escaping through the leak, so there was a risk of that some reactor components could melt. The dosimetrist told me to leave immediately. Later, however, he realized that someone had to do it. I knew that I could receive a dose of radiation that could kill me. Other people told me later that they thought I would not come back and that if I survived, I would have long-term effects. But when a person is at work and the rescue is up to them, they think differently than when they are at home on their couch. I couldn't tell myself to let someone else do it. Because there was no one else who could seal that pressure tube in the reactor. I also imagined how many people would be threatened by the disaster. I knew from my industrial training that radiation could endanger them. Among these people in my imagination were my wife and children. All this made me go further. I don't know how much dose I received. I didn't take my personal dosimeter with me in the rush. Measurements were taken later, but I didn't know the results. However, we can deduce a clue. For a year after this procedure, I was forbidden to go into the reactor hall so that I would not receive any further doses. When I left there, I didn't feel well. After a while, however, it passed. I also forgot that I had to get a checkup. It was a completely different time; the measuring instruments weren't that good, and a lot of things were kept secret. We also didn't have enough experience back then and we weren't well trained. Today, the requirements are much higher. My superiors and colleagues thanked me and shook my hand. That's all, no one gave me a bonus. Then, only on the fifteenth anniversary of the founding of Slovakia in 2008, President Ivan Gašparovič awarded me and Milan Antolik the Milan Rastislav Štefánik Cross (Photo 2.93). However, my great reward was that, although there was a lot of material damage, nothing worse happened».

On February 22, 1977, a severe accident occurred leading to fuel melting. The loading of a fuel that did not allow sufficient passage of CO_2, for reasons that are not well known, led to a local heat up that resulted in the melting of the fuel and

---

Fig. 2.71 (continued) fuel elements, 14- Control rods and emergency shutdown rods, 15- Hot carbon dioxide collection chamber, 16- Hot carbon dioxide evacuation pipe to the steam generators, 17- Injection of cooling CO_2, 18- "Cold" heavy water injection channel, 19- "Hot" heavy water extraction channel to a heat exchanger for cooling the heavy water (image adapted from the Slovakian company Javys)

Photo 2.93   Viliam Pačes (on the right of the left image) receives the Milan Rastislav Štefánik medal from the hands of Slovak President Ivan Gašparovič in 2008 for his courage during the accident at the A1 reactor in Bohunice (photo Slovak TV), Milan Antolik is also awarded (on the right of the right image). In 1987, both received a medal from the Prime Minister Lubomír Štrougal for services to construction during the Soviet era. Antolik reported that Štrougal was sweating profusely, his hands were shaking, and he seemed to be afraid of the contamination that the two former operators might have passed on to him 10 years later!

the piercing by the corium of the pressure tube separating the fuel and the CO_2 from the heavy water in the calandria. The mixture of heavy water brought to saturation by the corium and carbon dioxide contributed to the oxidation of the fuel cladding and the steam generator tubes. 132 assemblies (!) partially melted. The primary circuit, the secondary circuit and the reactor hall were contaminated. The contamination was such that the Czechoslovak government decided to close the reactor in 1979, which had produced a total of some 916 GWh in 5 years. 439 of the 571 spent fuel assemblies were evacuated to the Soviet Union from 1984 to 1990. The 132 badly damaged assemblies were sent to the Mayak site in Russia in 1999. The accident was classified 4 on the INES scale.

In June 1978, heavy rains at the site spread contamination to the Dudvah River, a tributary of the Vah River which itself flows into the Danube 90 km away, because no isolation action had been taken after the accident due to lack of funding. Dismantling operations did not really begin until 1995. The mass of deposits in the primary circuit was estimated at 14.3 tons, which is considerable. The gamma contamination is estimated to be between 10^{14} Bq and 10^{15} Bq. The alpha activity is between 10^{11} Bq and 10^{13} Bq. Starting in 1997, pits were dug around the reactor to pump tritium-contaminated water from the water table and limit leakage to the biotope (Fig. 2.72). ^{137}Cs activity (half-life 30.1 years) was measured at the site in 2004 (Fig. 2.73).

Constituyentes RA-2 (Argentina, 1983)

The Constituyentens accident falls into the category of criticality accidents in experimental reactors, like those of Vinča and SL-1 seen previously.

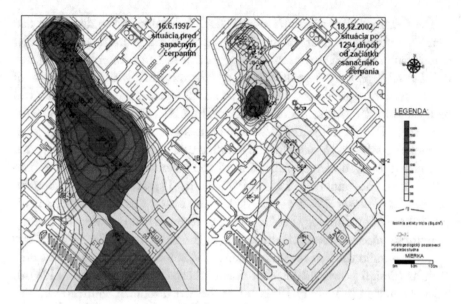

Fig. 2.72 Tritium contamination (in Bq/liter) of groundwater after the installation of pumping pits in 1997 around the reactor

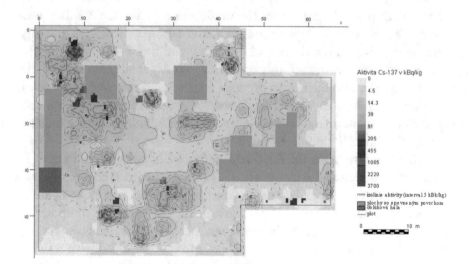

Fig. 2.73 Ground activity of ¹³⁷Cs on the site. A higher concentration is observed around the reactor building and the circular stack

The *Centro Atómico Constituyentes* (CAC) is located in the district of San Martín, Buenos Aires, Argentina (Photos 2.94 and 2.95). The center houses several facilities, such as the first nuclear reactor in Latin America (RA-1), RA for *Reactor Argentino*. A second RA-2 reactor was built with the objective of testing core arrangements for a more powerful reactor: the RA-3, which will be located at the Ezeiza Atomic Center. The CAC also has a heavy-ion gas accelerator TANDAR13. The CAC also houses a plant for the manufacture of uranium powder and a plant for the manufacture of fuel elements for research reactors, initially under the direction of the physicist Jorge Alberto Sabato (1924–1983. he founded in 1955 the Metallurgy Department of the *Comisión Nacional de Energía Atómica*—CNEA).[41] The center hosts laboratories dedicated to nanotechnology, solar energy, research and materials testing. Argentina's nuclear development began in the early 1950s. In 1957, it was decided to build the first research reactor. The RA-1 reactor, built in just nine months (design power 120 kWth, now authorized to operate at 40 kWth), began operation in January 1958. It was the first reactor in South America to diverge, which is a legitimate source of pride for Argentina. Originally, the RA-1 was an *Argonaut* reactor of American design operating with enriched uranium supplied by the Americans. The RA-1 is an open vessel reactor, reflected by graphite (imported from France), and whose moderator and coolant are demineralized light water. The maximum neutron flux is $2 \ 10^{12}$ neutron/cm^2/s against an average neutron flux of $3 \ 10^{14}$ neutron/cm^2/s in a large PWR. In the early 1960s, the core of RA-1 was modified. Fuel rods (20% ^{235}U enrichment) were introduced in place of the old Argonaut core design. The RA-1 facility was also the first to produce domestic radioisotopes for medical and industrial purposes. RA-1 is still used today for material activation testing, radiation damage and research onto new therapies in nuclear medicine, among other areas (Photos 2.96 and 2.97).

A critical facility called the RA-0 zero-power facility was first built at ACC and then transferred to the University of Cordoba. The RA-0 core has a circular ring geometry formed by two concentric and separable tanks made of anodized aluminum (Photo 2.98). It houses the fuel elements, composed of 20% enriched uranium in the form of rods, and demineralized water used as a moderator. The control rods are made of cadmium cladded with stainless steel. A rapid draining of the water from the vessel completes the safety measures. As this is a very low-power reactor (maximum neutron flux of 10^7 neutron/cm^2/s), no coolant is required and there is virtually no wear and tear on the fuel, so it does not need to be replaced. The RA-0 is used to train the

---

[41] The CNEA was created on May 31, 1950, by President Juan Domingo Perón to oversee Argentine work in the field of the peaceful use of the atom.

Photo 2.94   The *Centro Atomico Constituyentes*

Photo 2.95   The tower of Constituyentes is emblematic and "watches over" the site

Photo 2.96   The RA-1 reactor known as "Enrico Fermi" in honor of the builder of the first-ever reactor in Chicago

Photo 2.97   Operators working on the loading face of the RA-1 reactor

operators of the two power reactors Atucha-1 and Atucha-2. A digital reactimeter (numerical inversion of the Nordheim equations from a neutron flux measurement) was implemented at the start-up of the RA-0, which is the only reactor in the country to have such an instrument.

Photo 2.98 View of the Argentinean reactor RA-0

After that, the RA-3 project began to build a 5 MWth multipurpose nuclear reactor of the pool type Material Testing Reactor (MTR), for radioisotope production and research. For this reason and to define the characteristics of the RA-3 core, another critical zero-power facility was built, the RA-2. Initially, RA-3 was a 90% enriched fuel reactor, and its operation began in 1967 at the Ezeiza Atomic Center. The maximum fast neutron flux of RA-3 is 2.5 10^{14} fast neutron/cm^2/s and thermal neutron flux of 8 10^{13} thermal neutron/cm^2/s. RA-3 operates 4 days a week just for medical isotope production (Photo 2.99). When the Atucha-I nuclear plant project began, a German-designed power reactor, a small homogeneous reactor, was offered by the German government to Argentina (1969). It was the RA-4 reactor of the University of Rosario (20% enrichment, 1 W). In 1982, the pool reactor RA-6 of the Bariloche Atomic Center reached criticality. It is a 500 kW reactor with MTR fuel elements enriched to 90%. In 1990, the RA-3 began operating with 20% enriched fuel. In 1997, the RA-8 (a multi-purpose critical facility located at Pilcaniyeu) began operation. The RA-3 reactor is CNEA's most important reactor for the development of Argentine research reactors. It is the first of a series of Argentine MTRs built by CNEA (and INVAP Se) in Argentina and other countries: RA-6 (500 kW, Bariloche-Argentina), RP-10 (10 MW, Peru), NUR (500 kW, Algeria), MPR (22 MW, Egypt).

The RA-2 reactor, in charge of testing the RA-3 configurations, is much less documented (if none!) since it is the one on which the criticality accident occurred. Even today, government agencies and the CAC are more than

Photo 2.99 The RA-3 reactor whose main function is the production of medical isotopes such as metastable technetium 99

discreet about this reactor, which is never presented on official sites, as if they had wanted to erase this history so as not to harm the export effort. The RA-2 is a critical installation which diverged in July 1966 and of very low power (0.1 W) whose objective is the study of fuel lattices. The RA-2 reactor uses fuel in the form of enriched uranium plates clad in aluminum. The lattice is easily changeable for research in reactor physics. The core of RA-2 has a cross-section of 305 mm × 380 mm and an active height of 655 mm. In this geometry, different configurations of MTR fuel elements made of uranium enriched to 90% in ^{235}U are inserted, arranged in 19 uranium plates for the standard elements (width 75.5 mm× thickness 1.6 mm× height 655 mm), and 15 uranium plates interspersed with 2 cadmium plates for the control elements, both cladded in aluminum alloy. The power of the reactor is controlled by 4 cadmium control rods covered with stainless steel. The fuel casing is surrounded by a graphite reflector about 75 mm thick. The reactor vessel is entirely filled with demineralized light water, which acts as a coolant and moderator. Cooling is by convection and natural circulation of water inside the reactor core. On May 17, 1967, a mock-up core of RA-3 reached criticality in RA-2, in order to verify the configuration of the fuel assemblies. After the successful test, the work necessary for the inauguration of the RA-3 on December 20, 1967 was accelerated. Thereafter, the RA-2 continued to be used for various types of tests until the time of the accident. One can get an idea of the shape of the RA-2 by looking at the RA-1 pile.

On Friday, September 23, 1983, Osvaldo Rogulich, (Photo 2.100), chief operator in electro-mechanic of the RA-2 reactor of the CNEA with 14 years of experience, was waiting for the end of his shift at 5:00 p.m.. Since everything went well in the morning, he gave leave to his assistant around 2:00 p.m. since there was no more work, thus giving him an early weekend departure. However, around 3:00 p.m., Rogulich was asked to load a new core configuration for an experiment using a pulsed source and given his competence, he decided to do it alone. The procedure required a complete draining of the moderator fluid before any change of configuration of the fuel assemblies, to avoid any criticality risk. But complete draining means complete reflooding, which takes time. Probably voluntarily (?), Rogulich only half emptied the vessel of its water, convinced that the new half-full vessel geometry would be subcritical. At this level of progress in the change of configuration, he was right; but the fuel substitution operations that he was going to carry out will cruelly disabuse him of this belief. In direct view of the core, he could not ignore the presence of water. However, he violated the safety rules. Unfortunately, the partial removal of water from the moderator was not the only violation of safety procedures. Contrary to standard practice, two standard MTR fuel elements were left transiently near the graphite reflector but were not completely removed from the core. In addition, two control elements without their corresponding cadmium plates were inserted. The criticality of the lattice was reached at 4:10 p.m. when he tried to insert the second element. This second fuel element was found partially inserted afterwards, which suggests that it was at this moment that the power excursion took place. This consisted of a very short pulse of about 3×10^{17} fissions, which released about 10 MJ of power in the form of gamma and neutron radiation. This energy release occurred in about 50–70 milliseconds, long enough for Rogulich to see the flash of light emitted in the visible spectrum.

Since Rogulich was not wearing a dosimeter (?), the dose he received is estimated at 2000 rad (20 *Gray*) of gamma rays and 1700 rad (17 *Gray*) of neutrons, or a minimum dose equivalent of 37 *Sievert*, when the lethal dose is about 5 *Sievert*. The absence of a dosimeter shows how unaware the operator was of the risks he was running, especially alone. The other people on the site also did not have dosimeters. One of the conclusions of the investigation was that, probably because of several years of incident-free operation of the

Photo 2.100   Osvaldo Rogulich's official card as an agent of the CNEA. Rogulich was a methodical, cautious man who did not talk much. Married with three daughters, he lived in the working-class neighborhood of San José. He had joined CNEA as an electromechanical technician, and when one day, one of his daughters asked him what his job was on the RA-2 reactor, he replied, "*I turn a handle.*" (as reported by his daughter Marcela)

reactor, overconfidence may have played a role in simplifying steps and neglecting key safety factors. Thirty minutes after the irradiation, Rogulich experienced headaches, vomiting and diarrhea. Between 2 and 26 hours after the accident, "*the latency phase was observed, with no general clinical manifestations*" described scientists Dorval, Lestani and Marquez of the Balseiro Institute in a 2004 paper analyzing the accident. "*I saw him that night at the Policlínico Bancario de Caballito where he was hospitalized, and he was perfectly lucid*" recalled a retired operator who worked alongside Rogulich. A few hours after the radioactive accident, the president of the CNEA, the physicist and vice-admiral Carlos Castro Madero, visited Rogulich at the *Policlínico Bancario*, shortly before he lost consciousness. "*Workers use hammer and sometimes they get a hammer on their finger,*" the president reportedly told him cynically. This is probably the first act of scapegoating that Rogulich will be made to bear. 28 hours after the event, Rogulich went from the latency phase to the acute phase and began vomiting again. For the next 6 hours, he experienced anxiety and elation, although he remained lucid. Then the neurological syndrome started with loss of consciousness, and a symptom of vascular damage caused by radiation. He had convulsions, suffered three cardiac arrests, and finally died of acute radiation syndrome exactly 48 hours and 25 minutes after the nuclear accident at the RA-2 reactor. France, notified before his death, immediately offered to treat Rogulich in the department of Professor

Georges Mathé, where he could be given a bone marrow transplant like the Vinča accident victims, but when the French understood the level of radiation he had received, they declined: « *No need, he'll be dead before he gets there!* » they would have said prophetically. Moribund, he would not have been able to withstand the heavy operation of a transplant anyway. Eight other employees who were in the vicinity of the reactor at the time of the accident were affected by radiation, but at much lower doses that did not affect their health, according to evaluated dosimetry and subsequent follow-up (adapted and commented from an article of Facundo Di Genova in the Argentinean newspaper « *La Nacion* »). It was said that the alarms of the RA-2 reactor had not been triggered, unlike those of the RA-1 reactor, this prompted the operators of the neighboring reactor to take refuge in the RA-2 hall! If this is true, fortunately for them, the power peak was very short, and everything ended well before they had time to enter the hall.

Despite the seriousness of what happened, the 1984 annual report of the CAC (already in democracy, under the administration of President Raúl Alfonsín) did not mention the accident, and the only mention of the reactor is that the tasks of "updates" were going on. The RA-2, without giving reasons, was dismantled between 1984 and 1989. It will disappear completely from the history of Argentine nuclear power. Rogulich's daughter, who also worked for the CNEA, also felt that her father was being blamed, although the operational procedures were far from precise and their application was generally questionable. In 2007, the inventory of all the spent and unused fuel assemblies of this reactor was evaluated at 19 assemblies of highly enriched uranium and 91 plates of bent fuel, which had been made from highly enriched uranium (90% in ^{235}U supplied by the United States, the almost military enrichment made it possible to build a very small core). Fuel was sent back to the United States under the aegis of the US Department of Energy. These fuels had been kept until then in dry storage conditions on the site itself. Nowadays, it is very difficult to find detailed information about this accident, probably because Argentina had commercial interests in the sale of experimental reactors abroad. Although the remains of Osvaldo Rogulich lie in the cemetery of Lomas de Zamora, the RA-2 continues to haunt Argentina.

3

The Three-Mile-Island Accident

Abstract The core meltdown accident at Three Mile Island in 1979 in the United States was a major challenge to nuclear safety in the scientific community. An accident considered impossible by the specialists had nevertheless occurred, raising fundamental questions such as the understanding of the accident by the operators in the control room, the redundancy of the counter-measures and the efficiency of the equipment under accident conditions. A major plan of measures has been initiated by Western countries with the appearance of probabilistic safety studies and the increased development of defense in depth.

The accident of the second reactor of the American plant of Three Mile Island (without the s at the end of Mile, a frequent mistake!) sounded the death knell of the era of civil nuclear bliss. Until March 28, 1979, most of the public remained convinced of the infallibility of scientists. If the consequences of the accident on the public were more than modest, this accident created an

earthquake in the consciousness of the engineers. Mitchell Rogovin[1], responsible for the huge report[2] to the Nuclear Regulatory Commission's 1980
Commission of Inquiry, says himself: *"For years, the debate about nuclear power
in this country was the preserve of a handful of people, The TMI accident changed
all that. The fate of nuclear power has become ingrained in the American consciousness."* The terrible sequence of events that led to the loss of the reactor,
however, provided important real-world feedback.

The Three Mile Island plant (Photos 3.1, 3.2, and 3.3) is located on a small
island surrounded by the Susquehanna River, 16 kilometers from the city of
Harrisburg, Pennsylvania, and 180 kilometers from the capital, Washington.
Here are two similar reactors built by Babcock and Wilcox (B&W) for the
Metropolitan Edison Company. TMI-2 first reached criticality on March 28,
1978, exactly one year before the events we will describe. That is, the reactor
was completely new. The B&W reactors, whose core is very similar to those of
its competitor Westinghouse in the USA or France, have a primary circuit
that has some notable differences (Fig. 3.1). The containment is broadly similar to that of a Westinghouse reactor (Fig. 3.4).

The first point is the important difference in the design of the Steam
Generators (SGs). Indeed, the B&W SGs operate with forced convection fed
with hot water supply from the primary circuit entering from the top of the
SG (Fig. 3.2). The water flows downward through vertical straight Inconel

---

[1] Mitchell Rogovin (1930–1996). Famous American lawyer from Washington, D.C., government advisor
to the IRS between 1964 and 1966, and then attorney general. Very involved in civil rights, he defended
the New York Times in the case of the publication of the Pentagon Papers in 1971 and sued Richard
Nixon's re-election committee.

(photo University of Chicago)

[2] Mitchell Rogovin, George T. Frampton Jr., *Three Mile Island: a report to the commissioners and to the
public*, Nuclear Regulatory Commission special inquiry group, Janvier 1980. To ensure the impartiality
of the report, the NRC commissioned the firm of Rogovin, Stern and Huge to write the report for the
commission of inquiry and the public.

Volume I, NUREG/CR-1250, Vol. I, 183 pages, Volume II Part 1, NUREG/CR-1250, 1–306 pages,
Volume II Part 2, NUREG/CR-1250, 307–808 pages, Volume II Part 3, NUREG/CR-1250,
809–1272 pages.

Photo 3.1 The Three Mile Island plant is located on an island in the Susquehanna River, 10 miles from the city of Harrisburg. The No. 2 reactor is the building with the rounded dome that is located closest to the two non-functional cooling towers. Those in operation produce steam cloud

Photo 3.2 Unit-2 is in the foreground. The photograph is taken from one of the cooling towers of unit-2

Photo 3.3 Unit-1 in operation. Two cooling towers are required to cool the condenser. The auxiliary cold source comes from the river

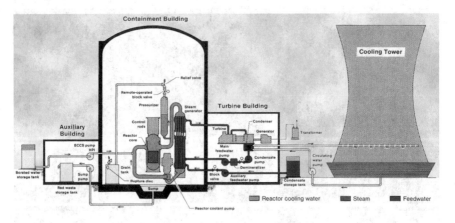

Fig. 3.1 Main components of the TMI-2 plant. *Réservoir d'eau borée* = Borated water tank, *Coeur* = Core, *Pressuriseur* = Pressurizer, *GV* = Steam Generator, *Puisard* = Sump, *Dispositif de chute des barres* = Control rod drive mechanism, *Soupape pilotable* = Power-operated relief valve, *Condenseur* = Condenser, *Transformateur* = Transformer, *Soupape de sûreté* = Safety valve, *Pompe de gavage principale* = SG water main feeding pump, *Pompe de gavage auxilaire* = Auxiliary feeding pump, *Aéroréfrigérant* = Cooling tower, *Circuit primaire* = Primary circuit, *vapeur* = steam, *eau secondaire liquide* = Liquid secondary water

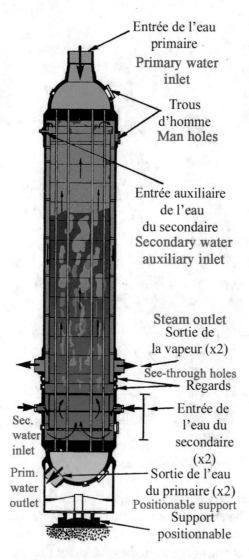

Entrée de l'eau
primaire
Primary water
inlet

Trous
d'homme
Man holes

Entrée auxiliaire
de l'eau
du secondaire
Secondary water
auxiliary inlet

Steam outlet
Sortie de
la vapeur (x2)

See-through holes
Regards

Sec.
water
inlet

Entrée de
l'eau du
secondaire
(x2)

Prim.
water
outlet

Sortie de l'eau
du primaire (x2)
Positionable support
Support
positionnable

Fig. 3.2 Babcock and Wilcox type steam generator (adapted from (M.F. Sankovich, B.N. McDonald: *One-through steam generator boosts PWR efficiency*, Nuclear Engineering International, July 1972)). The SG "OTSG" has a number of advantages: a model implemented on a 1150 MWe reactor provides a superheat of 28 °C at a pressure of 80 bars compared to the saturation temperature of the secondary due to the primary/secondary countercurrent flow (primary water enters from the top through a 91.4 cm diameter orifice and exits from the bottom through two 71.1 cm orifices, the secondary water enters from the bottom and starts boiling immediately, it is completely evaporated at two-thirds of the height, the steam produced flows into an annular down-comer and the steam comes out through two 61 cm diameter orifices) which improves the heat exchange performance. This reduces the kWh to 60.5 kcal of heating, or about 2.3% less than an inverted U-tube generator that operates as a boiler. The superheat produced allows to produce 100% dry steam, eliminating the need for droplet separators at the steam outlet. This saves considerable space as there is no separator/dryer at the SG outlet. The simplicity of the concept allows for infrequent cleaning, resulting in cost savings and increased availability. From a safety point of view, however, this type of SG drains much more quickly than a U-tube SG

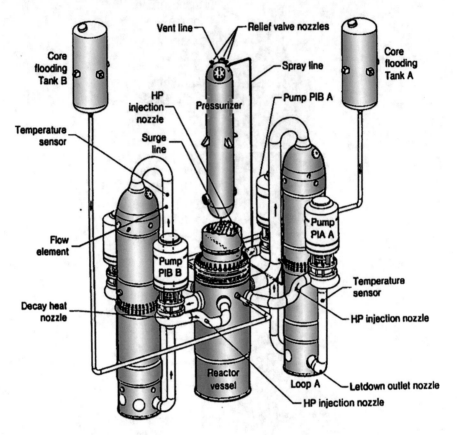

Fig. 3.3 Primary circuit of the TMI-2 reactor

tubes and transfers its heat to the secondary water fluid which flows vertically in counterflow. Therefore, these SGs are called "*once through*" steam generators. This type of circulation allows a great stability in operation, as well as an important superheating of the secondary steam (of the order of 30 °C), which allows better thermodynamic yields than in the case of Westinghouse SGs designed with inverted U-tubes, which operate as a basic evaporator, and whose steam temperature does not exceed the saturation temperature. Secondly, the B&W primary circuit has only two hot loops, which are split into four cold legs fed by four primary pumps at the outlet of the two SGs (Fig. 3.3). However, the vertical design (associated with a small volume of secondary water in the SG) means that SGs have very low thermal inertia compared to U-tube SGs, and they empty their secondary liquid very quickly (in less than a minute) in the event of an accident leading to a heat up the primary circuit when no auxiliary feed water is available from the secondary side. In the case of a U-tube SG, this time would be about fifteen minutes, giving the operator more time to recover the faulty situation. This is the main

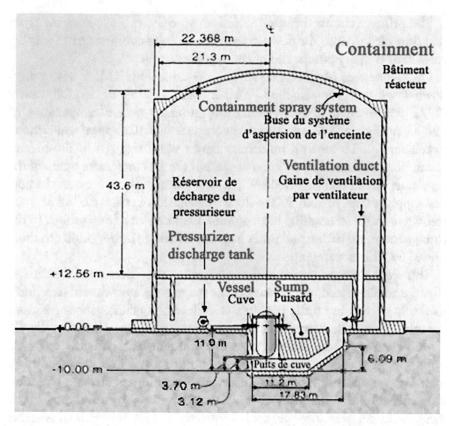

Fig. 3.4 Scale elevation view of the TMI-2 building. *Cuve* = Vessel, *Puits de cuve* = Vessel pit, *Puisard* = Sump, *Réservoir de décharge pressuriseur* = Pressurizer discharge tank, *Gaine de ventilation par ventilateur* = Fan cooler duct, *Buse du système d'aspersion de l'enceinte* = Containment spray system nozzle, *Bâtiment réacteur* = Reactor building

disadvantage of once through SGs, which leave little reaction time for operators to understand what happens. Another particularity is the possibility of cooling the containment (Fig. 3.4) by means of a Reactor Building Air Cooling System (= RBACS) called *"fan cooler"* i.e., a system of five fans inside the containment. The atmosphere of the containment passes in contact with air–water exchangers associated with an exchanger outside the containment. The water in question comes from the cold source, namely the air coolers[3] in normal situation or the river in case of LOCA. This active system was not retained on the French plants.

---

[3] Air coolers are huge-truncated cone-shaped towers from which steam sometimes escapes, and which can be seen from very far away, such as the Cruas air cooler in the Rhone valley (France) illustrated with a child seen from the A6 freeway.

The primary circuit therefore includes two SGs, two accumulator tanks that flow directly into the down-comer (the annular converter that feeds the core), four primary pumps and of course a pressurizer.

On Wednesday, March 28, 1979, at 4:00 a.m., while TMI-1 was in shutdown condition for refueling, TMI-2 was at 97% of its full power (2772 MWthermal, 905 MWelectrical, primary circuit temperature of 290 °C, pressure of 150 bars, boron concentration of 1026 ppm[4] and primary circuit flow of 16 tons/h), numerous alarms were triggered in the control room. The operators did not know it yet, but the 1A condensate pump of the condenser circuit had triggered.[5] When such pump triggers, it causes the normal supply pumps of both SGs to stop. This shutdown causes the loss of water feeding to the SGs, resulting in an automatic shutdown of the turbine. In the turbine room, cavitation and water hammer noises are heard, similar to those caused by air in a water pipe.

This event gives the time 0 of the accident scenario. Because of the turbine trip, the auxiliary feed water circuit of the steam generators should have automatically started, but it did not. In fact, following a maintenance operation, the valves located downstream of the three feeding pumps, which actually did start, were abnormally closed, in formal contradiction with the technical operating specifications. These valves had most probably been closed during a maintenance operation 2 weeks earlier. The operator restored the situation only after 8 min by manually opening the offending valves. But during this time, the reactor was no longer cooled correctly, and its pressure increased due to the heat up of the primary circuit. When there was a signal of high pressure in the pressurizer (153 bars), the discharge solenoid valve located at the head of the pressurizer, which protects the primary circuit from overpressure, opened between 3 and 6 seconds (Power-Operated Relief Valve (PORV), which is equivalent to the SEBIM valves on French reactors). At 8 seconds, the reactor was normally shutdown (SCRAM by rod drop) on a signal of very high pressure in the primary circuit (161 bars). The fluid leaking in the discharge line from the pressurizer to the discharge tank (RDP) allowed the pressure in the primary circuit to fall back to 148 bars, where the PORV solenoid valve should close again below 155 bars due to pressure difference. However, the valve remained stuck open. Very quickly (within 15 seconds), the operator realized that the valve had remained open and gave the closing command via

---

[4] *ppm = part per million.* This unit used in chemistry corresponds to the dissolution of 1 gram of a material in a ton of solvent (one million grams).

[5] The term "trip," also used, should be understood as the tripping of a circuit breaker. The system then shuts down. This term is used for a turbine, a pump, an electrical system...

the control console. The alarm check light (small light associated with the control button) indicated "*valve closed*."[6] But this alarm check light does not indicate the real state of the valve, but just the fact that the button is in the "*valve closed*" position and activates the current in the solenoid. In fact, even though the command was given to close it, the valve remained open. This error in the design of the alarms, linked to the fact that there was no real indication of the position of the relief valve stem, but only an indication of the presence of power on the control solenoid, will have a disastrous effect on the understanding of the accident by the plant's control team. The operators confirmed to the investigation committee that they firmly believed that the PORV valve was closed. The temperature rise at the valve was well measured, but recurrent leaks meant that the temperature was already high initially and this no longer worried the operator. As far as the alarms were concerned, we must admit that they all started on, both audibly and visually, causing great confusion in the control room. It was no longer possible to recognize the initiating alarms.

Even the printer in the control room refused to return the information contained in the data acquisition system during the two fateful hours, perhaps because of the saturation of the output buffer of the computer feeding it. From that moment on, the incident, which could have been controlled, became a LOCA (Lost Of Coolant Accident). At 60 seconds, the water level in the pressurizer rised rapidly. At the same time, both steam generators reached their low level. At 120 seconds, the automatic start of the high-pressure safety injection (HPSI) took place on low-pressure signal in the primary circuit (112 bars). From 4 to 11 min, the level of the pressurizer went out of its reading range to the great horror of the operators, whose obsession was the pressure rupture of the pressurizer or its pressurizer surge line. This phenomenon can happen quickly if the pressure compensation steam bubble at the head of the pressurizer has been lost, that is to say, in the jargon of the profession, if the pressurizer is "*solid*" (full of water). It should be noted that the water level in the pressurizer is measured by a differential pressure sensor (weight of a column of water). A control room reading of 33.5 feet indicates that the pressurizer is full of water with no vapor or bubble void. If a void fraction exists in the pressurizer, a level below 33.5 feet is indicated, the difference being directly proportional to the actual volume fraction of liquid water

---

[6] As a matter of fact, an alarm check light off means that there is no current applied to the valve opening solenoid, so the valve is supposed to be closed when the lamp is off.

in the pressurizer.[7] The commission of inquiry will show that the operating teams had been trained in the excessive fear of a pressurizer surge line rupture situation in case of "*solid*" pressurizer. It was in this context that the operators made a fateful decision: they decided to manually shutdown at 278 seconds the two safety injection pumps that could have saved the reactor.

In fact, the water level measurement system of the time was not reliable in a two-phase situation and the massive boiling from 6 min. Due to the drop in pressure (93 bars), it induced significant variations in the level of water swollen by steam. Still worried, the operator was also going to draw water from the primary circuit (!), which is already lacking, via the primary load/discharge circuit (RCV), always in the hope of reducing the (fictitious!) water level in the pressurizer. At the time of 8 min, the operator opened manually the valves of emergency water supply of the steam generators, finally aware that the steam generators were empty by lack of backed up water. From 11 min, the level of the pressurizer became readable again, unfortunately confirming the operator in his wrong behavior. The operator did not realize that a "small break LOCA" (SBLOCA) was occurring,[8] located at the head of the pressurizer. This induces a two-phase situation involving a high level of water in the pressurizer. The high-pressure injection was manually restarted but with a very low flow rate. At 15 min, the rupture of the rupture membrane in the pressurizer discharge tank (RDP), which had been completely filled with water leaking from the pressurizer, allowed radioactive water to flow from the primary circuit to the sumps located in the lower parts of the reactor building. The pressure indicator of the pressurizer discharge tank, into which the primary circuit water is discharged, indicated a rise in pressure. However, to make matters worse, this indicator was placed on a back panel of the control room, beyond the operator's vigilance.

The alarms then indicate the presence of radioactivity in the reactor building. From 20 to 70 min, the pressure and temperature of the reactor stabilized at boiling conditions of 72 bars and 287 °C. At 74 min, the operators shut down the two primary pumps on train B, then at 100 min, those on A-train, because of cavitation. This time initiates the second phase of the accident, namely the uncovering of the core (Fig. 3.5). Indeed, the pumps in motion were stirring a mixture of water and steam, ensuring the cooling of the core as best they could. After the shutdown of the pumps, the water and steam in the

---

[7] Between 5 min and 90 min, the pressurizer was full of a two-phase mixture, so the measured level can be considered as an indication of the void fraction. Note that 5 min. leave a very short reaction time to the operator.

[8] Since the pressurizer relief valve remained open, it was like having an accident where the fluid leaks through a small break.

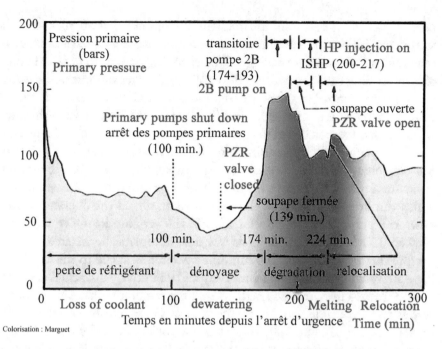

Fig. 3.5 Pressure of the primary circuit and description of the four main phases of the accident. *Perte de refrigerant* = loss of coolant, *dénoyage* = uncovery of the core, *dégradation* = melting of the core, *relocalisation* = meltdown to lower plenum, *soupape fermée* = valve closed, *arrêt des pompes primaires* = primary pumps shutdown

emulsion were to separate, and the residual water will end up in the lower parts of the primary circuit. No longer cooled, the core then heats up and its temperature reaches 327 °C at about 2 h from the onset on the accident, then goes out of the reading range for 14 min. The pressure rose to 148 bars. At 2 h20, isolation of steam generator B and discharge of the secondary steam to the atmosphere through the controlled relief valves. At 2 h36, the on-off isolation valve, located in series before the PORV valve, is manually closed, finally isolating the primary circuit (Fig. 3.5). We can consider that the operators have finally understood the problem and the fact that the core was dewatered (radioactivity appears in the radioactivity detectors of the sumps). During the next 5 h, the operator will try to cool the core, thanks to his SGs, by establishing a natural or forced circulation in the core, but the incondensable hydrogen produced by the oxidation of the zircaloy fuel cladding, trapped in the primary circuit, degrades the heat exchange towards the secondary circuit and blocks the convection. As for the residual power evacuation system (Low-Pressure Safety Injection, LPSI), it can only be operated at lower pressure (28 bars). It would have been necessary to depressurize the primary circuit in order to use it. At 174 min, the operator attempted a delicate maneuver, he started the 2B-pump to cool the core, initiating phase 3 of the accident: the reflooding of the core (quenching). This reflooding caused a rapid rise in pressure by sending colder water into the core, where water vaporizes. The operator will try to control this pressure by piloting the PORV valve, which despite its failures, remains controllable. This flooding fed the pressurizer with steam, which caused the water level to rise when the steam condensed in the pressurizer. The opening of the pressurizer spray line at 175 min. must have facilitated this condensation since the steam could escape by this way. But the terrible cavitation noise of the pump, which circulated a strongly two-phase fluid, and which can be heard clearly in the control room, urged the operator to shut down the pump after 19 min (at 3.2 h). The pressure in the containment reached 0.31 bar, the containment was then automatically isolated from an overpressure of 0.3 bar. At 200 min, the operator switched to the automatic high-pressure injection, which effectively and definitively quenched the core.

At time 224 min, it is conjectured that the corium, which formed in the core as a non-coolable liquid bath (the crust was impermeable to water), relocated (Figs. 3.6, 3.7, and 3.8). This relocation took place under water through the core bypass, the flow zone that surrounds the active core. Indeed, it was at this moment that the external neutron chambers measuring the neutron sources recorded a significant increase in signal.

At the same time, the internal chambers (Incore Self-Powered Neutron detectors) triggered an alarm, suggesting damage caused by the corium, which heated the vessel bottom penetrations by generating a thermoelectric current. This relocation was done under water and sent molten metals partially oxidized in the vessel bottom. At 10 h00 from the onset (Table 3.1), a containment pressure peak at 1.9 bar was attributed to a moderate explosion of hydrogen from the oxidation of the zirconium cladding. The automatic start of the containment spray (EAS) was shut down after having had time to inject about 20 m³ of water containing soda (soda favors the retention of volatile iodine in the sumps). At about 13.5 h, while maintaining the high-pressure injection to re-pressurize the primary circuit, residual power was finally released through the SG-A, because part of the hydrogen had been purged during the depressurization operations with the pressurizer valve. After 16 h, the plant returned to a stable state with the supposed presence of a bubble of non-condensable gas in the upper parts of the reactor, under the vessel cover. This gas, allegedly hydrogen, greatly worried the operators because of the risk of explosion, or even the risk of dewatering of the core in the event of a drop in primary circuit pressure. Throughout the first week, the operators tried to reduce this hydrogen bubble more or less dissolved in water, by heating it with

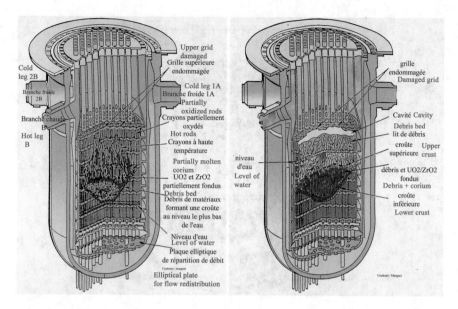

Fig. 3.6 State of the core before reflooding (before 174 min) and after reflooding (174 min to 189 min)

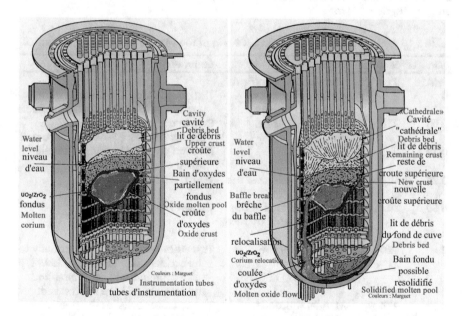

Fig. 3.7 State of the core before and after relocation (224 min)

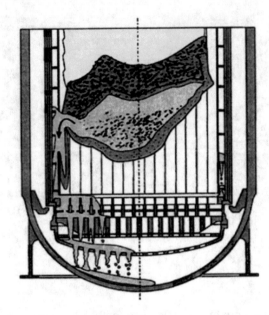

Fig. 3.8 Relocation of the corium from the core to the vessel bottom through the side bypass. Even though the reactor was full of water, this did not prevent the corium from progressing downwards

Table 3.1 Main chronology of the first 300 min of the TMI-2 accident (Figs. 3.6 and 3.7)

Minutes	Seconds	Event	Comment
PHASE 1: 0 to 100 minutes: Loss of coolant			
0	0	Untimely condenser pump shutdown	The turbine trips. The emergency power supply of the SGs is unavailable
0.05	3 to 6	PORV valve opens	Due to the increase in pressure because the core is no longer cooled.
0.13	8	Scram	On high pressure in primary circuit signal
0.20	12 to 13	The operator closes the PORV	**Which remains open!**
0.50	30	Attempt to launch the ASG (SG auxiliary feed water)	On low level signal in SGs
2	120	Activation of the high-pressure safety injection (112 bars)	This automatism responds normally to the accident due to system closure. If this action had not been inhibited, the core would have been saved!
4.63	278	Manual shutdown of the high-pressure injection	The operator believes the pressurizer is solid. **The accident will then get worse.**
5	300	Water withdrawal from primary circuit via RCV (primary circuit letdown)	The operator is afraid of a pressurizer rupture.

(continued)

Table 3.1 (continued)

Minutes	Seconds	Event	Comment
8	480	Opening of closed valves that inhibited ASG	The operator regains the use of his SGs. PORV still open.
15	900	RDP (pressurizer discharge tank) disk failure	Radioactivity in the reactor building
74	4400	Primary pumps B switched off	The core is badly cooled
20 to 70	1200 to 4200	Primary circuit stabilization at 72 bars	
PHASE 2: 100 to 174 minutes: heatup of the core			
100	6000	Primary pumps A switched off	The core is not cooled anymore
110	6600	Start of the uncovering of the core	Based on data from the external neutron chambers
142	8520	Manual closing of the PORV	
PHASE 3: 174 to 224 minutes: Reflooding of the core (quenching)			
174	10,440	Primary circuit pump 2B is restarted	The core is cooled again but the thermal shock causes the upper part of the core to collapse.
175	10,500	Pressurizer spray switched on	
192	11,520	The PORV is opened	Primary circuit pressure management
193	11,580	Primary circuit pump 2B is shutdown	Due to cavitation
		Pressurizer spray switched off	
197	11,820	PORV is closed	
200	12,000	The high-pressure safety injection is put back in automatic mode	Permanent reflooding of the core
220	13,200	PORV is opened	
PHASE 4: 224 to 300 minutes: Relocation of corium to the vessel bottom			
224	13,440	Relocation through bypass (29 tons)	
225	13,500	Pressurizer spray switched on	
262	15,720	Pressurizer spray switched off	
263	15,780	Heaters (11 out of 16) of the pressurizer in operation	Operators want to re-pressurize the primary circuit
267	16,020	High-pressure injection is activated at full speed	Permanent stabilization of the core

the heaters of the pressurizer to recover the hydrogen in the upper head of the pressurizer and degas it via the Pressurizer Discharge Tank (RDP). The nuclear auxiliary building (BAN) was contaminated by the spillage of about 40 m³ of radioactive water (which reached an activity of 800,000 Curie/m³ whereas in normal operation, it is less than one Curie/m³), which overflowed the liquid effluent treatment tanks. The radioactive gases were not filtered by the BAN filters and were the cause of the small release of radioactivity into the atmosphere out of the BAN, despite the presence of iodine traps upstream of the gas evacuation stack. This release of radioactivity caused a psychosis in the public.

The problem of hydrogen in the containment has been solved by the use of passive autocatalytic recombiners, which were installed on April 3. Their installation required the use of 400 tons of lead protection to limit the doses to the personnel.

The core was cooled for months using a single primary pump and the SG-A alone. The problem of the presence of a hydrogen bubble under the reactor cover was of great concern to the operator. Pessimistic calculations (up to 25 m³) had raised fears of a risk of uncovering the core if the pressure was lowered. The operator therefore limited the pressure drop and tried to degas the primary circuit. From April 5, no more bubbles were detected, which made some people doubt its existence. Its volume should never have exceeded 2 m³. On April 27, the natural thermosiphon circulation of water in the core was sufficient to ensure the cooling of the core. The accident was definitively stabilized.

From July 1980 onwards, Operators began to enter the reactor building in a cautious way and strongly protected against contamination (Photos 3.4 and 3.5). Decontamination (Photo 3.6) and dose mapping (Fig. 3.9) were performed. The objective was to insert a shielded video camera into the core via a control rod adapter (Photo 3.7). This visual inspection confirmed the presence of a cavity of 9.3 m³ in the core, resulting from the collapse of the assemblies. The camera allowed to visualize the debris bed resting at the bottom of this cavity.[9]

It is known that the core was not completely dewatered and that about 0.6 m of fuel height remained under water (a situation later called "*cold foot*").

[9] [Bandini et al., 1987] Bernard R. Bandini, Anthony J. Baratta, Victor R; Fricke, *Determination for the end state of the three Mile Island unit-2 accident using neutron transport analysis*, Nuclear Technology, Vol. 81, p 370–380 (1987).

[Duffy et al., 1986] L.P. Duffy, E.E. Kintner, R.H. Fillnow, J.W. Fisch: The *Three Mile Island accident and recovery*, Nuclear Energy Vol. 25 n°4, pp199–215 (1986).

[Knief, 1988] R.A. Knief: *Nuclear criticality for the TMI-2 recovery program*, Nuclear Safety Vol. 29–4, p 409–420 (1988).

[TMI PVIP, 1994] *Three Mile island reactor pressure vessel investigation project*, proceedings of an open forum sponsored by the OECD NEA and USNRC, Boston (USA), 20–22 October 1993, OECD, 1994, 402 pages.

Photo 3.4 Inside the reactor pool at an undetermined date. The control rod mechanism (RGL) is still upon the reactor vessel cover. One can see the studs closing the cover on the vessel. A ventilated tent allows you to momentarily escape the ambient radioactivity

This is confirmed by the position of the corium crust that formed at water level and served as a crucible for the molten bath. The collapse of the fuel was probably caused by the thermal shock due to the reflooding (the word "*quenching*" is used) and the terrible vibrations induced by the cavitation of the 2B-pump. A vault was formed by the collapse of the upper part of the assemblies.

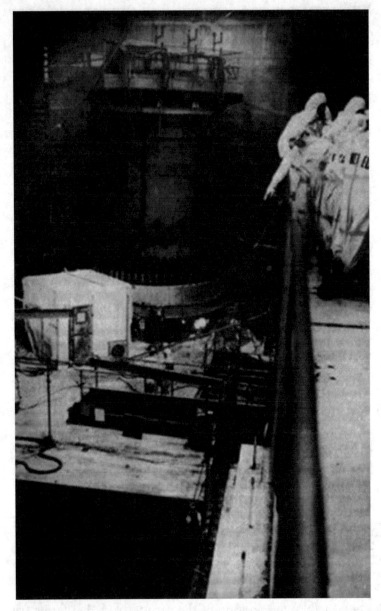

Photo 3.5   Operators walking along the empty reactor pool. They wear filtering mask, but no ventilated suits

From a sonar inspection (Photo 3.8), scientists were able to reconstitute a precise image of the cavity, which is crucial for understanding the core degradation scenario. We can distinguish the relocation of the corium in the bypass.

Once formed in the core, the molten bath, blocked vertically by its lower crust, progressed sideways by natural convection movements that pierced the

Photo 3.6 Cleaning the floor for decontamination

lateral ceramic crust, then going to pierce at about 224 min. the baffle at a height of 1.80 m. Approximately 29 tons relocated by this path through significant breaks (Fig. 3.14). The bypass trapped 9 tons of corium between the spacer plates and corium was even found at a higher altitude of the baffle tear (the corium would have risen due to the clogging of the bypass). Twenty tons were relocated in the vessel bottom and heated up the steel of the vessel, bringing about 2.5 MW of residual power at the time of relocation (224 min after the rod drop). By a phenomenon that is still not fully understood, a gap probably formed between the corium and the inner face of the vessel, allowing water to pass through, which cooled the corium. This water intrusion probably prevented a rapid attack by melting of the vessel, thus avoiding a bottom rupture.

From 1981 onwards, under the aegis of the NEA's Committee on the Safety of Nuclear Installations, 11 OECD member states joined the American DOE in analyzing samples from the core to understand the progression of the core damage. In 1983, in order to understand the anomalies in the measurements

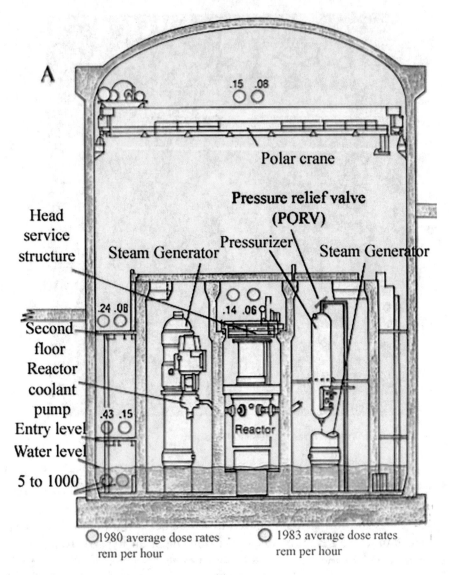

A

Polar crane

Pressure relief valve
(PORV)

Head
service
structure

Pressurizer

Steam Generator

Steam Generator

.15 .06

.24 .08

.14 .06

Reactor

Second
floor
Reactor
coolant
pump
Entry level

.43 .15

Water level

5 to 1000

◯1980 average dose rates
rem per hour

◯ 1983 average dose rates
rem per hour

Fig. 3.9 Dose map et evolution from 1980 to 1983

of the external chambers, which measured a much stronger signal even long after the accident (4 orders of magnitude of the signal expected for the postulated reactivity of the degraded core), wires were placed between the concrete pit of the vessel and the steel vessel, to which were attached dosimeters placed at different altitudes (Fig. 3.10) (Bandini et al. 1987). The dosimeters were exposed for 3 months and then analyzed. In 1984, the extent of the damage

Photo 3.7 Photos taken by the video camera of the debris bed of TMI-2 (1982) from (Duffy et al. 1986)

Photo 3.8 Sonar analysis of the cavity formed in the TMI-2 core (1983) from (Duffy et al. 1986)

was still unknown. Such an analysis allowed to show that some fuel was indeed present in the vessel bottom. The presence detected by the cameras and the sonar allowing the detection of the cathedral cavity at the top of the core, it was then possible to establish a neutronic model of the degraded core and the corium in the vessel bottom. This model showed that at least 10 tons of corium should have relocated in the vessel bottom. This was verified during dismantling (in fact about 20 tons).

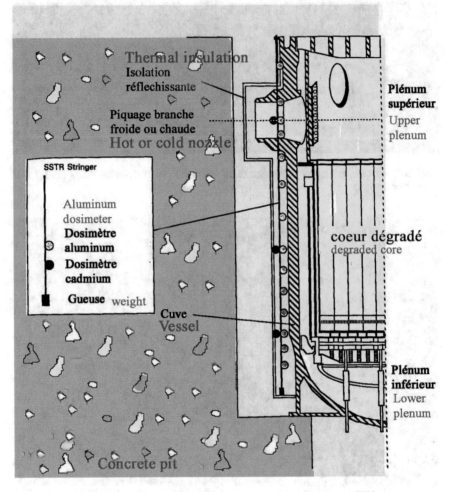

Fig. 3.10 Dose measurement in the vessel pit [from Bandini et al. 1987]

In 1985, an international inspection project[10] of the vessel (Vessel Inspection Project) was set up under the aegis of the OECD/NEA. It appeared that 19 tons of molten corium had been in contact with the vessel bottom of the reactor. A budget of $nine million between 1988 and 1993 allowed samples to be taken from the top of the reactor[11] thanks to an electric discharge

---

[10] [TMI PVIP (1994).

[11] The NRC prohibited breaching the integrity of the vessel by drilling from below.

cutting[12] under 12 meters of water (Metal Disintegration Machine or MDM), while the visibility was only 3–4 meters (Figs. 3.11 and 3.12).

The post-mortem analysis showed that the corium was divided equally between a solidified bath in the vessel bottom, topped by a debris bed, and a solidified bath that remained in the center of the reactor. The question of whether it was possible that the accident led to the downward ejection of some instrumentation tubes is relevant. It was found that some of the instrumentation tubes drowned in the corium bath were severely damaged, while some were curiously intact. Two phenomena are to be considered: on the one hand, the ejection of the tube after melting of the Inconel weld, and on the other hand, the rupture by thermal creep outside the vessel. Concerning the first point, the metallurgical examinations showed that the welds of the penetrations had not melted, suggesting that the temperature of the weld had never exceeded the temperature of *liquidus*[13] of Inconel 600, i.e., 1415 °C. Regarding the creep rupture, it should be noted that the pressure in the vessel is not perfectly known at the time the hot spot appeared at the vessel bottom. At a conservative pressure of 150 bars i.e., assuming that the system was re-pressurized, the thermal creep failure at the highest estimated temperature of the vessel (about 1100 °C) varies according to the hypotheses between 4 and 17 h, whereas this temperature did not finally exceed one hour. It was noted that some of the vessel penetrations (the taps where the instrumentation tubes pass through the vessel bottom) had failed without being ejected, but the corium had frozen inside of them (!) (Fig. 3.13).

The analysis of the samples showed the existence of an elliptical hot spot of about 1 m by 0.8 m, having reached a temperature of 1100 °C in the inner liner, and caused by an intimate contact. Around this spot, it is known that the temperature of 727 °C, which corresponds to the ferrite-austenite transition, was not reached (Fig. 3.13). Cracks of 5 mm were found in the stainless-steel buttering of the inner face of the vessel. They were located around three of the instrumentation tubes and had no extension in the steel of the vessel itself.

---

[12] Electric discharge cutting consists of a clean cut (no chips, preserves the integrity of the component). The principle is based on a graphite or copper-tungsten electrode, placed at a controlled distance from the part to be cut. This distance allows to modulate the energy on the cutting surface and is adjustable by the operator according to the potential difference between the electrode and the surface. The electrode is supplied with a pulsed current. Controlled arcs remove small molten particles that solidify on contact with the dielectric fluid that is interposed between the electrode and the surface: reactor water in the case of the TMI-2 vessel cutting. The shape of the electrode determines the shape of the cut sample. The roughness after cutting remains acceptable and the integrity of the component is not threatened.

[13] For a pure chemical body, we would speak of melting temperature.

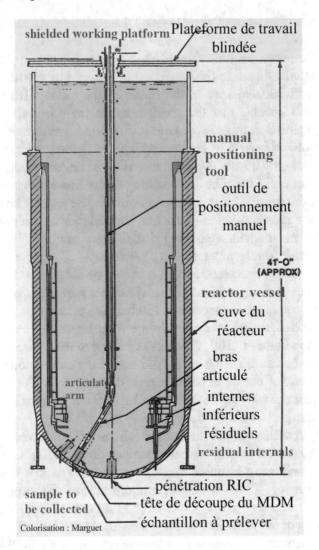

shielded working platform Plateforme de travail blindée

manual positioning tool

outil de positionnement manuel

41'-0" (APPROX)

reactor vessel

cuve du réacteur

bras articulé

internes inférieurs résiduels

articulated arm

residual internals

sample to be collected

pénétration RIC

tête de découpe du MDM

échantillon à prélever

Colorisation : Marguet

Fig. 3.11 View of the complete MDM system from the shielded work platform. The vessel is completely filled with water for effective biological protection

In all, 62 tons of core material melted and were distributed as we said in the bypass and vessel bottom, the rest "freezing" in the destroyed core. A "cathedral" cavity of 9.3 m³ appeared at the top of the core (Fig. 3.14). A molten bath of 33 tons at a temperature between 2800 K and 3100 K finally solidified in the center of the core. The oxidation of the core materials, in particular the zirconium of the cladding but also the steel of the upper internals produced about 460 kg of hydrogen gas. Reflooding calculations based on the hydrogen

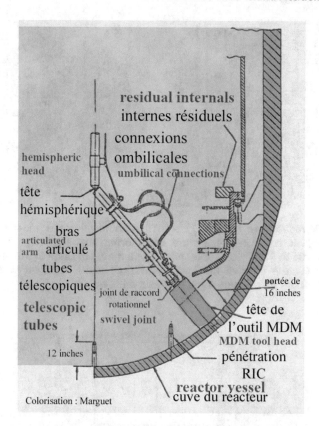

Fig. 3.12 Magnified view of the MDM tool head. Underwater television cameras allow to see in detail the cutting operations. A hemispherical articulated head allows the tool to be raised along the vessel bottom wall. The samples taken have a straight triangular section, 16.5 cm long, in the shape of an oblong boat. From these samples, specimens of standardized shape were extracted (Charpy impact test, etc.), which allowed access to the mechanical properties of the project vessel (TMI PVIP 1993, p. 88)

explosion and the ratio of oxidized zirconium in the core (45%), postulate that 300 kg could have been produced before reflooding and 160 kg during the reflooding phase at the start of the 2B-pump. These 160 kg are the subject of particular attention because they would have been produced very quickly (hence the impossibility of treating this sudden production with passive auto-catalytic recombiners). On the other hand, it is difficult to reproduce this phenomenon experimentally and computationally, which suggests progress on the physics of the reflooding phenomenon.

On the radioactive release side, the overflow of the tank of the TEP system (treatment of primary circuit effluents) for liquid effluent treatment resulted in the release into the BAN of approximately 40 m³ of highly contaminated

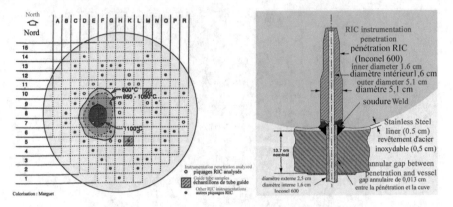

Fig. 3.13 Degradation of the TMI-2 core. The corium relocated under water in the vessel bottom, which rose in temperature. Post-mortem analysis showed that the vessel bottom steel rose to 1100 °C (red spot) without melting, but some of the vessel bottom penetrations (of the internal RIC instrumentation) were damaged without being ejected. The corium froze inside some perforated RIC tubes

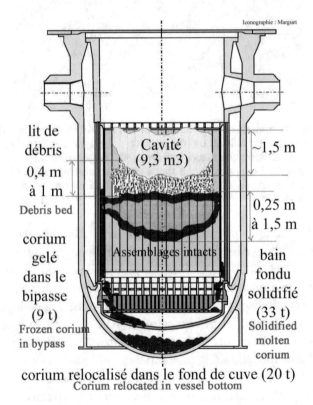

Fig. 3.14 Post-mortem assessment of the degradation of the TMI-2 core

water. The ventilation system released rare gases such as xenon and krypton after filtration ("absolute" filter for aerosols and iodine filter). It is estimated that the total activity released in rare gases is about 50,000 Curies (especially krypton 85 during the voluntary degassing phases in 1980 to enter the building) and less than 15 Curies in iodine 131 during the accidental phase. It is estimated that 99.9% of the cesium and iodine remained trapped in the water retained in the plant. The NRC, the regulatory agency in the United States, has produced a hypothetical value of 80 mrem as the maximum individual dose and an average of 9 mrem for the nearest 2000 inhabitants.[14]

The absolutely incredible timing of the release (12 days before the accident) of James Bridge's film "*The China Syndrome*" with Jane Fonda as the incorruptible journalist, Michael Douglas and Jack Lemmon will exacerbate public attention. Lemmon plays the role of an honest nuclear plant manager, a former Navy officer,[15] who was forced to testify to a commission of inquiry following an accident that occurred because of malpractice by the utility company that falsified the X-rays of the welds of the primary circuit. The Chinese syndrome is a paradoxical image meaning that nothing could utopically stop the radioactive corium, which would pierce the concrete raft down to China. The poster of the film presents an inoffensive air cooler curiously surrounded by aggressive chimneys (?). Let us underline that the film is well made, and the scenario remains credible (Photo 3.9).

On Friday, March 30, the operator decided to make significant releases: 1.2 rem/h from the stack of the Auxiliary Nuclear Building, which prompted the NRC to recommend to the governor the evacuation of pregnant women and young children within a 10-mile radius. Walter Cronkite (1916–2009), the famous American broadcast journalist for the CBS Evening News for 19 years. Often cited as "*the most trusted man in America*," could not elude the TMI-2 accident and presented it as a major event, concluding with his departing catch phrase "*And that's the way it is*" (Photo 3.10). The accident triggered an indescribable panic: gas stations were stormed, money was withdrawn from banks (more than $ten million in one day!), local religious authorities even authorized a general extreme unction by local radio station, a unique fact in the whole history of Church. One hundred and forty thousand people stayed away from their homes during the events. It was reported that some

---

[14] Recall the 100 mrem/year dose due to natural irradiation.

[15] In fact, many of the technicians operating on the real TMI-2 plant were former Nuclear Navy retirees under Admiral Hyman G. Rickover (1900–1986), the founding father of the US Nuclear Navy. Actually, the four operators on the fateful night, C. Faust, E. Frederick, W. Zewe and F. Sheimann, were all former Navy sailors (nuclear submarines).

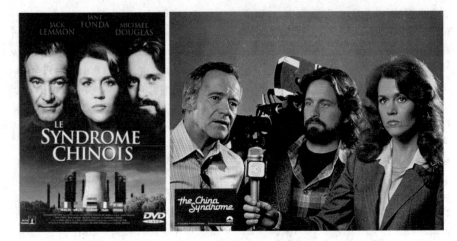

Photo 3.9 French poster of the film: *Le syndrome chinois* and still

Photo 3.10   Walter Cronkite presenting the TMI-2 accident (CBS Broadcasting)

pregnant women have decided to have an abortion for fear of malformation of the fetus.

The next day, the major French newspapers commented on the news. The *Association Française de Presse* relayed information that was either badly translated or unintentionally false ("*explosion of a valve in one of the pumps of the reactor cooling system?*"). Then a certain realism set in in the French press when it was understood that the release was very small. On April 1, the daily *Libération* headlined with a certain black humor "*A clean catastrophe, a mild panic.*" Brice Lalonde, a figure of French ecology, did not hesitate to say in the *Nouvel Observateur* of April 23 "-*I am not afraid of the atom, I am*

Fig. 3.15 The satirical press attacks the French utility's haughty communication « *American engineers are donkeys.* ». (drawing of Konk in a newspaper of 1979, DR)

afraid of technocrats, What is good for EDF is not good for the French!" The balanced newspaper *Le Monde* itself wrote "*- We are beginning to know more about the American plant than about its French counterparts.*" Satirical drawings flood the press (Fig. 3.15), and EDF was taking the fall. However, EDF was very concerned by the accident, both to learn from the experience and to get its own idea of the case. In April 1979, EDF executives were sent to the United States on a fact-finding mission. Finally, French scientists took up the cause on both sides. Maurice Tubiana explained "*We talk about nuclear power, but we think about bombs.*" The fact is that no operator will be able to hide behind a "*It's scientifically impossible*" or a very low frequency of occurrence.

From October 1985 onwards, the fuel and debris were removed under water with great care after opening the vessel cover. The risk of untimely re-criticality was a constant concern for the operator.

To guard against surprises, the absorbing boron concentration was held at 3500 ppm until early 1983, and conservative studies[16] were carried out by postulating an unfavorable "lens" geometry where the most enriched fuel batch (2.96%) is "coated" by the least enriched fuel (the two other batches) and by assuming the disappearance of the absorbing fission products and the control rods (Fig. 3.16). The concentration was even raised to 5000 ppm

---

[16] Described in Knief (1988).

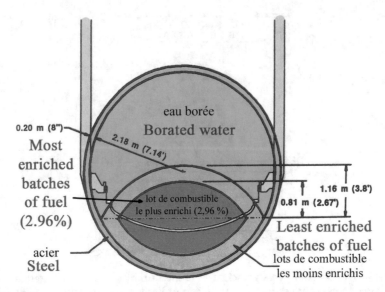

Fig. 3.16 "Lens" modeling to assess the risk of re-criticality of TMI-2 corium in the vessel bottom (adapted from Knief 1988)

before unloading to avoid any surprises. Have in mind that boron crystalizes over 7000 ppm.

The repercussions of the accident were considerable because it created a real intellectual earthquake in the scientific community. Indeed, the historical approach to safety has always been to consider bounding scenarios, which are supposed to cover less severe situations. For example, the "large break" LOCA scenario (case the double break of a primary cold leg), which is supposed to cover smaller breaks, has long been considered as the most penalizing, which is not necessarily true. This scenario leads to a massive depressurization of the primary circuit that a novice would detect without fail. The "*small break*" scenario, less impressive at first sight, is much more difficult to analyze. It is called "*weak signature*." In this idealized "*large break*" scenario, the operator is always assumed to be infallible, in the sense that he always responds perfectly to the needs of the plant. The emphasis is therefore only on the failure of the equipment, never on the failure of the man who pilots it. After TMI-2, many commissions of inquiry have tried to extract some truths from this affair. The lack of capitalization in the analysis of significant events was universally pointed out. On September 21, 1977, an event with the same signature as the first 30 min of the accident occurred at the Davis Besse plant, a reactor of the same type as TMI. The incident had no consequences insofar as the PORV valve was finally closed again after 20 min by the operator. Unfortunately, the feedback from this case did not reach the TMI teams. A loss of SG auxiliary feedwater occurred on the same Davis Plant on June 9, 1985, the same signature of TMI-2. The event started with a capacitor failure causing loss of main feedwater. This was followed by an operator pushing the wrong buttons during the transient. This error was multiplied in impact by steam feedwater rupture control system and auxiliary feedwater pump design deficiencies, equipment failures, and human factors problems. Other equipment failed to perform properly or was damaged as a result of the transient. Fortunately, the operator could close the block valve and stop the primary fluid to escape through the cycling PORV. Auxiliary feedwater was recovered 1150 s after the initiator and ended the incident. It is interesting to see that the 1000 first seconds are very similar to what happened on TMI-2 (Fig. 3.17).

The lack of standardization of reactors in the United States, where no two reactors out of the 75 in operation at that time are really identical, unlike in France, is also a remote cause of the accident. Another reason given was that the installation was also operating in a degraded mode with a leak of one ton of water per hour (!) through the pressurizer discharge line, inducing a high

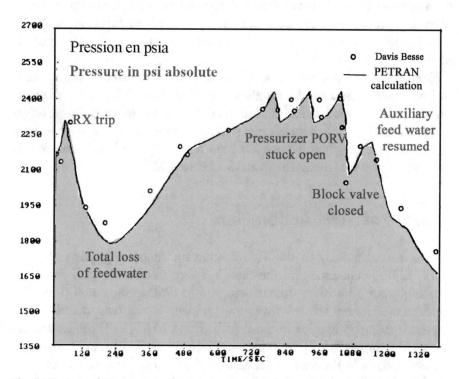

Fig. 3.17 Loss of SGs auxiliary feed water on Davis Besse on June 9, 1985. Fortunately, the fate of this plant was happier than TMI-2

temperature on this line and thus helping to mask the beginning of the accident. And what about the fact that the emergency power supply to the SGs was condemned, in absolute contradiction with the Technical Operating Specifications. The most significant technical cause that misled the operators was probably the alarm check light indicating the order and not the state of the pressurizer relief valve. On the organizational side, numerous failures were uncovered. The poorly defined responsibilities of those in charge contributed to the confusion in the management of the crisis. In the ultimate caricature, up to 60 people were present simultaneously in the control room, where even the governor of the State was invited by his own authority with his bodyguards! In France, the Internal Emergency Plan (PUI) and the Special Intervention Plan (PPI) specific to each site, clearly explain the role and prerogatives of each one.

The consequences of the TMI-2 accident on the improvement of safety are important. First, the principle of defense in depth and three barriers has been definitively imposed, silencing those who thought that "*too much was being done*" in terms of safety and that "*it was too expensive.*" Operator training has been improved, both in terms of knowledge and simulator training. Similarly, operating procedures have been completely revised, in particular with the introduction of specific procedures for severe accidents and the prioritization of alarms.

From the health point of view, it has been demonstrated that the accident had no consequences on the health of the inhabitants living near the plant, except for the mental trauma (Photo 3.11) during the uncontrolled leakage, which had unexpected consequences (voluntary interruption of pregnancy by choice of some pregnant women has been reported). The conclusion of this case can be read for free on the roadside (Photo 3.12).

French Post-TMI Action Plan

France reacted quickly to the TMI-2 accident. At the beginning of April 1979, EDF, Framatome and the French Safety Authorities (AS) formed a working group to analyze the accident and to develop an action plan. The fundamental lesson learned from the accident is that the overall safety approach currently applied to the French design of PWRs is fundamentally sound. The importance of the analyses and studies carried out in France since the early 1970s in the field of design safety and operational safety has been confirmed by the accident. The concept of defense in depth, which is the basis of the French approach to nuclear safety, has never been called into question

Photo 3.11 Sad joke on this house for sale

Photo 3.12 This sign on the road aptly sums up the whole affair

by TMI-2. As EDF is the only public utility with nuclear reactors in France, the company plays the role of architectural engineer, and all aspects of safety, including the design of the plant, as well as the construction and operation of the plant, are managed by EDF as a whole. In addition, the standardization of the plants allows EDF to efficiently provide generic analyses and studies. The method used by the French safety authorities and the technical support organization for safety analysis are based on "barrier analysis" which is of great value with regard to public health and safety. In April and August 1979, the AS requested EDF to provide additional studies and analyses of the experience gained from TMI-2. Consequently, the French post-TMI action plan was established in response to the requirements of the AS. (Photo 3.13). This plan includes 46 actions, each divided into specific items.

Technical Insert: "Details of EDF's Post-TMI Action Plan

1. *Plant design and man-machine interfacing*

TMI-2 focused on the area of operational safety. This includes the human–machine interface concept, operator training, and the structure of the operating team. Another important factor for improving operational safety is feedback. The human–machine interface concept covers all the hardware and software that the operator needs to operate the plant under normal conditions, as well as under incident and accident conditions. A man–machine interface was developed by EDF before TMI-2. However, the following additional analyses and studies were carried out in this area after April 1979. The review of the control room was carried out with the help of operating engineers specialized in the field of nuclear plant operation, as well as operating engineers specialized in the field of industrial plant operation, but also teams specialized in Human Factors who advocate that it is not up to Man to adapt to the Machine but the other way around and who introduce a new concept: ergonomics. This concept is relatively new because it has been said that even Gagarin was full of praise for the user-friendliness of the control panel of his Vostok-1 capsule, and that Russian engineers refused to install a small refrigerator on the first Russian atomic submarine: the K-3 (Marguet 2019, p. 30). However, it should be remembered that the first control rooms did not have chairs. A working group conducted a survey of nuclear plant operating personnel and simulator instructors. In addition, the behavior of the operators was recorded on video during several exercises performed on the 900 MWe simulator at the Bugey plant training center. After collecting the information from the survey and drawing conclusions from the simulator test recordings, EDF decided to build a full-scale mock-up of the control room as a working tool for the analysis of the changes. The main aspect of the change analysis was to improve the operator–machine interfaces. Two types of modifications were considered during the analysis conducted on the mock-up: First, the addition of information (alarms, valve states…) for certain safeguard

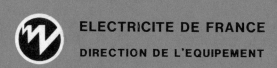

ELECTRICITE DE FRANCE

DIRECTION DE L'EQUIPEMENT

THE FRENCH POST TMI ACTION PLAN

Photo 3.13   EDF's post-MIT action plan

(continued)

Photo 3.14 Control panel improved by a "universal" color code on all French plants (EDF photo)

systems; Second, the improvement of the control panel layout. Based on the results of the modification analyses, the displays and controls in the control room are now arranged to improve the operator's capabilities: The new layout is based on better grouping of all function-related controls using demarcation lines; Use of colored functional areas; Improved labeling throughout the control room; Use of different types of symbols for rotary equipment controls and valve controls; Use of an active block diagram in conjunction with the passive block diagram (Photo 3.14).

Installation of a safety panel
There is always a potential risk of human error. TMI-2 has shown the importance of improving operator assistance to deal with this potential risk. In response to this problem, a computerized operator assistance, called a safety panel, was developed (Photo 3.15). The safety panel is designed to monitor the critical parameters of the plant in a concentrated way to give a systematic view of the plant safety state, mainly under accident conditions, and to assist the operators in their diagnosis and decision making. For this purpose, several functions are computerized: Identification of the cause of the first trip; Monitoring of the actuators; Assistance in diagnosis and selection of the accident procedure after the safeguard injection; Assistance in monitoring the safeguard injection; Monitoring of the residual power removal system; Display of the plant parameters, including saturation margin monitoring; Continuous monitoring of the

(continued)

(continued)

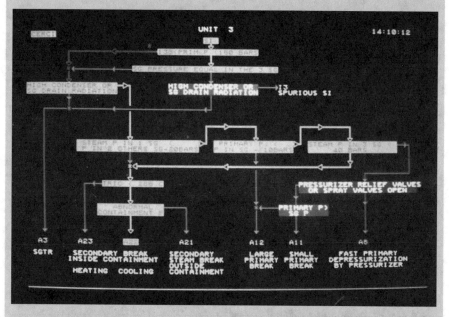

Photo 3.15 Example of a safety panel display (in this case, assistance in diagnosing and choosing procedures after a safety injection). This presentation dates from 1986. The more recent plants like the N4 have renovated Man–Machine Interfaces (MMIs)

plant state; Assistance in the U1 emergency procedure. In the event of an accident, and depending on the severity of the accident, three categories of personnel are involved in diagnosing the state of the plant: the operator in the control room, the Safety and Radiation Protection engineer (ISR in French) in the control room and the experts in the on-site technical support center. (Fig. 3.18). Therefore, the safety panel has three platens: two in the control room and one in the technical center. The safety panel is designed to complement the control room equipment normally used under normal and accidental conditions. In case of unavailability of the safety panel, the usual methods involving the control room equipment could be applied as a backup. In this context, the requirements for control room instrumentation are not necessary for the safety panel.

Design of advanced procedures for abnormal plant transients and crash recovery
In response to the Post-TM12 Action Plan, it was decided to organize an expert working group to review, analyze, and develop the existing incident (I) and accident (A) procedures. The operators of the plants were involved in order to help the reviewers benefit from the experience gained from accident analysis, operation, and training. Several exercises were carried out on simulators, to record the behavior of operators during simulated transients of the plant. These exercises were then analyzed to assess human factors. In addition to these actions

(continued)

(continued)

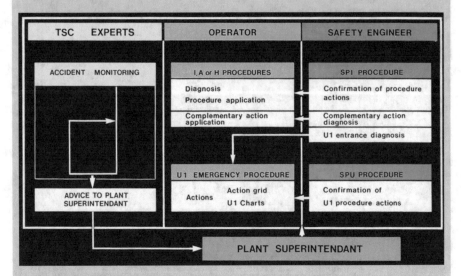

Fig. 3.18 Distribution of tasks between the teams on site. The I, A, and H procedures are event-driven procedures. U1, SPI (Permanent Post-Incident Surveillance) are State-oriented approach (APE) procedures. Event-driven procedures have been progressively abandoned in favor of the state-oriented approach (APE)

concerning incident and accident procedures, several actions were carried out to define the way in which the operators could deal with accidents outside the design basis. To this end, two approaches were followed: The State-oriented approach based on the physical state of the plant, and the event-oriented approach based on the historical triggering event. We have progressively moved from event-driven procedures, which require knowledge of the initiator of the accident to implement a pre-established response, to the state-oriented approach (APE) where the operator re-evaluates the effect of his actions periodically according to a predefined cycle (not according to his own free will) by scanning the vital state functions of the reactor. Thus, there is no need to know exactly what the current scenario is, and we can respond to multiple failures. The final product is a large set of very reliable and "ergonomic" procedures covering incidents, analyzed accidents and beyond-design basis accidents. Knowledge of the information contained in a procedure is necessary: during training, operators need detailed and explicit information in terms of "*how to do it,*" "*why to do it,*" and "*where to do it,*" bases for each operator action must be provided; during day-to-day operation, operators usually need guidance on "*what to do.*" Thus, a two-tiered procedure consisting of two documents is used. First, the procedure guide (or operating rule) which defines the purpose of the procedure and the operator or PLC actions that must be performed in order to achieve shutdown of the plant after the incident or accident has been diagnosed, this

(continued)

(continued)

document is written by the parties in charge of the design. Secondly, the procedure (or operating instruction), which includes only what the operator must do, is written by the EDF Nuclear Generation Department according to the above-mentioned rule and according to standard format guidelines (layout, colors…).

Operator's training
Since the beginning of the construction of PWRs, EDF has been committed to staff training. To determine the qualification of personnel, university education, experience, and training are taken into account. The main element in achieving the desired level of competence is training. To this end, the training program has been defined and includes courses on full plant simulators as well as on function simulators. This program was not fundamentally changed after TMI-2. However, the plant simulators have been improved to increase their representativeness in accident simulation.

Organization of the management: The Safety and Radioprotection Engineer
The structure of the operating team has been completed by a Safety and Radiation Protection Engineer (SRI) pre-positioned in an office adjoining the control room. He is called to the control room at the start of a sensitive plant transient (reactor trip, safety injection, etc.) to assist the operating team in recovery efforts. During the execution of a procedure, the unit manager remains in charge of the coordination of the plant while the safety engineer monitors the state of the plant according to the state-oriented approach (APE). This approach complements the Event-Driven Approach followed by the shift manager and operators. Thus, a redundant and diversified approach to plant surveillance results in superior performance of the operating team. Routine ISR tasks and assignments include issues involving the technical evaluation of the day-to-day operation of the plant from a safety perspective.

Experience feedback
An essential component of improving operational safety is learning from experience. Before TMI-2, EDF already had a feedback organization. The fact that the French Nuclear Fleet is highly standardized with only four models (900 MWe, 1300 MWe, 1450 MWe and EPR), is a considerable asset for feedback. Any problem detected on a reactor benefits the entire plant or even the entire Fleet. The organization of feedback has been improved following the post-TMI studies in order to rapidly assess each event discovered during pre-operational tests, as well as during operation. A feedback group, made up of experts from several EDF divisions in charge of design and operation, is dedicated to the analysis of each experience data coming from French nuclear plants, or from abroad when available. This committee gives its requirements to the EDF departments concerned in order to study effective solutions for each event.

2. Reactor core cooling modes

Post-accident studies
A major effort has been developed by EDF and Framatome in the field of post-accident studies to improve knowledge of post-accident conditions: Ability to eliminate a steam bubble located under the vessel cover of the reactor vessel;

(continued)

(continued)
Breaks in the pressurizer steam zone (such as untimely opening of a pressurizer valve), transients of small breaks, criteria for manual tripping of primary pumps; Effect of interruption of the safety injection system for 10 min in case of small breaks; Possibility of heat removal by steam generators in two-phase flow.

The State-Oriented Approach
One of the lessons learned from TMI-2 was the inability of the operators to perform a satisfactory diagnosis using the available procedures. The uncontrolled conditions of the plant led the operators to apply several different and inadequate accident procedures. As a result, the plant conditions progressively deteriorated, with the core being uncovered during the accident. TMI-2 demonstrates the limits of the event-driven approach based on the analysis of (almost) all conceivable accident sequences. To remedy this problem, the State-Oriented Approach (APE for *Approche Par Etats*) has been developed. It is based on measurements of physical parameters allowing the operator to recognize the thermal-hydraulic states of the boiler and to perform corrective actions according to these states. Indeed, the thermal-hydraulic states of the primary circuit can be enumerated in a finite way, whereas the accidental sequences can be multiplied ad infinitum without being sure to cover them all. The state-oriented approach (APE) has led to the improvement of procedures. Typical of the improvement is the support of management for safety injections. Typical of the state-oriented approach (APE) is the emergency procedure that allows post-accident operation by monitoring the physical state of the primary circuit.

3. Reactor cooling systems

Pressurizer relief valves
After TMI-2, an analysis and research program were developed by EDF and Framatome in the field of pressurizer safety and relief valves, while additional tests and improvements were carried out on the current safety and relief valves. A new approach to pressurizer overpressure protection has been proposed. As an alternative to the current protection, a solution using pilot-operated valves was analyzed and tested. This solution consists of three relief lines with two pilot valves in series. One of the valves acts as overpressure protection, while the other one, located downstream, acts as isolation. These valves can also be operated manually from the control room. New valves called SEBIM are now mounted in tandem (Figs. 3.19 and 3.20) and have replaced the older spring-loaded models. SEBIMs provide unparalleled pressurizer relief efficiency by eliminating the difficult problem of valve springs and valve flutter in the presence of two-phase fluids.

Release of non-condensable gases under accident conditions
A study was carried out by Framatome and EDF to determine the quantity of non-condensable gases that could be produced under accident conditions and how they could be released. It was concluded that non-condensable gases can be released from the reactor cooling system using existing equipment. Heat removal from the core would not be disrupted, despite the fact that these non-condensable gases can be stored at the head of the SGs inverted U-pins. For this purpose,

(continued)

(continued)

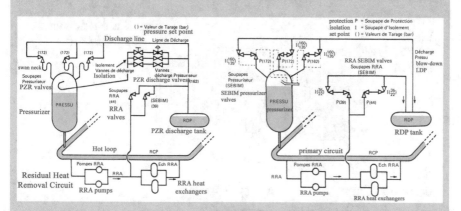

Fig. 3.19 Pre-1982 (left) and post-1982 (right) protection of the pressurizer

the primary coolant pumps are turned on or the feed and bleed process (safety injection plus discharge to the pressurizer relief valve) is used. This method does not require purging the reactor vessel.

4. Characteristics of active safety circuits

Steam Generator Safeguard System
The emergency feedwater system (ASG) is used to provide feedwater to the SGs under emergency conditions, involving loss of normal feedwater (ARE), as well as normal startup, normal shut down and hot standby conditions. After a full power reactor shutdown, the emergency feedwater system is automatically activated and the normal feedwater isolated. In order to limit the trip frequency and operating time of the SG Auxiliary Feed Water ASG, studies have been conducted to maintain a limited flow of the ARE through a predefined opening of the SG Normal Feed Water ARE control valve bypass. In addition, this solution limits temperature transients in the secondary side of the SG. As a result, the actuation of the ASG was modified. However, in case of very low SG level or if the ARE flow rate is lower than the required value, the ASG is automatically activated without delay.
Containment isolation system
In the event of a contamination accident inside the reactor building during the cold shutdown, with the safety injection signal inhibited, containment isolation is automatically provided upon receipt of an activity detection signal in the BR.

5. Nuclear auxiliary building and fuel building

Examination of radiation shielding
Several of the safeguard systems and auxiliary systems located outside containment could be required to operate in an accident with significant radioactive inventories in the fluids they handle. Some of these systems are located in the fuel building (BK) and in the nuclear auxiliary building (BAN). They include the

(continued)

(continued)

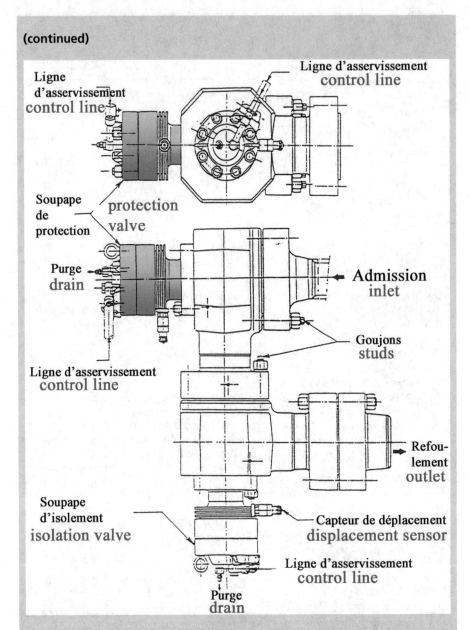

Ligne d'asservissement control line

Ligne d'asservissement control line

Soupape de protection / protection valve

Purge drain

Admission inlet

Goujons studs

Ligne d'asservissement control line

Refoulement outlet

Soupape d'isolement isolation valve

Capteur de déplacement displacement sensor

Ligne d'asservissement control line

Purge drain

Fig. 3.20 Tandem assembly of SEBIM valves

containment spray system (EAS) and the safeguard injection system (RIS). These systems are required to operate in the recirculation phase during an accident when the PTR tank from which they draw their water is empty. They then transfer water from the bottom of the containment (sumps) to the spray lines in the

(continued)

(continued)

reactor building or to the reactor vessel. Even if leakage from these systems is minimized, it is assumed that the premises housing the active components concerned may be contaminated. The radiological consequences inside the Auxiliary Building and the Fuel Building, resulting from an accident in which the reactor core is damaged, were estimated in accordance with the source terms for fission products that were updated in response to the post-TM12 action plan and studies. This estimate led to the conclusion that additional shielding was not necessary.

Ventilation
Additional ventilation tests were carried out in a standard plant using a simulation method to analyze contamination transport. These tests showed the need to improve the airtightness of the different rooms of the nuclear island; in addition, the air circulation inside some parts of the building was modified to avoid the spread of contamination in case of an accident.

6. *Radioactive effluents*

Transfer of highly radioactive leaks in the reactor building
As noted earlier, systems outside of containment may have to operate during an accident with significant radioactive levels in the fluids they process. Therefore, it would be necessary to collect and store leaking fluids for deactivation prior to treatment by the liquid waste treatment (TEP) system. In order to prevent the spread of contamination. The following principles are implemented: Detection and collection of highly radioactive leaks in the area where they are released; Transfer through the venting and draining system (RPE) pipes to storage capacity; Safe and radiation-protected storage of these highly radioactive liquids (reactor building containment); Installation of isolation devices between the venting and draining pipes used for the transfer of the highly radioactive liquids and the liquid waste treatment system The operator, from the plant control room, triggers the transfer of highly radioactive leaks into the reactor building on receipt of an activity alert signal in the reactor building (Fig. 3.21).

Flooding of the containment in accidental conditions
Following a LOCA or break in the main steam line inside containment, the lower portion of containment is flooded with water from the reactor coolant system, the safeguard injection system, and the containment spray system (EAS), or with water from the main steam lines and the ASG, as appropriate. The resulting maximum water depth was reassessed with an additional 15% margin. As a result, the locations of equipment likely to operate during an accident and which are located below the maximum water depth have been modified to allow their proper operation when the containment is flooded.

7. *Instrumentation and control*

Evaluation of the saturation margin and measurement of the water level in the vessel
An analysis was performed to define solutions that would allow the operator to better recognize inadequate core cooling. Two material modifications were

(continued)

(continued)

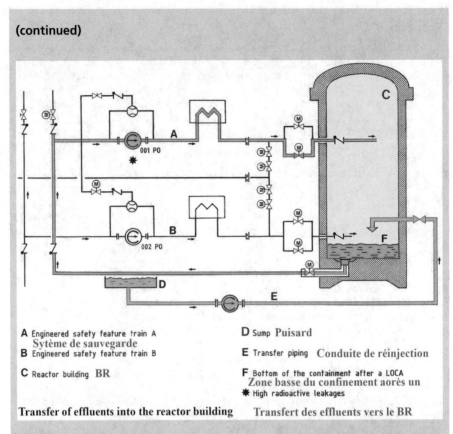

A Engineered safety feature train A
  Sytème de sauvegarde
B Engineered safety feature train B

C Reactor building BR

D Sump Puisard

E Transfer piping Conduite de réinjection

F Bottom of the containment after a LOCA
  Zone basse du confinement aorès un
  ✳ High radioactive leakages

Transfer of effluents into the reactor building   Transfert des effluents vers le BR

Fig. 3.21 Transfer of contaminated effluents to the reactor building

defined: The evaluation of the water saturation margin in the vessel (implemented in the 900 MWe and 1300 MWe plants), this system, called *"ebullio-meter"*, includes a computer device that processes the measurements of the core thermocouples as well as the temperature and pressure of the reactor cooling system, the saturation margin is displayed on the safety panel in the control room; A system for measuring the water level in the reactor vessel (implemented in the 1300 MWe plant), this system is based on the measurement of the differential pressure between the top and bottom of the vessel. The differential pressure system uses cells of different ranges to cover various flow behaviors with and without operation of the primary pumps. The reactor vessel level is displayed in the control room and provides the operators with reliable information, even in two-phase situations (Fig. 3.22).

Sampling of the primary circuit water
A review of the nuclear sampling system was conducted to determine the ability of personnel to obtain a sample of the reactor coolant under accident conditions. The fission product source term, updated in response to the post-TMI2

(continued)

(continued)

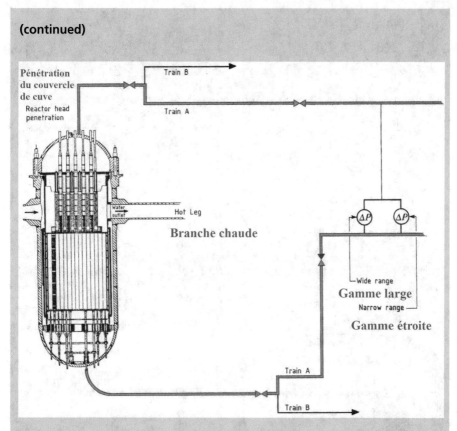

Fig. 3.22   Measurement of the water level in the vessel

analysis and studies, was considered to review the effectiveness of the radiation shielding. Based on the results of this review, a post-accident sampling cabinet was installed with additional specific radiation shielding. In addition, if the reactor coolant system sample lines are not available, additional sample lines, connected to the EAS recirculation pipes, allow for alternative post-accident sampling.

Monitoring of the activity in the containment
The radiation level inside the containment is a parameter closely related to the amount of gaseous fission products released into the reactor building. The monitoring range has been extended in the upper part from a dose rate of 10^5 rad/h to 10^7 rad/h.

8. *Equipment Qualification*

Equipment qualification makes it possible to demonstrate that the plant equipment can perform its intended safety functions, despite the unfavorable condi-

(*continued*)

(continued)

tions of a design basis accident during which the equipment must operate. Since the beginning of the PWR program, EDF, in collaboration with the CEA and the nuclear industry, has undertaken a vast program of equipment qualification. This program includes analyses and tests. In response to the post-TMI2 action plan, the qualification program has been improved at two levels. The first level is related to the revision of qualification requirements (equipment performance, analyses that have been performed to define more precisely the environmental conditions of containment resulting from an accident. The second level concerns the development and construction of new test facilities. These can simulate a wide range of accident conditions.

9. *Beyond-design events*

Beyond-design-basis accident procedures
Prior to TMI-2, the complete loss of some safety-related redundant systems was already analyzed by EDF. The systems analyzed for complete loss were those that are normally used continuously (e.g., the component cooling system, example in Fig. 3.23) or whose frequency of use could be significant (e.g., the ASG system). Since such an event occurred at TMI-2, EDF's analysis seems well founded. This

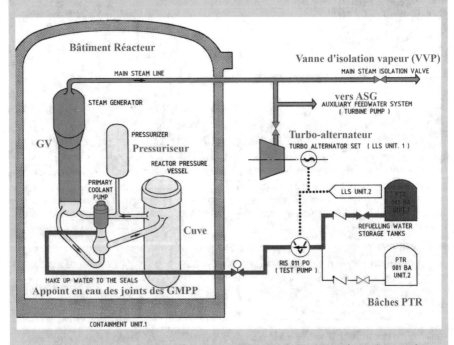

Fig. 3.23 Emergency cooling of primary pump seals in case of total loss of AC power

(*continued*)

(continued)

analysis has led to the installation of additional equipment and to the development of *"beyond design basis procedures"* (H procedures in the Event Approach) to enable the operator to cope with such events. An example of additional equipment is the turbo-alternator assembly that supplies power to the safety injection test pump and injection to the primary pump seals in preparation for a total loss of power (if the pump seals are ineffective, a small LOCA break is encountered). Each unit is equipped with a turbo-generator set powered by the steam produced by the SGs. Until the power supply is restored, the plant can be safely maintained under hot shutdown conditions, without affecting the tightness of the primary pump seals or the core.

U1 Emergency Procedure
If the operator encounters a situation that has not yet been analyzed or is unable to make a satisfactory diagnosis of the plant transient, the safety engineer instructs the team to abandon the event-oriented procedures and apply the state-oriented approach (APE) U1 emergency procedure. The U1 emergency procedure is initiated to prevent or delay potential core damage resulting from degraded plant conditions. Basically, most of the operator's accident actions when applying accident procedures are based on the event-driven approach. This approach is very effective for most transients. However, it is not possible to pre-analyze and formulate a predetermined response to every conceivable situation. To overcome this problem, in the case where control by the event-oriented approach is lost, the state-oriented approach (APE) was developed and introduced into the continuous monitoring of the plant state procedures (SPI (Permanent Post-Incident Monitoring), SPU (Permanent Ultimate Monitoring) and the U1 emergency procedure. This approach complements the event-driven approach when the operator encounters a situation that has not yet been analyzed or when he is not able to diagnose the plant state satisfactorily.

10. *Emergency Preparedness*

A nuclear plant is designed to operate within safety margins that guarantee very limited radiological risks for plant personnel and the public. Nevertheless, despite all the precautions taken at each stage, from design to operation of the plant, accidental conditions leading to a nuclear emergency cannot be excluded. This emergency situation is distinguished from other emergencies by the fact that it is likely to lead to significant radioactive releases into the environment. Therefore, adequate preparation must be made in collaboration with government authorities (national and local) and other organizations to deal with such a situation. The overall emergency state as defined for a nuclear plant accident includes both on-site and off-site emergency preparedness. Emergency preparedness has been improved in France following TMI-2. Among these improvements, the emergency organization is now equipped with a team of on-site experts specialized in nuclear safety and accident analysis. To this end, the on-site technical support center, which houses the experts, is designed to have the same habitability as the control room. Plant information can be displayed and recorded in the technical support center, where a safety panel dialogue console is located. The on-site technical support center provides internal support that complements the off-site technical support centers of EDF and the French safety authorities (National Crisis Center at the IRSN premises in Fontenay-aux-Roses.

Conclusion

The TMI-2 accident is an "earthquake" in the nuclear community because it puts the safety of nuclear reactors into perspective. If engineers admitted that experimental reactors could be fallible, the occurrence of a severe accident of a power reactor in the USA, in the most industrialized country, at the head of technological progress in the field, is more than a surprise, it is a painful questioning. As the saying goes, "*Every cloud has a silver lining,*" civil nuclear power has learned a lot from TMI-2. In addition to the technological advances and system improvements I mentioned earlier, I will especially remember the paradigm shift introduced by the State Approach.

What is revolutionary in this approach is that we no longer seek to know about the initiator, but only to analyze its consequences. The reduction in procedures is considerable. Studies on the state-oriented approach (APE) began in 1980 with tests on simulator, which led to new procedures for accident management of the containment spray (EAS) and the primary pumps around 1982. As an example, the start/stop of the high-pressure injection became based on a grid of the water level in the pressurizer according to the temperature difference at saturation. Around 1984, the SPI-U1 procedure emerged, based on a diagram between the RIC temperature (hot spot of the reactor) and always the difference to saturation. From 1990, the ECP (primary) and ECS (secondary) procedures were introduced on the P'4 plant, which use the water level in the vessel. From 1995, the second generation of APE procedures, known as APE*, was applied to the standardized plant P4 and N4. Indeed, the feedback from the application of the APE at the P'4 level has allowed the development of state-oriented approach (APE) procedures for older levels. The APE contains fewer procedures (Fig. 3.24), which have been grouped and standardized, i.e., five procedures for the whole domain in power or unconnected RRA, against the 40 or so event-driven procedures. The state-oriented approach (APE) is then generalized to the 900 plants by including situations where the primary circuit is open. The "*RRA connected, full and vented primary circuit*" domain during shutdown is covered by two specific procedures The RRA is the circuit that cools the reactor below 32 bars when the SGs can no longer extract power (Marguet 2019, p. 872). Only the "*open primary circuit shutdown*" state is still covered by event-driven procedures. The entry in the Severe Accidents Intervention Guide (GIAG for *Guide d'Intervention des Accidents Graves*)), strongly expanded after TMI-2, is carried out on quantified criteria.

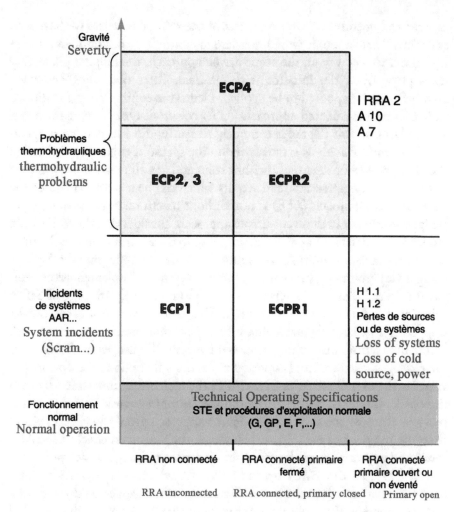

Fig. 3.24 State-oriented approach: second generation of procedures (APE*)

The implementation of the state-oriented approach (APE) was evaluated as early as 1991 using human factors techniques in order to identify its advantages and disadvantages. The state-oriented approach (APE) has the advantage of eliminating what has been called the "*contradiction between logic and rule.*" This contradiction appears during an event-driven procedure. Strictly applying the rule can come into opposition with the commonsense logic that appears when one no longer understands what he is doing or when the procedure is in fact not the right one. This can lead to significant stress in the consultation phase of the choice of the initiating event and a violent feeling of panic. It is said that panic is communicative, especially if it takes hold of an experienced operator on whom the rest of the shift crew relies. We have seen situations where a rookie would not dare to contradict a senior operator who was in the wrong way. In the event-driven approach, the looping phase only takes place when the initiator is determined. After that, the *"no-strings-attached"* rule is supposed to be applied. Hence the contradiction mentioned earlier. The state-oriented approach (APE) covers all types of situations and accumulations. This dogma is extremely reassuring for inexperienced operators. The work of analysis is transferred to the specialist engineers who design the method, well upstream of the shift team and at a time when there is time to think, because upstream of an activity which can be feverish. However, the state-oriented approach (APE) is not without its critics. Some people, especially professionals in the event-driven approach, consider that the APE would reduce the understanding of the actions required and would reduce the margin of the operators, limiting their initiative. Others consider that the APE, by being a very bounding procedure, is very heavy in its implementation compared to certain "simple" scenarios, typically the untimely tripping of Safety Injection (*"a hammer to crush a fly!"*). This type of spurious event can be stopped immediately by switching off the injection pumps concerned, well before the continuous looping proposed by the APE takes effect. The APE is therefore criticized for not proposing a diagnosis that allows the operator to visualize the overall scenario. In the APE, the incident is understood by continuous looping, i.e., by delta between the current situation and that of the previous scan, whereas the event-oriented approach gives a long-term vision from the moment of the initiator. However, the anti-stress effect of the APE is real and appreciated, as shown by the simulator trainings carried out at the end of 1991. As time went by, the criticisms, which came essentially from teams that had practiced the event-based approach, disappeared through natural rejuvenation. The only real question that remains is that of the universal covering of the APE. Taking into account the accumulation of failures, namely the art of *cyndinics*.[17] Is it exhaustive in the context of nuclear reactor accidents, given the complexity of the industrial object? Up to now, the state-oriented approach (APE) has always proved effective, but we have never had any "major incidents" in France.

---

[17] *Cyndinics* is the science of danger. This neologism was invented at the Sorbonne University in 1987. It appeared in the press for the first time in the newspaper Le Monde on December 10, 1987.

The implementation of the state-oriented approach (APE) was concomitant with a new structuring of the shift teams. The Radiation Safety Engineer (ISR), who is solely in charge of safety aspects and has no driving duties, was created in the shift team following the TMI-2 accident. The ISR will be removed from the shift at the same time as the generalized implementation of the APE. The position of Assistant Shift Supervisor, more specifically in charge of consignments, is also being removed. The establishment of the ISR, whose competences were to be used as a makeup for the decisions of the shift supervisor, seemed to be an adequate response to the feverishness (one could even speak of hysteria) in the control room during the disastrous accident of TMI-2 in 1979. However, the feedback from an ISR in a shift team is mixed. This one has no driving action. As in "*The Desert of the Tartars*" (*Il deserto dei tartari*) by the author Dino Buzzati, the ISR waits for the accident (the war in the novel), an unlikely event which, fortunately, never happens, but creates a real psychological tension and a heavy routine to support. In addition, this job was rewarded with substantial bonuses, making the recipient reluctant to transfer to another job. The situation of the specialized firemen on nuclear sites raises the same problem. How to keep the motivation of the agents in these positions of perpetual waiting? Most sites therefore rely on traditional firefighters in the nearby town, but with dedicated nuclear training, rather than on firefighters pre-positioned on site and often with nothing to do. Finally, the Chief Operating Officer (CO) has hierarchical powers, but also powers related to safety monitoring. Some people consider that the constraints of production and the constraints of safety resting on one man are incompatible (the ISR on shift, who had no hierarchical role, could serve as a counterweight).

Man has the ambiguous ability to desire change and to be reluctant to change at the same time. This ambiguity is also reflected in his acceptance of the state-oriented approach (APE). However, the state-oriented approach is a management tool that transcends the problems of competence. An accident will always be better managed by a very competent operator, but the stakes are such that we cannot rely solely on the random competence of the shift teams. The best option is of course the combination of state-oriented approach (APE) and competence. Future will bring its share of answers.

And to close this very serious chapter on a major nuclear accident, I cannot resist ending on a humorous note, which I hope will not be too out of place in the context, by showing you the visit of President Jimmy Carter[18] in the TMI-2 control room and in rather ridiculous yellow over-boots (Photo 3.16), dubious in front of a control panel (Photo 3.17), and on the corrosive but tender humor of the talented French cartoonist Jacques Faizant (1918–2006), who sketches in one page all the difficulty of informing the public (Photo 3.18) about this affair ("*the plumbers of Pennsylvania*"). After all, "*The only absolute thing in a world like ours is humor!*"- Albert Einstein.

Finally, if you think you can do better than the real operators in 1979, you can play on the Apple-II+ game "*Three Mile Island*" (Photos 3.19 and 3.20), a curious spin-off of the real story.

---

[18] Jimmy Carter. Born in 1924, the 39th president of the United States was a young US Navy lieutenant in 1952, in nearby Schenectady, New York, training to work aboard America's first nuclear submarine at the time of the accident of the NRX reactor in Chalk River, Ontario, just 180 km from Ottawa. Chalk River officials turned to the United States for help in dismantling the NRX reactor just after the accident of December 12, 1952 (INES 5). A total of 26 Americans, including several volunteers, rushed to Chalk River to help with the hazardous job. Carter led a team of men who, after formulating a plan, descended each one into the highly radioactive site for 90 seconds to perform specialized tasks. Carter's job, according to the CBC recounting, was simply to turn a single screw. But even that limited time exposure carried serious risks because of very high radiation surrounding. Carter was told that he might never be able to have children again, though in fact his daughter Amy was born years later. Carter told CNN in 2008: "*We were fairly well-instructed then on what nuclear power was, but for about six months after that, I had radio-activity in my urine. They let us get probably a thousand times more radiation than they would now. It was in the early stages, and they didn't know.*". Although short, his exposure was still very significant. So, although Carter was often mocked as a peanut farmer, he was not "*a sucker for the first brood*" during the TMI-2 accident, and he perfectly knew what a severe accident meant.

Lieutenant James Earl (Jimmy) Carter Jr. in main control room of USS K-1 (US Navy)

Photo 3.16 US President Jimmy Carter visits the TMI-2 Control Room in yellow overboots

Photo 3.17 Jimmy Carter looks doubtful at a control panel in the Command Room

Photo 3.18 The talented French humorist Jacques Faizant sketches a moment in the lives of French people who are visibly concerned about nuclear energy! Jean Elleinstein (1927–2002) was a French historian specializing in communism and a member of the French Communist Party (PCF). Georges Marchais (1920–1997) was the first secretary of the PCF from 1972 to 1997, renowned for his outspoken popular views (DR)

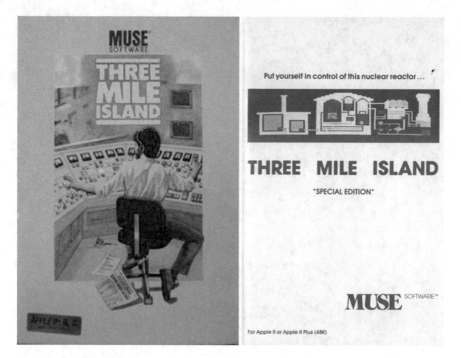

Photo 3.19 The Three Mile Island game from Muse software for APPLE-II+ (48 *ko* of Random-Access Memory, 1980)

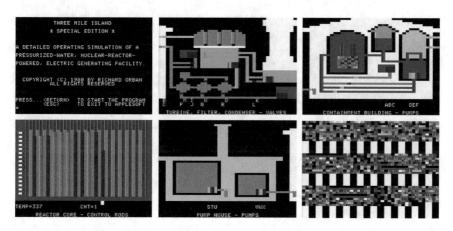

Photo 3.20 Detailed operating simulation? The designer Richard Orban (Richard Orban is a developer who was credited for video games at MicroProse Software and Riverbank Software in the 1980s. He was responsible for the 1988 C-64 game Red Storm Rising) tried his best on that poorly pixellized game but be indulgent for this game dating from the beginning of personal computer. However, you can simulate the secondary circuit, turbine and cooling tower (first row, middle); Core vessel, pressurizer (the little house with pink steam!) and the steam generator with green steam (obviously a U-tube SG instead of a once through SG) (first row, right); degradation of the core and the position of the control rods (second row, left). Auxiliary building with stack (second row, middle). Please, write to me if you can understand the second row- right picture?

4

The Chernobyl Accident

Abstract The 1986 Chernobyl accident in Ukraine spread a very large quantity of radioactivity over the European continent. Reactor 4 literally exploded, and firefighters had great difficulty in stopping the fire. Many people died in the ranks of the personnel who intervened to limit the consequences of the accident. A largely unprepared test and a particularly dangerous type of reactor are the causes of this accident that has left its mark on the collective consciousness. This chapter details this terrible day with scientific elements allowing a true understanding of the accident.

On April 26, 1986 (the day after a holiday and just before the four non-working days related to May 1rst, an unavoidable holiday in the USSR), at 1:23 a.m. (11:23 p.m. in France), the No. 4 reactor at the Chernobyl plant (Fig. 4.1) located in Ukraine exploded following an uncontrolled power excursion. About 300 million curies of radioactivity were released, 6000 times more than the Windscale accident of 1957, nearly 100 times more than the radioactivity of the all the aerial atomic tests.

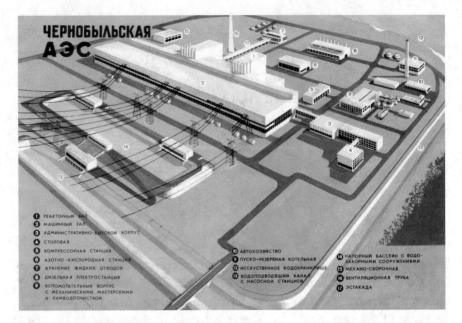

Fig. 4.1 An artist's drawing presenting an idyllic vision of the Chernobyl site from a 1978 presentation brochure. The turbine building (noted 2) is common to two twin plants. The two twin reactors are noted 1

The RBMK Reactor Type

The RBMK[1] reactor is of Soviet design and construction, the prototype of which is the Obninsk reactor already mentioned. The USSR's industrial development plans called for a steady increase in electricity production by accelerating the commissioning of nuclear reactors. The existing WWERs pressurized water reactor vessel design was not up to this task, as the country's heavy industry was unable to produce the required number of reactor vessels. In this context, the country's leaders assigned the Ministry of Medium Machine Building the task of creating a power reactor without any vessel, whose main equipment could be mass-produced. Thus, the RBMK reactor was born in a constrained economic context (Fig. 4.2). The starting point of the RBMK construction process was a technical meeting held in Leningrad on January 12, 1965, chaired by the first deputy minister of Medium-sized Machine Building A. I. Churin, during which the first organizational decisions were taken. On April 15, 1966, Minister E.P. Slavsky signed the planning mission

---

[1] RBMK is the Russian anagram for *Reaktor Bolshoi Mochnosti Kanalnii*, literally high-power channel reactor.

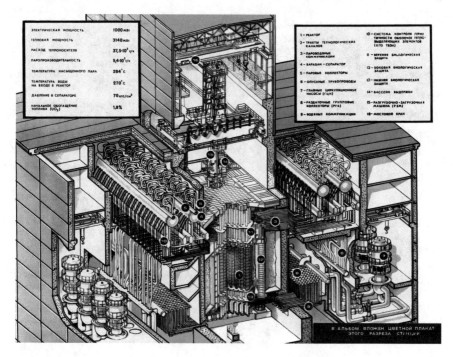

Fig. 4.2 Cut view of a RBMK reactor of the Chernobyl type

for the Leningrad nuclear power plant. The scientific supervisor of the project was the Kurchatov Institute. The development of the technical design of the reactor was initially entrusted to the Design Bureau of the Bolshevik plant in Leningrad and the reactor type to NII-8 (NIKIET). In 1967, the technical project of the B-190 reactor designed by the Design Bureau was presented to the Scientific and Technical Council, where the experts issued a negative opinion. As a result, the decision was made to transfer the duties of the chief designer of the entire power unit to NII-8, where the project was given the name RBMK-1000 (1000 MWe power reactor). At the beginning of June 1967, the RBMK-1000 project developed by NIKIET had received a positive response from the Scientific and Technical Council, which gave the start of the new reactor.

The RBMK is a pressure tube reactor with a graphite moderator (Fig. 4.3). The graphite stack is 8 m high. Each channel, consisting of a pressure tube, must resist a pressure of 70 bars of light water (saturation temperature 287 °C). The fuel is uranium oxide slightly enriched in ^{235}U (about 2% at the beginning, 2.6% at the end), which is in the form of a rod 3.5 m high and 13.5 mm in diameter. Two fuel rod clusters are placed one above the other, joined by a hook and a locking plug, for a total height of 7 m of active core. Each

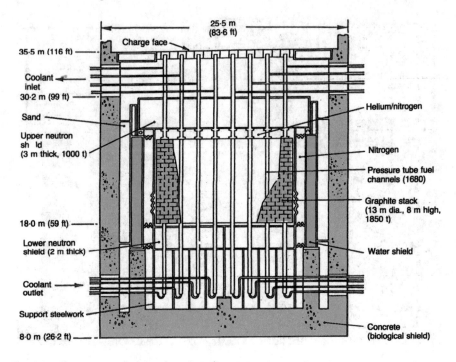

Fig. 4.3 The concrete caisson housing the core

assembly contains 18 fuel rods containing a stack of uranium oxide pellets (Figs. 4.4 and 4.5). The coolant is ordinary light water. The upper and lower parts of the channels are made of stainless steel, and the active part is made of zirconium alloy with 2.5% niobium, which has rather resistant mechanical and corrosion properties and good transparency to neutrons (low capture). The zirconium part of the channel is connected to the steel part by special welded adapters.

In addition to the fuel channels, the RBMK core has 179 protection system channels. These channels are designed for automatic power control, fast reactor shutdown, and control of the radial field and axial power sheet of the core (height 3050 mm). The shutdown control rods are 512 mm long. To control the energy distribution along the height of the core, there are 12 channels with seven axial section detectors (fission chambers), which are uniformly installed in the active core of the reactor outside the grid of the fuel channels and the channels of the Protection System CPS. The energy distribution along the core radius is monitored by means of detectors installed in the center tubes of the fuel assemblies in 117 fuel channels. At the joints of the graphite columns in the reactor masonry are 20 vertical holes, 45 mm in diameter, in which three-zone thermometers are installed to monitor the graphite

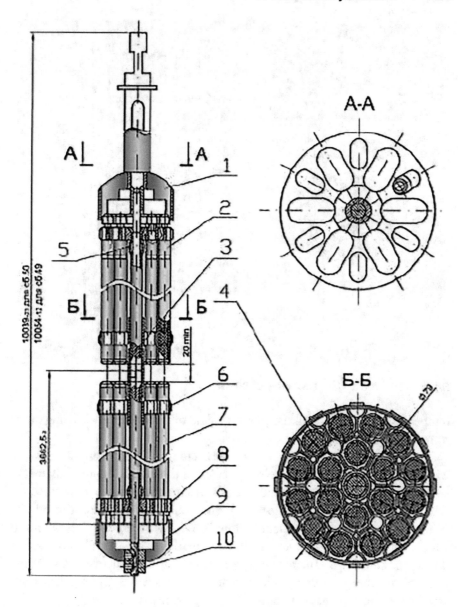

Fig. 4.4 Configuration of an RBMK-1000 fuel assembly: the assembly comprises two clusters of 18 rods distributed around a central core. (1) Assembly head with tie-down sleeve, (2) Upper control rod cluster, (3) Central core, (4) Fuel rod, (5) Central tube, (6) Spacer grid, (7) Lower control rod cluster, (8) Lower grid, (9) Assembly foot, (10) Holding nut

R

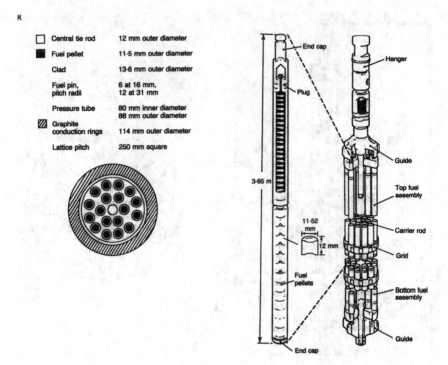

Fig. 4.5 Fuel rod and fuel assembly of RBMK

temperature. The Windscale accident has paid off, and graphite is being monitored much more.

The reactor is housed in a concrete pit with dimensions of 21.6 m × 21.6 m × 25.5 m. The lower slab, 2 m thick and 14.5 m in diameter, consists of a cylindrical shell and two plates into which the penetrations for the fuel and control channel pipes are hermetically welded. The entire volume inside the plate between the penetrations is filled with serpentinite, thanks to which, being a biological shield, it allows to work in the gap under the reactor during the shutdown. The reactor is surrounded by a lateral protection in the form of an annular water tank, which is mounted on support structures attached to the concrete base of the reactor pit (Figs. 4.6 and 4.7).

The axial holes in the graphite columns of the core are for fuel channels and the control and protection system channels. The peripheral openings of the reflector column are used for the cooling channels of the graphite reflector that heat up due to neutrons and gamma rays (Figs. 4.8 and 4.9). The channels are welded to the inner surface of the top plate risers and are connected to the bottom plate risers by bellows compensators, which ensure compensation for the thermal expansion of the channels during heating. Thus, inside

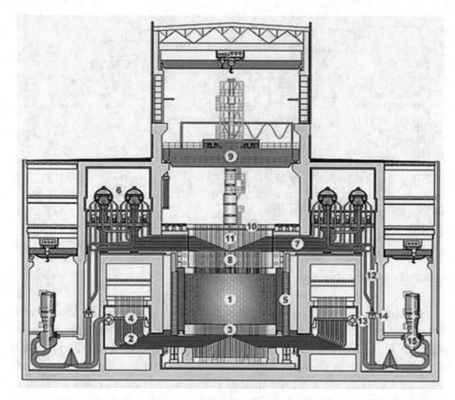

Fig. 4.6 Axial section of an RBMK-1000: (1) Active area of the graphite stack (2488 graphite columns, 18 m diameter 8 m height, 250 mm × 250 mm × 600 mm graphite block). (2) Demineralized light water pipes for cooling the core. (3) Biological shielding. (4) Pressure tube water distribution manifold. (5) Lateral biological protection. (6) Steam separator drum. (7) Communication pipes for steam and water to the drums. (8) Upper biological containment. (9) Fuel loading and unloading machine. (10) Removable slab decking. (11) Fuel channel ducts accessible after unscrewing the plug at the fuel loading machine. (12) Primary water drainage pipes. (13) Pressure manifold. (14) Suction manifold. (15) Primary circuit circulation pumps

the reactor, a cooling duct is formed by the pressure tube containing the light water itself and a part of the risers of the upper plate above the welding joint of the channels to these risers. But the swelling under irradiation of the graphite, which can see a neutron fluence up to $40 \cdot 10^{21}$ n/cm2 (!), causes the disappearance of the initial construction gap between the tube and the graphite, which will crack by compression (Photo 4.1). The deformation of the tubes of the control rod clusters (Fig. 4.10) is also a very important safety issue. The Russians estimate that a deflection of 50 mm in relation to the perfect straightness remains admissible to ensure a "free" passage of the control rod cluster.

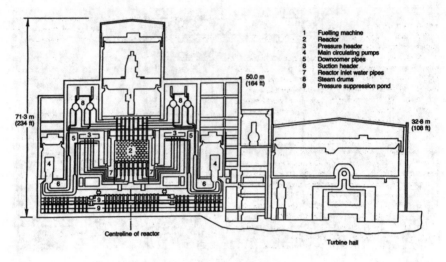

Fig. 4.7 Longitudinal section of an RBMK-1000

On September 10, 1973, the core of Liningrad-1 began to be loaded with fuel cartridges without water in the pressure tubes. When the pressure tubes began to be filled with water, the reactor began to behave in disagreement with the theory then developed by the Soviets, in that the reactivity suddenly began to increase, while all the rods of the standard and start-up protection control systems were in the core. The reactivity continued to increase until the middle of the core was filled with water and approached a critical value. The effect of the cold water proved to be positive. The deviation from the calculation was so large that the decision was made to suspend the reactor loading. After parametric calculations, the number of control rods in the core was increased, after which the physical start-up was successfully performed. This anecdote shows an imperfect mastery of reactor physics calculations in real geometry and a poor consideration of the moderating effect of water, because it is well known that the reactivity of a design increases with the density of single-phase water, neutrons being better slowed down. We have seen previously that the water level is played on to reach criticality in experimental reactors (Zoé, Vinča). The first critical state of the reactor was reached on September 12, 1973.

In these reactors, there is no steam generator as in pressurized water reactors (PWR). The water is vaporized in the core and the exit temperature of the core is 270 °C with a humidity level of 14.5%. The steam from the primary circuit, after being dried, drives the turbines directly (5400 tons/h of steam at 65 bars). The water pumped (37,500 tons/h) into the lower part of the reactor core boils as it flows up the pressure tubes. The flow rate through the channels

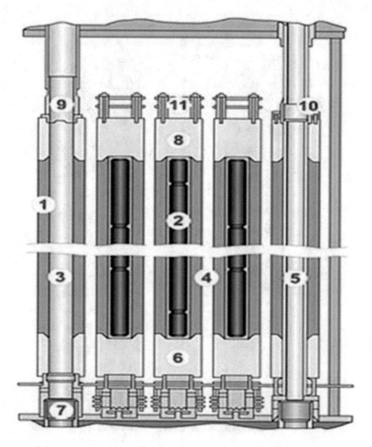

Fig. 4.8 View of a graphite block made of graphite columns. (1) graphite blocks. (2) graphite rods. (3) Plant column. (4) Columns of reflectors. (5) Peripheral reflector column. (6) Base plates. (7) Support sockets. (8) Protection plates. (9) Flanges. (10) Guide. (11) Heat shields

is regulated according to the pressure tube requirements by means of control valves installed in the feeder pipes. The steam quality is approximately 15%. These vertical pressure tubes contain the fuel surrounded by an alloy of zirconium and niobium (Zr-2%Nb). The Russians have already cleverly eliminated from the concept the magnesium and aluminum that caused so many problems in England, France, and Switzerland. The moderator consists of a stack of graphite bricks with a square cross-section, that slows down neutrons and promotes fission reactions. About 1700 pressure tubes of 88 mm diameter (thickness 4 mm, height 7 m) made of zirconium-niobium alloy are placed in this graphite stack, along with fuel elements or control rods. A mixture of nitrogen and helium is circulated between the graphite blocks to prevent its

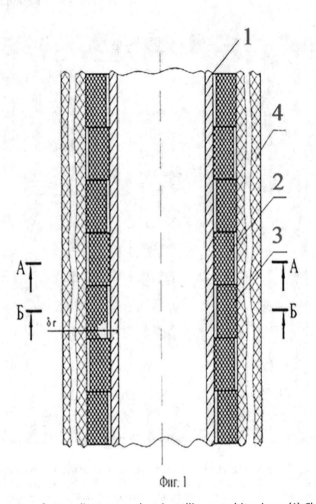

Фиг. 1

Fig. 4.9 Diagram of gap adjustment when installing graphite rings: (1) Channel tube. (2) Graphite ring installed in masonry. (3) Graphite ring installed in pipe. (4) Graphite masonry. Split graphite rings are mounted on the zirconium portion of the process channel. These rings adapt tightly around the channel tube or are pressed against the surface of the hole in the graphite masonry body. The split rings provide heat transfer from the graphite masonry to the coolant flowing through the channel and allow for changes in channel dimensions. Under the influence of irradiation, temperature and, in the case of channels, coolant pressure, the shape of the channel tubes, graphite blocks and rings changes. This in turn leads to the disappearance of the expected gap between the zirconium tube of the channel and the outer graphite ring, and to the appearance of a strong contact between the channel and the graphite pile. Consequently, this leads to the "pinching" of the tube in the channel of the graphite pile. In case of total blockage of the channel, deformations occur and, consequently, tensions appear in the graphite blocks, leading to the cracking of the blocks and the deformation of the graphite masonry. These circumstances lead to a reduction in the lifespan of the reactor. The filling of the expected diametrical gap between the channel zirconium tube and the graphite outer ring, added to the filling of the expected diametrical gap between the channel and the graphite pile, result in their premature cracking and deformation of the graphite stack. The methods of measuring and controlling the gap size are indirect and labor intensive

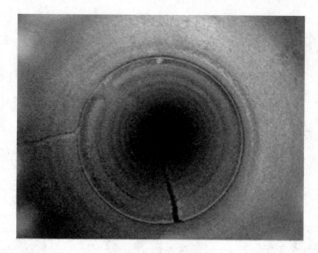

Photo 4.1 Impressive crack in a graphite block following the visual inspection during the 2008 examination of the plant 1 of the Leningrad nuclear power plant. The cracks are caused by the swelling of the graphite under irradiation which "pinches" the central channel of the image

oxidation, but also to improve the cooling of the stack. Two independent loops ensure the cooling of the core by light water and boiling is allowed in the pressure tube at 290 °C, in the same way as in the Boiling Water Reactor. The water-steam mixture is taken up by collectors and directed to the separators. Twelve collectors feed each separator. Each loop has two separators and four recirculation pumps (three in operation and one in reserve). The steam drives two turbines coupled to two 500 MWe alternators. The large number of channels in the RBMK reactors (1681 pressure tubes) makes it necessary to change the fuel during operation. The shutdown time would otherwise be too penalizing. Thus, to change the spent fuel into new fuel, each reactor has a handling machine which unloads the fuel from the channel which has reached a burn-up of 22,500 MWd/t on average, an unloading burn-up rate which is rather low because of a specific power which is lower than in PWRs (about 28 MWd/t instead of 38 MWd/t), and an even lower power density (with the same power, the core of an RBMK is 20 times bigger than that of a PWR!) Note that about 6% of the power is produced by gamma and neutron heat up in the graphite blocks (compared to 2.6% of power in the water of PWRs). Attention is therefore drawn to the imperative need to cool the graphite blocks with a mixture of chemically neutral helium-nitrogen gases. The spent fuel is then unloaded into a temporary storage pool. The reactor block rests on a mechanically welded structure that is contained in a concrete cavity. This is called a caisson, but unlike PWRs in case of accident, the caisson is not a

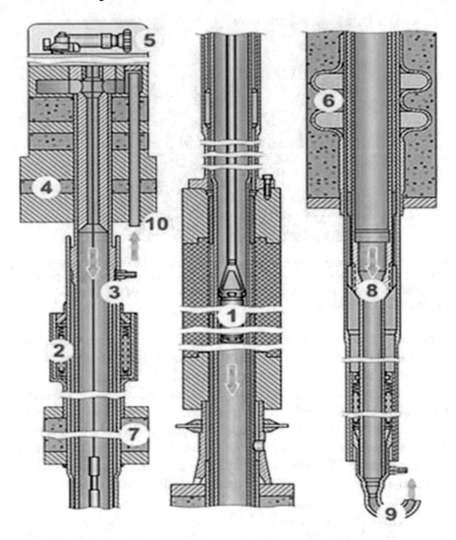

Fig. 4.10 Schematic diagram of the control and protection system channel: (1) Control rod drive for the sorbent cluster. (2) Bellows expansion joint. (3) CPS top duct. (4) Upper cap. (5) Control rod drive mechanism (6) Lower biological shield. (7) Upper biological shield. (8) Throttling device. (9) Channel drainpipe. (10) Water supply pipe in the channel. It must be ensured that there will be no blockage of the rods in the channels of the control and protection system, taking into account the possible bending of the graphite masonry columns. The control and protection cladding is designed to contain the control rods of the control system and to provide water circulation to cool the electrical actuators of the control system that release Joule energy. The control and protection duct is a welded tubular structure made of zirconium alloy and corrosion-resistant steel. Graphite sleeves are attached to the channel to provide the required temperature control for the graphite column. The upper part of the channel houses the actuator mounting heads and the channel cooling water supply. In the lower part of the channel is a constriction that ensures that the entire channel cavity is filled with water. The calculated permissible deflection of the columns is 50 mm, above which the free insertion of the control rod clusters is compromised

perfectly tight containment for fission products. Above the reactor, a machine allows the continuous loading and unloading of the fuel without shutting down the reactor. It has a special sealed cooling circuit. After fit-up of the machine on the head of a fuel channel, all the two assemblies of the channel are withdrawn in one block then, after rotation of a barrel, two fresh fuel assemblies are lowered into the channel and this one is closed, the machine will then deposit the spent fuel assemblies in a deactivation pool. The machine is surrounded by a biological shield (cask), the interior of which is pressurized. The cask is equipped with a rotating barrel with four slots for fuel assemblies and other devices. We had already presented this type of machine in the French (Natural Uranium-Graphite-Gas (UNGG)) and British (MAGNOX) concepts. In nominal operation, the RBMK-1000 is reloaded with 1–2 fuel assemblies per day. In addition to the replacement of spent fuel by new fuel, the purpose of the reloading is to increase the burn-up rate of the fuel and to equalize the power distribution in the core by rearranging the fuel cartridges. An automated refueling planning system is installed in the unit to determine the order of refueling and the type of fuel assemblies to be loaded.

The control of the reactor is ensured by 211 control rods that absorb neutrons and occupy pressure tubes similar to those that contain the fuel, distributed radially in the core. These rods are operated by mechanisms located above the core under the protective floor of the hall and require motors to be able to enter the core where they are pushed, unlike the control rods of PWRs, which fall by gravity. The question of why the designers did not choose a gravity system is explained when one sees the congestion of the light water pipes connected to the pressure tubes. The clusters move very slowly: 14 s for the complete introduction of the manual and automatic rods and 10 s for the short absorbents. Only the 24 shutdown rods have a shorter insertion time (2.5 s), but the first- and second-generation reactors were not originally equipped with them. Each pressure tube has its own connection at the core outlet, which flows into a steam collector (Fig. 4.11).

The cooling of the reactor is carried out by two loops, each evacuating the energy produced by half of the core. Each loop includes two steam separators and four recirculation pumps (3 in operation and 1 in reserve). The mixture of water and steam coming out of each pressure tube arrives through a pipe in one of these separator tanks, 30 m long and 2.30 m in diameter, in which the water and steam are separated. These separators are necessary because the quality of the steam leaving the core is poor (15%) and it is not possible to send too much moisture to the turbine (impact of

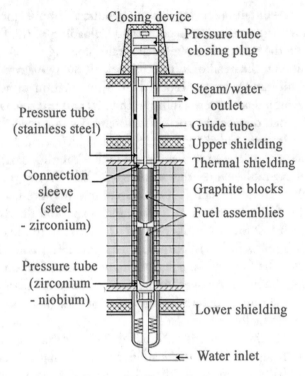

Fig. 4.11 A pressure tube from RBMK (DR)

water drops on the turbine blades). The steam is sent to the turbine and the water returns through 12 pipes to the collectors and recirculation pumps that feed the pressure tubes through a system of collectors, sub-collectors, and pipes. On each loop, there are 22 sub-collectors of 300 mm diameter. These headers are all weak points in this welded structure (it has been said that this reactor is a "plumber's reactor") because they considerably increase the risk of primary circuit breakage with possible dewatering of a pressure tube in the event of a break at the foot of the tube, or even of several pressure tubes in the event of a break in a particular header. However, an emergency cooling circuit allows the core to be cooled in the event of a break in the main cooling circuit (break in a steam line or a water supply pipe) in the least penalizing cases. Since neutron moderation is essentially due to static graphite elements (and not to the light water contained in the pressure tubes, an increase in light water boiling results in a decrease in cooling (steam cools less than liquid water) and in parasitic capture of neutrons by the hydrogen in the light water without inhibiting the fission reaction in the reactor. This results in a strongly positive void coefficient, which is particularly dangerous for the safety of RBMKs in case of power excursion. In a water moderated reactor, the heat up of the water and the appearance of steam (PWRs, BWRs) reduces the density of the moderator. As a result, the neutrons are less well slowed down, and the fission reaction decreases. We speak of a negative moderator coefficient, which is favorable to the safety of PWRs and intrinsically safe (taking into account the fact that if the water is strongly borated by addition of boric acid to compensate for the reactivity of the fresh fuel assemblies), the steam created does not carry boron, which is not good for safety in the steam zone. In any case, a positive void coefficient makes the RBMK reactor very vulnerable to a fast neutron reactivity accident with power increase. We will have the opportunity to come back to this point later.

On the biological protection side, the reactor core is surrounded, over its entire height, by an annular water tank, itself surrounded by a containment containing sand. The sand has a low density (1300 kg/m³) but provides some aerosol filtration and thermal protection. The entire graphite masonry cavity is enclosed in a lightweight cylindrical core barrel (shroud). This vertical wall of the reactor core is designed to provide good flexibility against thermal expansion caused in particular by the expansion of the graphite, but its containment role is questionable and it does not replace a 20 cm steel vessel as in the case of PWRs. The RBMKs are equipped with various systems, such as the emergency core cooling system, which consists in countering the eventuality of a double-ended guillotine break of a pipe, with total loss of electricity off site. This involves the rupture of the pressure manifolds or the suction manifold of the main circulation pump. In this eventuality, this system provides both immediate core cooling and long-term residual power removal with six pumps fed from the accident location system to cool the damaged half of the core and three pumps fed from the clean condensate tanks to cool the undamaged half of the reactor in case of an accident or incident. There have been three generations of RBMK reactors. The first generation was designed in the early 1970s and included six RBMK reactors. The second generation built between the end of the 1970s and the beginning of the 80s includes ten units, and finally, the third generation only two. The essential difference comes from the protection of the core cavity in the event of a pressure tube rupture. For the first-generation reactors, a valve allows the steam to be discharged directly into the atmosphere. The reactor cavity is not designed to withstand a pressure of more than 3.1 bars. The second- and third-generation reactors are equipped with an accident location system which consists in condensing the steam. There are still valves opening to the atmosphere in case more pressure tubes break. However, the third generation can withstand the rupture of three pressure tubes. This accident location system detects a drop in pressure in the primary circuit to identify the half of the core concerned. It can relieve the pressure by degassing in sealed compartments. An overpressure protection system in the reactor vessel protects against overpressure that could occur in the event of a pressure tube rupture inside the caisson, the letdown being ensured by tubes that connect the caisson to the detection and depressurization system via a hydraulic guard that isolates the two systems in normal operation. The guard is removed by overpressure in case of accident. This assembly makes it possible to manage the rupture of two or three pressure tubes (for the reactors of the first- and second-generation, respectively). This system has been improved to be able to support, today, the simultaneous rupture of a maximum of nine pressure tubes. A larger break is beyond the scope of the system. The RBMK concept has undergone technical evolutions and improvements, especially after the Chernobyl accident.

In 2006, Rosatom stated that it was considering a 15-year lifespan extension for RBMKs through the modernization of its 11 remaining RBMK reactors. Ten of these had had their licenses extended by mid-2016. Following significant design changes, as well as extensive refurbishment, including pressure tube replacement, a 45-year lifespan was considered realistic for most 1000 MWe (first generation) RBMK units. The limitation comes from graphite irradiation. For older RBMK units, lifetime performance restoration operations consist in correcting the deformation of the graphite pile due to graphite irradiation, which can deform the pressure tubes. After disassembly of the pressure tubes, longitudinal cutting of a limited number of graphite columns, which were no longer axially straight, restored the geometry of the graphite pile to a state consistent with the original design requirements. This procedure will allow each of these old reactors to operate for at least 3 more years and can then be repeated. Leningrad-1 was the first reactor to undergo this procedure in 2012–2013, followed by the Kursk units, and then Smolensk in 2017. In early 2012, Rosatom announced a 45 billion ruble ($691 million) program to modernize and extend the lifespan of the Smolensk 1–3 RBMK units. In 2012, Smolensk-1 was licensed until December 2022, a 10-year extension after this "*large refitting*" or "*Grand carénage*" as they say in France. The modernization of Smolensk-2 began in 2013, for a return to service in 2015, and included replacement of the fuel channels and modernization of the reactor control and protection system, as

Photo 4.2   Smolensk-3 loading face being upgraded (photo Rosatom)

well as the radiation monitoring system, and strengthening of the building structure. In April 2015, an application for a 15-year license extension was filed. The upgrade of Smolensk-3 (Photo 4.2) followed. In 2017, work began to restore the lifespan performance of the graphite stacks. The three Smolensk units are planned to operate for 45 years. Eventually, WWERs will replace the old RBMKs (adapted from a Rosatom communication, relayed by World Nuclear News).

In 2011, the inspection of the reactor of Unit-1 of the Lenigrad nuclear power plant also revealed premature deformation of the graphite pile caused by the swelling of graphite under the effect of radiation and its subsequent cracking. Vladimir Asmolov, first deputy general director of Rosenergoatom Concern, gave an interview to AtomInfo.Ru in 2012 where he states, "*It is correct that the power capacity of RBMK will not increase. The problem comes from the graphite. At the design stage, graphite expansion under irradiation was expected. It was supposed to start after 40–45 years of operation. The degradation we are witnessing cannot be called catastrophic. But there is a trend. And we have realized that if we increase the power, this will accelerate the degradation. In the first unit of the Leningrad nuclear plant, we have reduced the power to 80 percent to allow the unit to operate until a replacement unit is available. We will continue to measure pressure tube bends. If we see that the trends are bad, we will reduce the power further. I can't rule out the most radical solution. After a certain dose of irradiation, cracks appear in the graphite masonry. It is a complicated process, but at some point, the masonry loses its original geometry. The most unpleasant part of this is the bending by deformation of the channels. In terms of heat transfer, this is still fine, as it will remain at a sufficient level even if the curvature range is large. But there are measuring chains in the channels of the control rods, and for some, the bending is unacceptable. Therefore, we pay particular attention to the issue of graphite degradation."* In fact, these cracks and deformations in the graphite were endangering its further operation. In 2012–2013, work was done on the reactor that reduced the deformation of the masonry by making cuts in the graphite to compensate for the swelling and shaping. In 2013, the reactor was restarted, but the increasing rate of defect graphite required near-yearly masonry corrections. Nevertheless, the reactor was able to remain operational until the end of its lifespan in 2018. In 2014, similar work was required at Leningrad Unit 2. On December 21, 2018, at 11:30 p.m., after 45 years of operation, Unit-1 of the Leningrad nuclear power plant, the first 1000 MWe high-power reactor in the USSR, was shut down permanently. Since its connection to the grid on December 21, 1973, it had produced 264.9 billion kWh of electricity.

However, conclusion on the embedding of pressure tubes and rod insertion channels in the graphite columns arose a fundamental problem of RBMK construction. The swelling of the graphite under irradiation remains the main parameter conditioning the lifespan of the reactor, whereas it is rather the aging of the vessel under irradiation that will determine the lifespan of a PWR reactor (assuming that this component cannot be reasonably changed on site).

Analysis of the RBMK-1000's performance showed that its design had significant reserves for cost-effective power increases. Several reactor parameters, such as the temperature of the metal structures and the graphite pile, were lower than expected, and the main circulating pumps had a significant power reserve. It was therefore decided to increase the power to 4800 MWth i.e. 1500 MWe. Such an increase in power allows a significant reduction in the specific investment costs (20–30%) and, consequently, in the cost of the electricity produced. Specialists from the scientific project team (INPP) and the lead designer organization (NIKIET) began to verify the idea of increased heat transfer in the core, engaging in its design calculation and experimental justification. The main task was to increase the critical power of the fuel channel, i.e., the power at which a boiling crisis occurs on the surface of the fuel element leading to a drying of the cladding/fluid wall, accompanied by an inadmissible increase in the temperature of the fuel element. This problem was successfully solved by introducing heat exchange intensifiers, i.e., grids with axial mixing of the coolant flow (water). These grids were mounted on the upper fuel assembly of the fuel cartridge with a pitch of 80 mm, which allowed a 1.5-fold increase in the exchange capacity of the cartridge. On this basis, the technical design of the RBMK-1500 reactor was published in July 1975. The Scientific and Technical Council of the *Minsredmash* approved the technical design, stressing the need for certain additional studies, in particular the performance of vibration and wear tests on the fuel assemblies with a mixing grid. After testing the new fuels in Leningrad-2, it was confirmed that the performance of the cartridges containing 6.5% enriched uranium left a sufficient margin for the boiling crisis. Compared to the RBMK-1000 reactor, the layout and design of the RBMK-1500 reactor core were modified to improve the circulation of coolant in case of an emergency, as well as a number of design innovations such as increasing the length of the steam separator and the diameter of the condensate feed pipes and the steam circuit. The increase in reactor capacity also resulted in an increase in the number of feed, condensate and drain pumps, ejectors and other thermal engineering equipment, an increase in water consumption in the technical water supply system, and an increase in the auxiliary power system. In order to reduce the magnitude of radioactive noble gas emissions, a two-stage treatment scheme has been considered for gaseous-aerosol releases to the atmosphere through a ventilation stack up to a height of 150 m. In addition, to reduce the values of radioactive aerosol emissions, filter cleaning stations were planned, capturing aerosols on special filters. The new reactors were deployed in Ignalina, Lithuania (Photos 4.3 and 4.4), the general designers being VNIPIET and Gidroproekt of the Ministry of Energy. The site has two RBMK-1500 units, each using 189 tons of 2% uranium oxide with four K-750-65/300 turbines, each with a capacity of 750 MWe. Ingalina-1 was commissioned in 1983 and closed in 2005, and Ignalina-2 in 1987 and 2009. There remains the problem of the positive void coefficient and the instability of xenon, which require rigorous operating conditions.

During the start-up of the Ignalina-1 unit, while measuring the calibration characteristics of the automatic protection and manual control rods, a positive drift in the responsiveness at the initial moment of rod movement was detected. This was caused by water displacement in the lower part of the channel, which required a change in the rod design. This phenomenon

Photo 4.3 Construction of the two Ignalina reactors in Lithuania in the early 1980s

Photo 4.4 Service desk of the Ignalina plant in Lithuania

of reactivity insertion at the beginning of the rod insertion (by insertion of the graphite rod follower) will be found in the case of the Chernobyl accident. Large water leaks in the channels of the measurement systems caused by defective pipes were detected at the beginning of operation, requiring their replacement. At the same time, the operation of the reactors revealed some uncertainties that increased significantly as they approached their nominal capacity, particularly with regard to releases of noble radioactive gases and volatile fission products. In addition, it was found that due to irregularities in power distribution, periodic power peaks occurred in some channels resulting in fuel element cladding cracking. Therefore, a special commission recommended, and the Ministry approved, the reduction of the long-term power of the RBMK-1500 reactor to 1250 MWe. In terms of fuel, the Russians have introduced a burnable poison[2]: erbium introduced homogeneously at 0.41% into uranium oxide enriched to 2.4%. The erbium is consumed during burn-up and produces, by neutron capture, isotopes that are less absorbent than those initially present in natural erbium. This tactic avoids a too strong enrichment at the beginning of the life cycle, which must be compensated by fixed heterogeneous absorbers, while lengthening the natural cycle length. In 1996, after the theoretical studies had been verified experimentally, the RBMK-1500 reactors were converted to uranium-erbium fuel in a progressive manner, during which additional absorbers were gradually removed from the core. The use of uranium-erbium fuel has significantly increased the average burn-up of discharged spent fuel and the average energy production of the fuel assembly in the reactor. Compared to the situation before the introduction of the new fuel, the average burn-up of fuel assemblies in Unit-1 rose to 41% (from 850 to 1200 MWd/t), in Unit'2 - 47% (from 850 to 1250 MWd/t). Beginning in 2005, Unit'2 saw an increase in fuel enrichment with a first pilot batch of uranium 235-erbium fuel enriched to 2.8% and containing 0.6% erbium (adapted from http://www.biblioatom.ru/evolution/istoriya-osnovnyh-sistem/istoriya-reactorov/rbmk/). The two Ignalina reactors are now being dismantled (Photo 4.5).

---

[2] In France, gadolinium is used as a burnable poison.

Photo 4.5 The dismantling of the two Ignalina reactors involved the evacuation of 190 spent fuel transport casks CONSTOR/RBMK-1500 M2 from the reactor pools to an intermediate storage building

Chernobyl (1986, Ukraine)

The plant is located 14.5 km northwest of a small provincial town, Chernobyl, and 110 km northwest of Kiev, the capital of Ukraine. The plant is very close to the border with Belarus. The city of Chernobyl had about 12,500 inhabitants. The town had an economic boom thanks to the installation of the nuclear plant. Three kilometers from the plant was the new town of Pripyat, with a population of 49,000. Thus, there were between 115,000 and 135,000 inhabitants within a 30 km radius of the plant. The plant was built from 1977 for reactor-1 to 1983 for reactor-4. In 1986, reactors-5 and -6 were still under construction. An artificial lake with an area of 22 km² was built on the banks of the Pripyat River to provide cooling water for the reactors.

Chernobyl Reactor-4 is an RBMK type reactor (Photo 4.6) of second generation of 1000 MWe (3200 MWth). Chernobyl-4 started up for the first time on November 11, 1983. Reactor 4 shares with Reactor 3 the same fuel building and the same turbine building. The core consists of a stack of 190.3 tons of uranium oxide enriched to 1.8%, contained in 1696 assemblies cladded with a zirconium alloy, and 600 tons of graphite. The core is cooled by 1693 pressure tubes in zirconium alloy with 2.5% niobium (88 mm in diameter and 4 mm thick). These pressure tubes penetrate the reactor vertically and through which boiling light water flows. Each pressure tube houses

Photo 4.6   Cover of the presentation brochure of Chernobyl

two assemblies, each 3.64 m high, bolted together on a central tie rod. Each sub-assembly contains 18 fuel rods, arranged in two concentric rings around the central tie rod. The fuel consists of pellets of 2% enriched UO2, 11.5 mm in diameter and 15 mm high, clad in a zirconium-niobium alloy (1%). The pressure tubes are placed in holes that penetrate stacks (columns) of graphite bricks (25 cm × 25 cm × 60 cm) (Figs. 4.12 and 4.13).

The core is cooled by two loops and two steam separators (drum 2.3 m in diameter and 30 m long) filled with the water-steam mixture (average quality 14.5%), which leaves the core (Figs. 4.14 and 4.15). Radially, there is then a biological protection against neutrons (water compartments), then an annular compartment of sand.

In the RBMK concept, there is no steam generator, which allows savings in components, but does not physically isolate the Turbo-Alternator Unit from the core. The water pressure in the core is low compared to a PWR (about 81 bars at the core inlet and 69 bars in the steam separators). The steam produced is sent to two turbines of 500 MWe each. After condensation in a condenser, the water is returned via 12 collectors and then 22 sub-collectors per loop to supply the pressure tubes (Fig. 4.16). The multiplicity of piping in

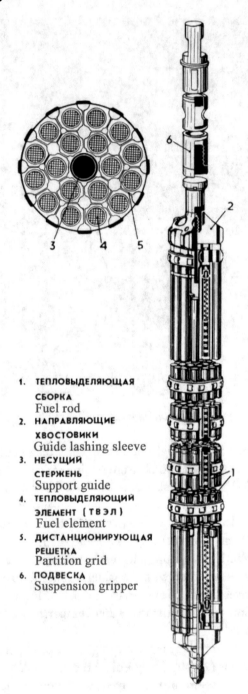

1. **ТЕПЛОВЫДЕЛЯЮЩАЯ СБОРКА**
   Fuel rod
2. **НАПРАВЛЯЮЩИЕ ХВОСТОВИКИ**
   Guide lashing sleeve
3. **НЕСУЩИЙ СТЕРЖЕНЬ**
   Support guide
4. **ТЕПЛОВЫДЕЛЯЮЩИЙ ЭЛЕМЕНТ (ТВЭЛ)**
   Fuel element
5. **ДИСТАНЦИОНИРУЮЩАЯ РЕШЕТКА**
   Partition grid
6. **ПОДВЕСКА**
   Suspension gripper

Fig. 4.12   A fuel rod cluster from Chernobyl

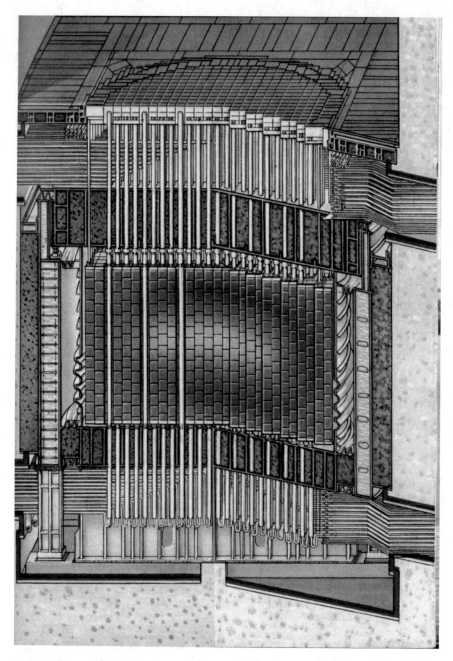

Fig. 4.13 Graphite moderator penetrated by the pressure tubes. One can see the complexity of the water supply (at the foot of the massive graphite block) and the steam collectors (above the core). Drawing from the presentation brochure of the reactor (1977)

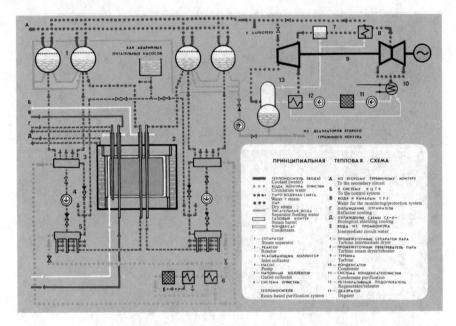

Fig. 4.14 A rare document from the presentation brochure of the plant (1977): the main circuits of Chernobyl

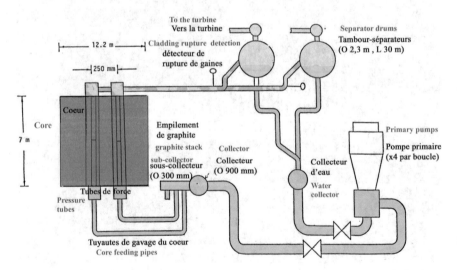

Fig. 4.15 Primary circuit of Chernobyl (x2 loops). There is a certain resemblance with the first boiling water reactors as regards the steam separation system, and Natural Uranium-Graphite-Gas (UNGG) because of the constitution of the core

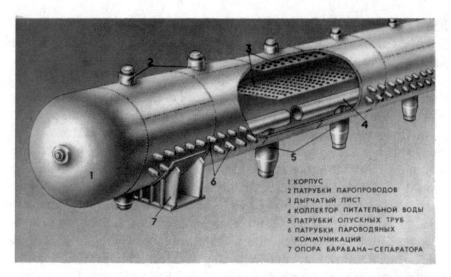

I КОРПУС
2 ПАТРУБКИ ПАРОПРОВОДОВ
3 ДЫРЧАТЫЙ ЛИСТ
4 КОЛЛЕКТОР ПИТАТЕЛЬНОЙ ВОДЫ
5 ПАТРУБКИ ОПУСКНЫХ ТРУБ
6 ПАТРУБКИ ПАРОВОДЯНЫХ
   КОММУНИКАЦИЙ
7 ОПОРА БАРАБАНА—СЕПАРАТОРА

I НАГНЕТАТЕЛЬНЫЙ КОЛЛЕКТОР
2 РАЗДАТОЧНО—ГРУППОВЫЕ КОЛЛЕКТОРЫ
3 НАПОРНЫЕ ТРУБОПРОВОДЫ
4 РАСХОДОМЕР
5 КЛАПАН РЕГУЛИРУЮЩИЙ
6 ТРУБОПРОВОДЫ ВОДЯНЫХ
   КОММУНИКАЦИЙ

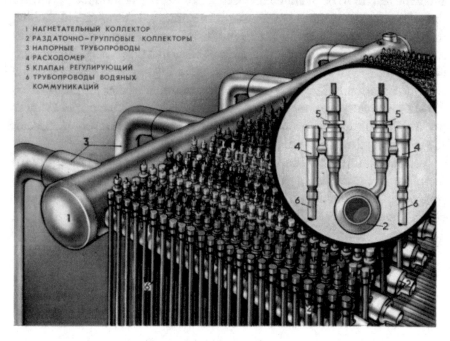

Fig. 4.16   Drawings of a steam separator drum and a steam header from the presentation brochure of the reactor (1977). The RBMK concept induces an important complexity of piping and ducts because each pressure tube is isolated

RBMK concepts is a weak point that multiplies the possibilities of a Loss of Primary Coolant Accident, which technologically prohibits increasing the pressure.

The circulation of water in the pressure tubes allows for on-going refueling since it is sufficient to isolate only one channel, the others continuing to flow. This makes it possible to produce plutonium continuously for military purposes. The core has imposing dimensions (11.8 m × 8 m), which requires many rods to control the reactivity and spatial instability of the neutron flux: 4 groups of rods for a total of 211 B_4C boron carbide rods. The first group consists of 139 rods that are inserted by manual control from the top of the reactor. A second group consists of 24 "short" rods that are inserted manually from the bottom of the reactor for axial power control. A third group of 24 rods constitutes the shutdown group. The fourth control group consists of the last 24 rods, 12 of which are controlled by the average power of the core and allow the reactor to be operated between 20 and 100% of nominal power, and the other 12 at local power. The reactor building is a conventional building that is not designed to retain fission products, nor to resist internal overpressure, because of its relatively small volume (1000 m³) (Fig. 4.17). This absence of resistant external building is expressed by the will of the Soviets to limit the costs of construction. It happens that the nominal operating point of the RBMKs is slightly over-moderate,[3] which makes the void coefficient slightly positive. The situation gets worse if the core is poisoned with xenon 135[4] and the void coefficient becomes more strongly positive.

RBMKs exhibit instability when the core operates below 25% of the nominal power, an instability aggravated by the large size of the core, and by the presence of void fraction. The void coefficient depends strongly on the concentration of xenon 135, a gas produced by fissions in the fuel. Indeed, xenon 135 strongly absorbs slow neutrons and a hardening of the neutron flux

---

[3] The k_{eff} curve as a function of the water volume to fuel volume ratio (called moderation ratio) presents a maximum for a value called moderation optimum. Above this optimum (resp. below), the reactor is said to be over-moderated (resp. under-moderated). In an over-moderated situation, the appearance of additional steam decreases the moderation ratio which thus tends towards the optimum, hence an increase in reactivity.

[4] Xenon is a gaseous fission product that appears in the fuel. Its isotope 135, radioactive with a half-life of 9 h14, is a formidable absorber of "thermal" neutrons, i.e., neutrons strongly slowed down by the carbon atoms of graphite. This absorber is so powerful that it is referred to as "neutron poison" which "suffocates" the reactor.

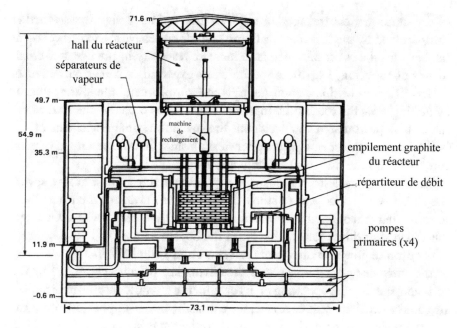

Fig. 4.17 General plan of Chernobyl-4

spectrum[5] due to the presence of vacuum causes a decrease of the neutron absorption of xenon. In fact, the void coefficient is more strongly positive at high xenon concentration. This concentration increases if the reactor is operated at low power because xenon 135, produced mainly by radioactivity of iodine 135 of half-life 6.57 h, disappears less by capture if the neutron flux decreases. This problem was known to the designers, whose major concern was full power operation in a country with a chronic shortage of electricity production. It is therefore foreseen in the Technical Operating Specifications not to remain below 700 MWth for too long, in any case without a rod inserted in the core. The design of the core has therefore accommodated a positive void coefficient of about $\partial k/(k\partial\alpha) = 2000$ pcm at nominal power, and which increases when the power decreases! On the other hand, the more the rods are extracted, the more the void coefficient is positive.

The electromechanical control rods insertion system takes about 20 s to insert the shutdown rods, which is more than ten times the gravity drop time

---

[5] A neutron flux spectrum is said to harden when the average speed of neutrons increases, which means that the proportion of fast neutrons increases compared to slow neutrons. A hardening of the spectrum can have several causes: disappearance of the moderator whose function is to slow down the neutrons, increase of poisons absorbing in the thermal domain (case of xenon 135), appearance of fissile isotopes with a harder neutron emission spectrum (plutonium)…

of the shutdown control rods of a PWR! This serious design flaw is further compounded by the fact that the boron carbide rods have a 1 m carbon tip at the bottom, the effect of which is to increase reactivity in the first few centimeters of insertion, i.e., the opposite effect expected of a reactivity control device. The idea of this carbon meter is to "complement" the lower reflector by nesting once the rod is fully inserted. Thus, it was probably the start of the scram that precipitated the accident! The maximum design accident of the RBMK was a rupture of a large upper steam header, in no case a massive pressure tubes rupture due to a reactivity accident.

The reactor was destroyed during an experimental test phase at low power, the objective of which was to study the unstable behavior of the reactor and to show that it remained controllable without using safety devices. During a scheduled shut down for maintenance, the Soviets wanted to take advantage of the drop in power to carry out a test at around 1000 MWth. This was a test of the emergency power supply to the circulation pumps (Fig. 4.18) by the turbine-generator set operating on its own inertia after the turbine has tripped. In certain accident sequences where external power supplies are lost, the plant's diesels start-up and supply power to the safety systems.

In the design of the RBMKs, the latency time for the recovery by the diesels is about 40–50 s. However, the recirculation pumps cannot wait for such a long period of time without being supplied with electricity. The experiment was therefore intended to test a new alternator voltage regulus. In fact, three safeguard systems were voluntarily inhibited (i.e., the safeguard injection plus two automatic scram devices) to carry out this test. The accidental transient was initiated while the reactor was operating for a long time at reduced power, with less than 30 rods inserted in the core, i.e., two violations of the operating instructions. On April 25, at 1 a.m., the operators began to reduce the load and at 1 p.m. reached half power (1600 MWth). They shut down Turbo-Alternator Unit No. 7 and switched off the safety injection according to the experimental procedure (otherwise it would have triggered automatically and thwarted the test). At 2 p.m., the national grid asked them to remain at 50% of nominal power to cover the demand, and it was not until 11.10 p.m. that the dispatching office authorized the continuation of the load reduction. The reactor therefore remained without a safeguard system for 9 h, in formal violation of the safety regulations. At this point, the operator made a significant error in the execution of the procedure. The reactor control system requires that at low power levels, the LAC (Local Automatic Control) system be switched to a global regulation system for average power. This switchover is manual and requires the operator to synchronize the control parameters to switch from one system to the other. The operator missed the switchover

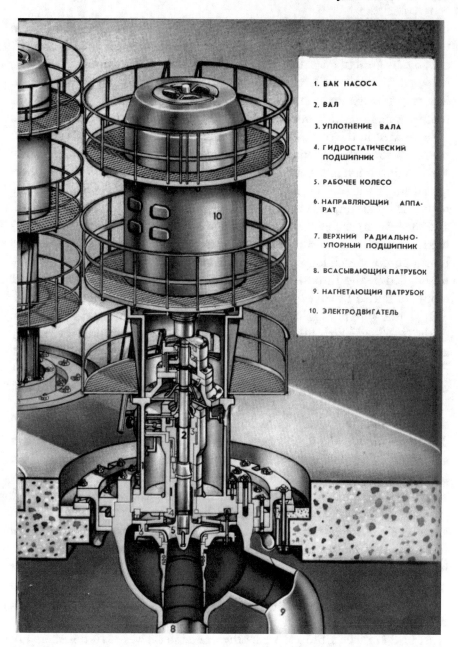

1. БАК НАСОСА

2. ВАЛ

3. УПЛОТНЕНИЕ ВАЛА

4. ГИДРОСТАТИЧЕСКИЙ
   ПОДШИПНИК

5. РАБОЧЕЕ КОЛЕСО

6. НАПРАВЛЯЮЩИЙ АППА-
   РАТ

7. ВЕРХНИЙ РАДИАЛЬНО-
   УПОРНЫЙ ПОДШИПНИК

8. ВСАСЫВАЮЩИЙ ПАТРУБОК

9. НАГНЕТАЮЩИЙ ПАТРУБОК

10. ЭЛЕКТРОДВИГАТЕЛЬ

Fig. 4.18 Circulation pump (drawing from the Chernobyl presentation brochure—1977)

maneuver, and the reactor was choked by the misdialed global regulation. The reactor power dropped to 30 MWth. The operator therefore tried at about 1 a.m. on 26 April to raise the power to 200 MWth to start the test. To do this, he raised the rods to a level above the Technical Operating Specifications. In fact, they provide that the reactor must always operate with a sufficient anti-reactivity margin. This margin is defined by an equivalent number of rods fully inserted in the core. This number must be at least equal to 30 rods, and below 15 rods, the reactor must be shut down immediately. This technical specification is not automated, and the operator can violate it without suffering an automatic shutdown. The operation at intermediate power, previously requested by the dispatching, had strongly poisoned the reactor in xenon 135, and the void coefficient was then positive of the order of +30 pcm/% volumetric steam. At 1 h 03 min, the operators started the test and started the fourth pump of each loop so that during the test, 4 pumps (2 per loop) were re-supplied by the Turbo-Alternator Unit n°8, and the 4 other pumps (2 per loop) continued to cool the core by being supplied electrically by the grid. This pseudo-symmetrical configuration feeds the core with colder water due to the overflow, which initially reduces the quantity of steam in the core by condensation and the pressure in the separators, the level of which falls below the alarm threshold. The operator compensates by raising the last rods in the core, losing even more of his anti-reactivity margin. The level and pressure in the steam separator drums (the steam that comes out of the core is very wet, since it is actually a water/steam mixture containing about 15% steam) fall dangerously low, reaching the shutdown threshold. The operators inhibit this automatic safety device so as not to be disturbed during the test (!), and increase (by tripling it!) the feed water level to raise the water level in the separators. The automatic control rods are now completely removed from the core, and the operator raises the so-called "manual" rods (not controlled by an automatic control system). At 1 h 22 min 30 s, the operator in the control room notes on the monitoring printer that the available anti-reactivity margin has reached such a level that the emergency shutdown must be immediate because the reactivity margin is only 6–8 rods. He informs the test manager, who nevertheless decides to launch the test, a decision with serious consequences.

At 1 h 23 min 04 s, the test began: the steam inlet valves of Turbo-Alternator Unit (GTA) n°8 were closed (this is the idea of the test). The closing of the valves increases the pressure in the primary circuit, resulting in a contraction of the steam and a decrease in reactivity (because the void coefficient is positive), this leads to a new rise in the control rods. The diesels start-up but only reach their nominal speed after 40 s. The emergency shutdown on the "*2 units unavailable"* signal comes on without any consequence on the reactor, because the operator has inhibited this protection to let the reactor run in order to be able to restart the test in the event of an unsuccessful first test (for a single test or a real loss of GTA incident situation, the reactor would stop on automatic scram). The pumps are then supplied by the inertia of the GTA and the flow of primary water decreases according to the declining speed of the GTA. We find ourselves in a situation of a test where we are not sure to be able to ensure the supply of the pumps with a reactor still running! The slowing down of the four pumps leads to a drop in the flow of primary circuit water and a decrease in the sub-cooling of the circulating water. It is therefore water at the limit of saturation that enters the core. The temperature of the core and of the water increases, as well as the steam quality (void fraction). The reactivity increases and at 1 h 23 min 40 s, the operator presses the scram button because all the indicators are in the red. But because of the positive void coefficient, the power increases exponentially and reaches 100 times the nominal power in less than 4 s (Fig. 4.19). The power will only shutdown because of the Doppler effect. The safety control rods, which are inserted very slowly into the core, can do nothing, especially since at the beginning of insertion, the end of the graphite rods flushes out water by penetrating the core, contributing to the void effect, and over-increasing the reactivity by the moderating effect of the

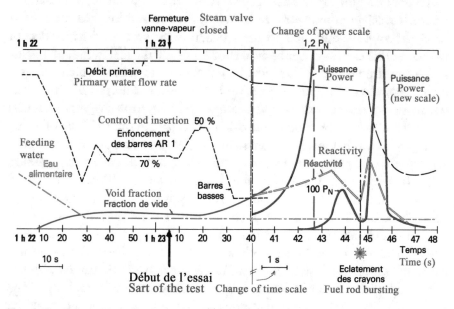

Fig. 4.19 The accidental transient of Chernobyl on April 26, 1986, 1 h 23 min 40 s (change of time scale from this time, change of power scale from 1 h 23 min 42 s, adapted from report IPSN 2-86)

Photo 4.7   Reactor slab inverted almost vertically in the vessel pit (DR)

graphite during the first insertion steps. As the reactor is already very hot, the channels receiving the control rods were deformed and blocked the rods at barely 1.5 m instead of the 7 m active height of the core. Under the effect of the power that can no longer be evacuated by steam, the fuel swells and bursts, dispersing droplets of molten fuel in the water and steam, which causes a steam explosion. The power flash probably causes radiolysis of the water, which dissociates into oxygen and hydrogen, a highly explosive gas that is also produced massively by exothermic oxidation of zirconium (83.7 kg of zirconium per active channel, i.e., more than 141.7 tons of Zr, potentially producing 6.5 tons of hydrogen), whose combustion contributes to the pressure spike. It is estimated that 30% of the channels exploded. If the contribution of hydrogen combustion to the explosion cannot be excluded (one can even speak of detonation, which appears under certain conditions during a deflagration to detonation transition in mixtures of hydrogen and air), the steam explosion alone can justify the mechanical power that pulverized the core and raised the containment slab (Photo 4.7) since the pulverization/micronization of the molten fuel at some 3000 °C in water presented a considerable exchange surface with the creation of an insulating surface of steam on the surface of the drops of corium made up of molten fuel and molten cladding. This almost instantaneous transfer of heat created a compression wave which itself

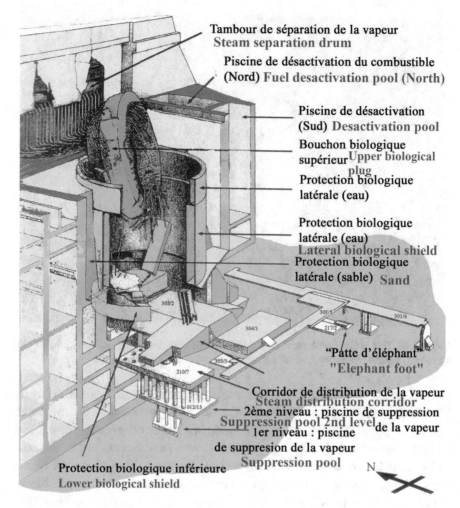

Tambour de séparation de la vapeur
Steam separation drum

Piscine de désactivation du combustible
(Nord) **Fuel desactivation pool (North)**

Piscine de désactivation
(Sud) **Desactivation pool**

Bouchon biologique
supérieur **Upper biological plug**

Protection biologique
latérale (eau)

Protection biologique
latérale (eau)
Lateral biological shield

Protection biologique
latérale (sable) **Sand**

"Patte d'éléphant"
"Elephant foot"

Corridor de distribution de la vapeur
Steam distribution corridor
2ème niveau : piscine de suppression
Suppression pool 2nd level de la vapeur
1er niveau : piscine
de suppresion de la vapeur
Suppression pool   N

Protection biologique inférieure
Lower biological shield

Fig. 4.20 Inner view of the Chernobyl reactor building after the explosion (adapted from Borovoi and Sich 1995)

micronized larger drops of corium, resulting in a chain reaction of thermal and mechanical effects, unrelated to neutron production.

This explosion broke new pressure tubes causing an overpressure wave of about 10 bars. This overpressure raised the upper plug (2000 tons!), which serves as a slab (resisting a pressure of 2 bars), destroying the other pressure tubes while dragging the rods. The cylindrical plug fell back vertically (Photo 4.7) and got stuck in an inclined position in the reactor pit (Fig. 4.20). The explosion also destroyed the light structures of the conventional building, including the core, the crane, and the fuel loading machine (Photo 4.8).

Photo 4.8 The building of reactor-4 is torn apart. On the right the building of the reactor-3, where the teams remained at their post during the accident

This explosion also claimed its first victim: the patrol auxiliary operator Valery Khodemtchouk, who must have been close to the service desk on the upper side of the pile and who did not even have time to alert the control room. His body was never found. Very hot graphite blocks were thrown out of the core and caught fire on contact with air, generating several fires that took heroic firefighters several hours to bring under control. Part of the core was blown out of the broken building and a large quantity of radioactive products were released into the atmosphere without any filtration (Photos 4.9 and 4.10).

The building was only designed for lateral pressure tube breaks, but not in the core. The side bunkers are rated at 4.5 bars, and the pump and manifold area were only rated at 0.8 bar overpressure. The bunkers are connected to the steam suppression pool[6] located in the lower parts of the reactor by connection pipes that are equipped with rupture discs. In the accident, the reinforcement of the lateral structures only had the effect of channeling the blast upwards, nevertheless protecting the turbine building, which continued to operate for the twin unit 3, before the shift supervisor of the plant 3 decided

---

[6] The function of this steam suppression pool is to condense the steam from the core and to percolate it into the pool. This component also exists in the Boiling Water Reactors .

Photo 4.9   Aerial view of the reactor-4 or rather what remains of it. This view allows to see inside the pit of the reactor

to return to a cold shutdown at 5:00 a.m., against the orders of the director of the plant V.P. Brioukhanov.[7] Two agents of the plant were killed on the spot because they were present in the reactor building at the time of the explosion.

[7] Viktor Petrovich Brioukhanov (1935–2021). Ukrainian hydraulic engineer. After his studies at the Tashkent Polytechnic Institute, he worked from 1959 onwards in several Ukrainian hydroelectric plants where he rose through the ranks. From 1970 he worked on the construction of the Chernobyl plant and became the plant manager when it started up in 1977. He has been described as a zealous civil servant who did not pay much attention to safety rules. He was not present during the test that led to the accident. He was sentenced to 10 years in prison in a closed trial of the case in 1987. He was released from prison in 1992 and became director of the state-owned power plant import-export company Ukrinterenergo, which is particularly active in the hydroelectric sector. His rehabilitation to this position shows that he was used as a scapegoat. The causes of the disaster must be sought in the Soviet system as a whole, rather than in the individual responsibility of a single engineer, because such an accident cannot be the work of a single man.

V. P. Brioukhanov

Photo 4.10 An impressive photo taken through the open door of a helicopter, during the construction of the first sarcophagus (photo DR). A truck, which one wonders how it arrived there, gives the measure of the whole picture

From a neutronic point of view, based on a $\beta_{core} \approx 500$ pcm for the core and taking a fast neutron lifetime $\ell \approx 1$ ms, the reactor behavior if the inserted reactivity ρ is large, is driven by prompt neutrons, i.e.

$$\frac{dP(t)}{dt} = P(t)\frac{\rho - \beta}{\ell} \quad \text{thus} \quad P(t) = P_{(t=0\,s)}e^{\frac{\rho - \beta}{\ell}t}$$

With a void coefficient of 30 pcm/% void, a change in void fraction from 10% (nominal value) to 60% injects 1500 pcm of reactivity, or $\rho - \beta \approx 1000$ pcm. In one second, the power is multiplied by more than 20,000 times, whereas the Doppler effect, which is only –1 pcm/K, will only be effective after about 5 s. It is very likely that there was a steam explosion when the molten fuel sprayed into the water. The oxidation of the fuel cladding heated the graphite rings which surround them, setting them on fire, then the fire spread after one hour to the moderator graphite block.

The energy deposited in the fuel was sufficient to fragment the fuel into fine particles, generating an intense corium-water reaction. If all the pressure tubes had emptied, a reactivity insertion of $2.7 would have been obtained according to the Soviet analysis. It seems that about 200 kg of hydrogen were produced during the accident. Borovoï and Sich estimate that 135 tons of fuel out of the initial 190.3 tons melted and flowed into the lower parts of the reactor[8] (Photo 4.11). Six point seven tons were expelled into the atmosphere in the form of particles and about 30 tons in the form of dust fell back into the upper levels of the reactor. The remaining 11 tons have yet to be precisely located in the reactor building or the sarcophagus. The reactor building was heavily damaged above the service desk, and the upper roof was blown off, reducing the height of the building from 71.5 to 46 m. The blast destroyed or displaced the support columns of the building threatening its integrity, hence the difficulty of constructing the sarcophagus on top of it.

By 1:28 a.m. on April 26, the first firefighters were already working on the site. human reinforcements, consisting of new firefighters and technicians, arrived until 4 a.m. to try to contain the various fires that had appeared in the building of plant-4, on the roof of the adjacent turbine hall, and in the diesel fuel and flammable materials storage areas. The largest fires in the turbine building were extinguished by 2:10 a.m. and by 2:30 a.m. the main fires on the roofs of the reactor building were under control. At 4:50 am, most of the fires were extinguished. A graphite fire broke out about 20 h after the start of

---

[8] Borovoï and Sich (1995).

Photo 4.11 Flow of solidified corium in the shape of an "elephant's foot." Even after partial solidification, the residual power can melt this corium until the surface in contact with the air is large enough to evacuate the power by natural convection. The activity of this corium is such that the remotely operated inspection robots that approached it the first time failed due to the intense radiation (photo DR)

the accident. This fire was formed from gases formed by the action of the steam on the graphite and by oxidation of the zirconium in the fuel cladding. The oxidation reaction of the zirconium is very exothermic, which contributes to the temperature increase and feeds the graphite fire, which is itself exothermic. The firefighters were unable to control it and called in helicopters and the army. This fire did not cease definitively until May 7 and was largely responsible for the dispersion of radionuclides in the wind, whereas a simple explosion without fire would probably have contaminated only an area near the plant.

In order to extinguish the graphite fire and to reduce the release of radioactive products, helicopters dumped 5000 tons of various materials, in the form of 80 kg bags, between April 27 and May 2. These materials included boron carbide, to absorb neutrons; dolomite, which was to serve as a cold source and as a source of carbon dioxide to smother the fire; lead, to absorb gamma radiation; sand and clay, which were supposed to prevent the release of particles. However, these materials may have acted as thermal insulators, thereby increasing the temperature of the core, which led to a new release of radionu-clides a week later. In addition, because of the lack of visibility, and because the high irradiation rate forbade helicopters to remain over the area for a long time, the materials were pierced a little on the spot, instead of being concen-trated on the cavity and the burning roofs. The bags of material falling in the wrong place caused further damage to the building structure and increased the dispersion of contaminated dust in the atmosphere. It is known that the melted fuel and structural elements, forming corium, melted through the lower biological shield, and flowed to the ground, intensifying the release of radionuclides.

The explosion of the reactor released a very large quantity of radioactivity of the order of 300 Mega-Curie (i.e., 12 Exa-becquerel,[9] or about 20% of the total initial radioactivity of the core, including 0.1 Exa-becquerel of ^{137}Cs) which significantly contaminated an area close to 150,000 km², and a good part of Western Europe, depending on the wind. It was in Sweden that the first radioactivity measurements alerted the West on April 28, 1986,[10] report-ing a massive release of radioactivity. It was noted that sites located 2000 km away received more radioactivity than nearby areas 200 km away. In France, Corsica and the southeast of France were the most exposed. A controversy arose over the fact that the French government had minimized the impor-tance of the deposits, which the press seized upon by ironically stating that *"the cloud had stopped at the borders!"* It is estimated that about one-third of the radioactivity was deposited in the USSR (Fig. 4.21), another third in the rest of Europe, and the last third in the northern hemisphere outside Europe in a more or less uniform way (Figs. 4.22 and 4.23).

Initially, the main concern of the authorities was to ensure the subcriticality of the destroyed core by massive helicopter dumping of solid materials, 5000 tons of sand, clay, boron, dolomite and lead, to smother the fire and control a possible return to criticality. *"Liquidators,"* as the term is currently used, were sent on the roof to release the expelled radioactive debris with shovels into the reactor hole. A medal honors their sacrifice as most of them suffered heavily from radiation (Photo 4.12).

From May 6, the Soviets succeeded in injecting nitrogen to definitively stop the graphite fires, while cooling the corium. On May 14, the idea of a slab with a built-in liquid nitrogen cooling system was proposed, to be installed under the reactor, in order to slow down the breakthrough of the raft

---

[9] 1 Exa-Becquerel = 10^{18} Becquerels.

[10] N.J. Pattenden: *A review of long-term studies of radioactivity in the environment from the Chernobyl acci-dent by AEA technology*, Nuclear Energy, Vol. 30, n°6 pp 341–359 (1991).

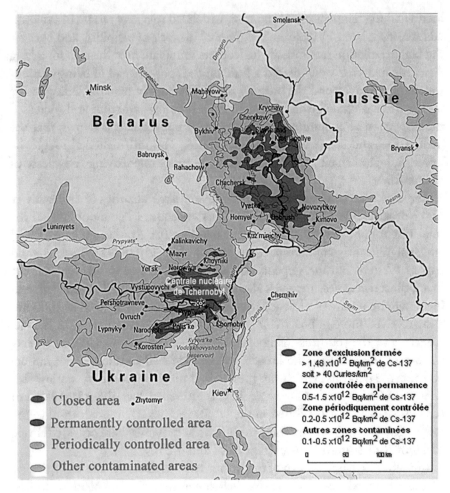

Fig. 4.21 Deposits measured in 1996 around Chernobyl in Belarus and Ukraine (source http://www.astrosurf.com/luxorion/Physique/tchernobyl-contamination-map.jpg)

that could contaminate the groundwater with molten radioactive material, and to cool the core. To do this, a tunnel was dug by miners called in as reinforcements for 15 days from the base of plant-3. They dug a gallery under the reactor with the aim of installing this liquid nitrogen exchanger. Concrete was poured there instead to slow down any vertical progression of the corium.

In order to contain the radioactivity, the Soviet state called upon a large number of civilian and military personnel. These liquidators carried out, among other things, decontamination of the site and roads, storage of waste, and the construction of dams. However, their most important task was the construction of the sarcophagus around the reactor. The sarcophagus was

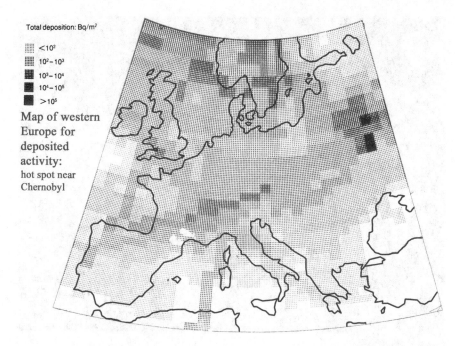

Total deposition: Bq/m²

      <10²
      10²–10³
      10³–10⁴
      10⁴–10⁵
      >10⁵

Map of western
Europe for
deposited
activity:
hot spot near
Chernobyl

Fig. 4.22   Total deposit of ^{137}Cs between April 25 and May 8 (according to H.M. Simpson, J.J.N. Wilson, S. Guirguis, P.A. Scott: *Assessment of the Chernobyl release in the immediate aftermath of the accident*, Nuclear Energy Vol. 26, n°5, pp 295–301 (1987))

built in a hurry from July to November 1986 (Fig. 4.24). The purpose of this sarcophagus was to prevent the radioactivity remaining in the reactor from dispersing into the environment, to limit the entry of rainwater, which could contaminate the ground, and to continue the operation of reactor no. 3. Indeed, this reactor was adjacent to reactor No. 4 and had common facilities such as the turbine hall and the auxiliary building. This sarcophagus (Photo 4.13) was made of beams and large metal plates, as well as concrete. This 300,000-ton building is 175 m × 70 m square and 66 m high. However, because of the very high dose rates, the beams and metal plates had to be installed with remote cranes, and thus without the possibility of fixing them precisely together. The surface area of the openings in this sarcophagus was estimated to be about 1000 m². Voluntary or not, these openings contributed to cool the structures thanks to the circulation of air, but left escape routes for the radioactive particles, partially stabilized on the ground by deposited resin. These gaps were reduced by half following work carried out from 1995 to 1997.

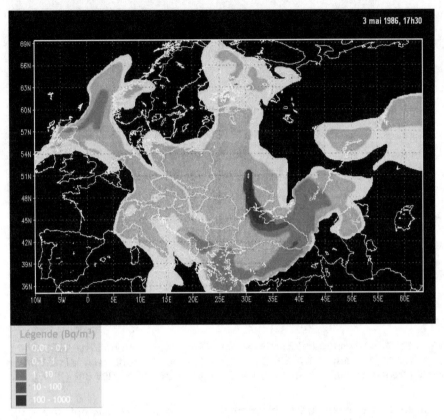

3 mai 1986, 17h30

Légende (Bq/m³)

0.01 - 0.1
0.1 - 1
1 - 10
10 - 100
100 - 1000

Fig. 4.23 Dissemination of the radioactive cloud on May 3, 1986. France will be particularly affected in the South-East and in Corsica

But this new structure is not designed to withstand a major earthquake. However, this area of Ukraine is at risk of suffering an earthquake of magnitude 6 on the Richter scale[11] every hundred years. It will thus be necessary in the long term to rebuild a reinforced structure. The roof of the sarcophagus is also composed of sheet metal plates that were installed remotely because of the radiation (by cranes and remote handling), hence the difficulty of making them perfectly joined. Finally, a polyvinyl solution is regularly sprayed inside the sarcophagus in order to fix the radioactive particles to the ground as much as possible.

---

[11] Charles Francis Richter (1900–1985), American seismologist. After studying at Stanford University, then a doctorate at Caltech where he worked from 1936, he proposed in 1935 the scale that bears his name, based on the energy of the earthquake. In 1956, he calculated with Beno Gutenberg the correspondence between the magnitude of an earthquake and its developed energy.

Photo 4.12 Soviet medal of the *"liquidators."* Alpha, beta, and gamma radiation penetrate a drop of blood

In 1997, an international fund for Chernobyl was set up under the control of the European Bank for Reconstruction and Development (EBRD) to contribute to the maintenance of the sarcophagus and its reconstruction. The principle of building a second sarcophagus over the current one has been shut down, 720–760 million dollars needed have been raised by the EBRD. In addition, the donor countries of the International Chernobyl Shelter Fund pledged in May 2005 to provide an additional $200 million for the construction of a new sarcophagus: a steel protective arch built 180 m from the reactor and slid on rails over the current sarcophagus (Photo 4.14). The largest new contribution came from the G8, with $185 million, and Russia for the first time pledged to participate in the international community's financial

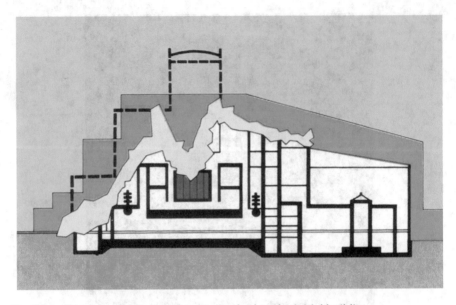

Fig. 4.24 Principle of the sarcophagus attached to the initial building

effort. Ukraine is to contribute $22 million. In September 2007, a contract was signed with the Novarka consortium (Vinci and Bouygues) for the construction of a new accessible containment that will overlap the old sarcophagus. The new arch-shaped structure will have several objectives: to protect the first sarcophagus against external aggression, to ensure a perfect seal between the radioactive ruins of the destroyed reactor and the environment, and to allow the dismantling of the old sarcophagus and the removal of radioactive materials under safe conditions. Bouygues Travaux Publics and Vinci built the new containment for Reactor 4 between 2007 and 2018. This sarcophagus of titanic dimensions will have an estimated lifespan of 100 years. In particular, it will allow the dismantling of the accident reactor. The workers on the site, intensely decontaminated to limit the doses received, have benefited from reinforced medical monitoring and specific protective equipment. Composed of two metal structures, with a total weight of 25,000 tons, this containment arch is 162 m long, 108 m high and has a span of 257 m. It alone could overlap the Stade de France or the Statue of Liberty! The containment was designed to withstand temperatures ranging from −43 to +45 °C, as well as a class-3 tornado or an earthquake measuring VI on the Mercalli scale. This project involved international expertise. The arch was built in Italy, the siding by a German-Turkish company, and the lifting was outsourced to a Dutch company. On the site, we met Ukrainian site managers, Italian mountaineers for the work at height, and Azeris used to offshore oil platforms

Photo 4.13   The first sarcophagus of Chernobyl 4

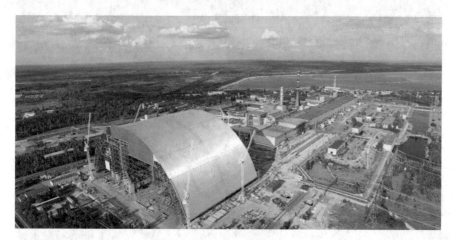

Photo 4.14   The new containment arch built by the French companies Bouygues and Vinci (source site internet Bouygues)

for the lifting part (adapted from https://www.bouygues-construction.com/projet-emblematique/tchernobyl).

The impact on the population was very important. 200,000 people were evacuated and the town of Pripyat (Photos 4.15 and 4.16) closest to the plant was permanently abandoned (45,000 inhabitants). About a hundred people died of acute burn-up syndrome, but it is estimated that several thousand people died of induced cancers. The accident was classified at the highest level of the INES scale, i.e., 7. It can be considered as the most serious reactor accident ever, in terms of radioactive releases and mortality.

Following the accident, many international programs analyzed the causes and consequences of the destruction of the reactor. On other reactors of the same type, 80 additional absorber rods were installed to counter the void effect and bring it back below β. The anti-reactivity margin of spare rods (which should not be inserted) was increased to more than 43 rods. A new rod insertion system that reduces the insertion time from 20 to 2 s was installed on all RBMKs. Water displacement induced by the start of shutdown cluster insertion has been geometrically eliminated by lowering the position of the fuel rods to ensure anti-reactivity insertion as soon as the rods are inserted. The number of shutdown control rods to be inserted from below was increased from 24 to 32. The command control was modified to make it more difficult

Photo 4.15   The Pripyat cinema theater in 1970

Photo 4.16   The abandoned carousel in Pripyat, an iconic image of the effects of Chernobyl

to inhibit the automatic safeties. The Soviets themselves noticed the fact that the plant had been operating since its launch with a very high rate of availability had made the operators lose all notion of risk, and the obligation of results to an order received, in a political system where orders were not contested, had amplified the disempowerment of the operators.

The first modification to be made concerns the positive void coefficient. Two measures have been adopted to decrease this positive reactivity coefficient: decrease the graphite mass and increase the amount of absorbing material in the core. The decrease in the mass of graphite makes it possible to slow down the neutrons less well. Water plays a significant role in the moderating effect. Thus, when the water vaporizes because of an increase in temperature, the neutrons are less slowed down and the fission rate increases less rapidly. This solution can only be implemented in the new RBMK reactors. The increase of absorbing materials can be done in already existing RBMK reactors. This involves replacing fuel elements with fixed absorbing elements made of boron steel. Eighty additional absorber assemblies have been introduced in the 1000 MWe RBMK reactors and fifty in the 1500 MWe RBMK reactors. The loss of reactivity generated is compensated by an increase in the enrichment of the U235 fuel from 2 to 2.4% in the RBMK-1000 reactors. Enrichment remained at 2% on the 1500 MWe reactors. The void coefficient remains positive but has been reduced from $4.5 \times \beta$ to $0.7 \times \beta$. Indeed, a negative or zero void coefficient would have rendered the safeguard systems inoperative. The control rods also underwent modifications. The graphite follower on the rods was reduced because it increased the reactivity of the control rod cluster insertion. In addition, the first two generations of RBMK reactors were not equipped with a shutdown system. They only had control rods that went from the mid-core position to the bottom position in 14 s. Twenty-four fast control rods were added, capable of shutting down the reactor in 2.5 s. In general, the speed of the control rods has been increased. The radial division into nine or twelve zones instead of seven will therefore be finer. The number of so-called "short" control rods has been increased, as well as the number of manual control rods. The first two generations of RBMK reactors had a depressurization system of eight 300 mm diameter pipes allowing the evacuation of steam to the atmosphere to cope with a pressure tube rupture. Larger diameter pipes have been added to most of the existing RBMK reactors to cope with the rupture of about ten pressure tubes. It is planned to build recondensation pools, to avoid releasing steam to the atmosphere in case of an accident, on the RBMK reactors that are not yet equipped with them. The emergency cooling system of second- and third-generation reactors is designed to ensure cooling of the core in the event of a rupture of a 900 mm diameter pipe. In the first-generation reactors, the pumps of this circuit were only capable of coping with the rupture of a 300 mm pipe. The modifications, already carried out or planned for the Russian reactors Nos. 1 and 2 at Leningrad and Kursk, benefit from this improvement.

The consequences in terms of releases and radioactivity are terrible. Many books have described the courage and self-sacrifice of the first firemen who intervened to control the fire, with the fear of a propagation towards the twin reactor n°3. The reactor building was completely split open, with an open view of the gutted reactor core. Pieces of the graphite columns of the moderator were propelled on fire all around the reactor and glowed to melt the bitumen. The twenty or so first firemen were trying to get close to the blaze with poor hoses. Army soldiers, armed with shovels and loosely protected by masks and aprons, were sent to shovel the graphite pieces off the roof and release them into the reactor core. Most of these "liquidators," as they are called today, will suffer lethal doses of more than 6 Sieverts. A plume of radioactivity spread to the north-west of the plant and rapidly expanded to become a cloud 15 km wide. On 27 April, the order was given to spray the cloud with silver iodide from Tupolev-16 bombers in order to provoke rain by condensation on the silver iodide. These artificial rains by seeding were intended to prevent the clouds from reaching the large cities like Moscow or Kiev [operation Cyclone-N (Brown 2021, p. 71)], but these rains concentrated of course the activity in the zones where they were established. It is reported that the pilots of these planes, led by the commander Alexander Grushin who was decorated by Vladimir Putin in 2007 for his action, suffered multiple serious disorders related to radioactivity causing permanent disabilities. It is not within the scope of this book to develop[12] the physical and moral suffering endured by the population (radiation sickness, leukemia, cancer), evacuated too late from an exclusion zone of 30 km around the site and insufficiently protected by a rapid distribution of stable iodine tablets. Valery Legasov, a renowned scientist and one of the main designers of the RBMKs, was warned in Moscow and arrived at the site around noon. He said that he had begun to realize the seriousness of the accident when he saw the glow of the fire in the distance. Two years after the catastrophe, Legasov,[13] a member of the USSR Academy of Sciences (he had claimed that these reactors were so safe that they could have been built on Red Square!), hanged himself by denouncing the pressure he had been under not to divulge the ins and outs of the Chernobyl case.

---

[12] Let's mention Kate Brown's very detailed study published in 2019 in English language, 2021 in French language (Brown 2021). Kate Brown had access to Russian sources and state documents in Russia and Ukraine at first hand. She recorded testimonies of people present at the time of the events which give a real value on the aftermath of the accident. On the other hand, one can regret a rather conventional description of the accident itself and its real causes which are treated in half a page. On page 31, it says: *"But by the time of the shutdown, the atomic chain reaction in the reactor core had become "critical" (in other words, uncontrollable) and the power of the reactor had suddenly increased."* Although any reactor in normal operation is critical (the word is a false friend after all), I will not be taken too seriously if I say that this sentence does not contribute much to the technical debate at least. On the other hand, if the "technical" part is not there, the real-life testimonies are extremely interesting to get an idea of the context and I recommend reading them. I could be criticized for not having paid enough attention to health issues in my book to concentrate on the scientific and technological parts, but it is precisely this field that I was aiming at, since most of the other books concentrate on the medical or psycho-social aspects.

[13] Valery Legasov (1936–1988). Soviet physicist. Member of the USSR Academy of Sciences. After a doctorate in chemistry in 1972, he became a specialist in inorganic chemistry. Professor at the Institute of Chemistry and Physics of Moscow, he then directed the prestigious Kourtchatov Institute of nuclear physics. In this capacity, he participated in the commission of inquiry on the Chernobyl accident. He presented the final report to the IAEA, the International Atomic Energy Agency in August 1986. He hanged himself in April 1988, leaving a tape recording in which he explained his vision of the affair, thwarted by the blackout of the Soviet government at the time. Many saw it as a political murder in disguise.

Legasov just graduated in the early 1960s.

Chernpbyl Forever?

"*Chernobyl*" is a five-episode HBO television miniseries co-produced with Britain's Sky channel, and broadcast in France by OCS. This series, considered extremely realistic by experts, has been an unexpected success since the beginning of its broadcast on May 6, 2019. On the IMDb website, the largest online database of films and series, it has become the highest-rated series in history. According to Reuters, tour operators who offer to visit the sites of the disaster in Ukraine, have seen even a 40% increase in bookings in the weeks following its broadcast. Jared Harris and Stellan Skarsgard play the roles of Valery Legasov, the dedicated academician, and Boris Shcherbina, the gruff but sincere apparatchik, who lead the operation (Photo 4.17). The beautiful role that is given to them should not make us forget that their functions in the Soviet State apparatus did not fit well with a protesting behavior. Having seen the series myself, I can confirm that the scientific explanations given to the commission of inquiry are very satisfactory, with the exception of a few rare errors in translation into French. The series brings this drama even more into the collective consciousness.

At a highly confidential meeting of the PolitBuro on July 3, 1986, in the presence of Mikhail Gorbachev, Boris Shcherbina, the head of the special commission in charge of managing the Chernobyl disaster, said, "*The accident is attributable to flagrant violations of safety regulations, which were aggravated by serious flaws in the reactor design. However, these two causes are not of equal importance. The commission believes that it was operator error that triggered the accident. The employees of the plant had eyes only for the production of electricity (salary bonuses obliged) and privileged it to the detriment of safety.*" This statement therefore incriminates the operators, which suits the whole audience and will send the engineers in charge of the test to court. However, he added: "*The mistakes made by the operators are one thing. This accident, the largest in nuclear history, also highlights the shortcomings of the construction and design of the RBMK reactor. In the last five years, there have been 1,042 accidents in our nuclear plants, and yet since 1983; the Ministry of Energy and Electrification has never met to discuss reactor safety. There were 104 accidents in the Chernobyl plant alone. Let's face it, the RBMK reactor is potentially dangerous. It does not meet modern safety standards. No other country would want to use it. We must therefore, as difficult as it may seem, question whether it should continue to be built.*" That says it all. Shcherbina himself died 4 years later of a radiation-induced disease (Brown 2021, pp. 82–84).

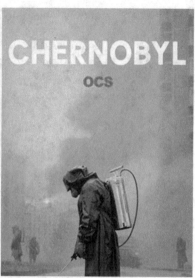

Photo 4.17 Stellan Skarsgard and Jared Harris, the two main actors of the excellent series "*Chernobyl*" produced by HBO. On the right the not very reassuring cover of the series

Sosnovy Bor: The Precursor Accident of Chernobyl (Leningrad, Russia, 1975)

Sosnovy Bor (*Сосновый Бор*) is a city in Leningrad Oblast, 70 km west of St. Petersburg (formerly Leningrad), on Koporskaya Bay on the southern shore of the Gulf of Finland at the mouth of a small coastal river, the Kovachi (Photo 4.18). The name of the city means "*pine forest.*"

The nuclear plant, intended to supply the city of Leningrad, began to be built in 1967 when the site was selected. Sosnovy Bor was granted the status of a city in 1973 and in the same year the first nuclear reactor went into operation. Note that the city hosts the statue of academician Anatoli Alexandrov (1903–1994, founder of the Sosnovy Bor Institute of Scientific and Technological Research in 1962, an offshoot of the Kurchatov Institute devoted to nuclear research) sculpted by Albert Charkin. Today the site includes two 925 MWe RBMK reactors in final shutdown (shutdown in 2018 and 2020), four reactors in service (two 925 MWe RBMK reactors and two 1085 MWe WWER-1200 type pressurized water reactors). Reactor 3, which interests us here, was commissioned at the end of 1973, connected to the grid on December 7, 1979, and put into industrial service on June 29, 1980. The first two shutdown reactors are first-generation RBMKs, the other two

Photo 4.18 Location of the city of Sosnovy bor near Saint Petersburg, located very close to Finland

RBMKs belong to the second generation as was the one at Chernobyl (Photo 4.19).

The nuclear plant, operated by Rosenergoatom, is the main industry in the region. Rosenergoatom is subject to the authority of the Russian Federal Agency for Atomic Energy (Rosatom), which is the equivalent of the French Nuclear Safety Authority or the NRC in the United States. The reactor is of the RBMK type whose ancestor is the Obninsk reactor already mentioned above.

The operation of the Leningrad RBMKs has not been without accidents. At the initial stage, the blocking devices (plugs) of the fuel channels became massive problem and had to be urgently redesigned. On February 6, 1974, as a result of water boiling, followed by water hammering, the intermediate cooling water circuit of Leningrad-1 was ruptured. On November 30, 1975, the first severe accident also occurred on unit 1, accompanied by the destruction (melting) of some fuel channels, resulting in radioactive releases. This accident, which highlighted the design flaws of the RBMK-1000 reactor, was in fact a precursor to the major accident that occurred in 1986 at the RBMK-1000 reactor of the Chernobyl nuclear plant. On that day, while the reactor was operating at 20% of its rated power, power fluctuations began to occur due to the instability of xenon at low power. The greater the active height of a reactor, the more unstable it is with respect to an axial-offset oscillation. The

Photo 4.19 Two reactors of Leningrad (the highest cubic blocks contain one reactor each)

axial-offset is defined as the axial power imbalance of the core (difference between the power of the top of the core and the bottom of the core normalized to the total power and expressed in %):

$$\text{Axial} - \text{Offset}_{[\%]} = (\text{Upper Power} - \text{Lower Power})$$
$$/ (\text{Upper Power} + \text{Lower Power}) \times 100\%$$

It turns out that the axial-offset (AO) is highly dependent on the axial temporal oscillation of the concentration of xenon-135, the strongest neutron absorbing gaseous fission product produced during reactor operation. This oscillation appears when the control rods are moved rapidly and attenuates in small reactors (<3 m) (Fig. 4.25). They sometimes appear in a divergent way in PWRs with an active height of 4.27 m (French P4 type), which requires corrective piloting actions by the operators who are on the lookout for this type of behavior, in particular when the power level is increased rapidly. These oscillations will be all the more important in a 7 m reactor like the RBMKs. It should be understood that a large value of the axial-offset, either positive or negative, means an overpower in the top of the core (AO > 0) or in the bottom of the core (AO < 0). This overpower will lead to a stronger vaporization and, in the case of a positive void coefficient as it is the case of RBMKs, a vicious circle contributing to increase the power locally.

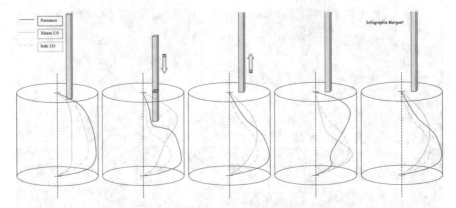

Fig. 4.25 Xenon oscillation following a rod insertion and withdrawal. The power distortion created by the introduction of the control rod will affect the production of xenon 135, which is produced by the decay of iodine 135 (half-life of 6 h) and which disappears by capturing neutrons to produce xenon 136, but also by β^- radioactivity with a half-life of 9 h. Under certain conditions, xenon starts to oscillate axially as a function of time, and if the oscillation is divergent, the risk of runaway is possible. This behavior is particularly feared by operators with large axial size (source Marguet 2017, p. 1275)

The power fluctuations intensified and became threatening in nature. On the morning of November 30, 1975, the head of measurements at the plant received a phone call from the nearby NITI Scientific and Technological Research Institute: *"Are you okay? Our dosimeters are off the scale. But everything is clear at the Institute's premises."* This is how the NITI (Scientific Research Institute named after A.P. Aleksandrov), located 3 km from the first unit in Leningrad, reacted to the emission of radioactive aerosols, carried by the winds from the plant. This was the first signal of an accident detected outside its area. According to Vitaly Abakumov, an actor of the events who worked during this shift as a reactor control engineer, on November 30 at 6:33 a.m., *"Several emergency signals appeared at the same time in the reactor control room, indicating a loss of containment of pressure tubes."* This was the moment of the accident. Senior reactor control engineers then closely monitored the power release throughout the core volume and inserted control rods into the parts of the core where the fuel was overheating and removed them from the parts where the chain reaction was shutting down. An Emergency Shutdown eventually prevented an explosion. Nevertheless, there was a local overpower in the plant, and the cladding of ten neighboring assemblies deteriorated, so that some of the radionuclides they had accumulated went into the cooling system and then into the atmosphere. The backup automatic controllers succeeded in shutting down the reactor. But the information about the accident was immediately classified. Neither the country, nor the city, nor even the plant's employees knew about it. "*At the time, I was working as a senior engineer in charge of turbine control,*" says Valery Koptyaev, a former employee of the plant. "*On November 30, my team was on vacation. When I arrived in the main control room on December 1, I saw my co-worker Mikhail Khudyakov wearing a breathing mask. I already knew that the unit was shut down, but I had no idea why. Usually, the management, from the director and chief engineer to the managers and their deputies, showed up at our morning briefing in suits, ties and casual shoes. That day, I saw management in white overalls and special shoes. I ask Michael*", -"*Why are they wearing respirators, what is the activity level in the air?*", - "*I don't know exactly, but over 200 times the norm, according to the dosimetry guys!*" he replied. -"*Then we found out how much 'junk' had been dispersed not only on the plant, but throughout the city.*" So, what really happened in 1975? Vitaly Abakumov tells us about it in detail. On the night of November 30, one of the two working alternators had to be decoupled from its turbine and its rotor removed for repair. The operators unloaded the required generator. But by mistake, the operating one was disconnected from the grid instead of the unloaded one. This triggered the emergency protection tripping the turbine and then the automatic shutdown of the reactor. Without the turbine, the steam is vented to the atmosphere. Up to this point, we are in an incidental phase but not yet an accidental phase. These conditions are foreseen in the normal operation of reactors. In this way, the reactor no longer produces electricity for the grid, but also for itself. As a general rule, it is always possible to reduce the power without initiating the scram to produce at least the power required by the nuclear auxiliaries. This is known in the industry as house-load operation. The plant is "*house-load operating*" when it can still produce the 10% necessary for its own operation. This is important from the point of view of safety because the plant does not then need to call on an external source of electricity; or even a backup (diesels). -"*Realizing that the staff had made a mistake, the station foreman gave the order to restart the mistakenly disconnected turbine as soon as possible,*" recalls Abakumov. -"*All the preparation for conditioning and loading the turbine-generator unit took place in a nervous atmosphere, facing the real threat of unacceptable xenon poisoning of the reactor during the shutdown, entry into the "black hole" of the xenon pike, and the resulting load following on the unit. The operators had to remove almost all of the control rods from the reactor to bring it up to power to compensate for the xenon poisoning, and bringing the reactor back to the minimum controllable power level became a dangerous and difficult task for the senior reactor control engineer, prohibited by regulation. However, the shift supervisor and the control engineer proceeded with this violation of the Technical Operating Specification without hesitation. They sought to compensate for the consequences of the operator's error, as the main indicator at the time was the power generation plan. The shutdown of a reactor means the loss of megawatt-hours unproduced!"*

This explanation deserves some physics explanations on the poisoning of xenon 135. Xenon is a radioactive gas produced in the fuel by fission, but also by the decay of another fission product, radioactive iodine 135, which is produced in large quantities (about 6% of fissions), Fig. 4.26). It remains trapped in the fuel, even forming bubbles or in the gap between the fuel pellet and the inner face of the cladding.

The concentrations of iodine 135 and xenon 135 are governed by differential equations involving the nuclear properties of the fuel (Σ_f), neutron flux in the core (Φ) and nuclear properties of iodine and xenon (σ_I, σ_{Xe}, λ_I, λ_{Xe}):

$$\begin{cases} \dfrac{d\left[{}^{135}_{53}I\right]}{dt} = \gamma_I \Sigma_f \Phi - \lambda_I \left[{}^{135}_{53}I\right] \\[2mm] \dfrac{d\left[{}^{135}_{54}Xe\right]}{dt} = \gamma_{Xe} \Sigma_f \Phi + \lambda_I \left[{}^{135}_{53}I\right] - \lambda_{Xe}\left[{}^{135}_{54}Xe\right] - \sigma_{Xe}\Phi\left[{}^{135}_{54}Xe\right] \end{cases}$$

When the reactor is shutdown, the iodine 135 produced in greater quantity, decays (half-life about 6 h) into xenon 135 from which a strong increase in the concentration of xenon 135. This xenon absorbs very strongly the thermal neutrons, stifling the chain reaction. But this production is limited by the

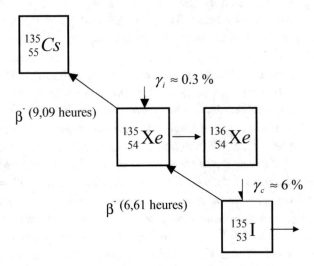

Fig. 4.26 Diagram of xenon production 135 (Marguet 2017, p. 1143), *heures* = hours

decay of xenon 135 itself radioactive (half-life about 9 h). As the reactor is critical just before the shutdown at a certain position of the control rods, it will be necessary to raise the rods if one wants to restart the reactor during the Xenon pike and all the more so as the concentration of Xenon 135 is important. This period with a xenon overconcentration is about twenty hours and, after a short delay after the shutdown, can even prevent restarting if the rods are in the extracted positions (Fig. 4.27). One speaks then about "*black hole*" in the shift teams, in reference to the colossal absorption of the black holes. It is then necessary to wait for the disappearance of xenon, which induces shortages of electricity production (and shortages of exploitation) that Russia cannot afford at that time.

To restart the reactor quickly after the turbine trip, the operators have to pull out the reactor control rods a lot, and in addition, at low power, the reactor is chronically unstable due to xenon oscillations. "*The RBMK reactor is large not only in terms of design parameters, but also in terms of reactor physics, which means that it is possible to achieve criticality not only for the reactor "in general", but also in local areas of the reactor core*," Abakumov continues. In other words, the reactor is so large that neutron decoupling occurs between areas of the reactor that begin to behave independently. This phenomenon is also called the "mini-core effect" and is specific to large cores. "- *With total "poisoning" of the core and no practical means of influencing reactivity (all control rods were removed), the chief engineer was able to keep the reactor at a minimal*

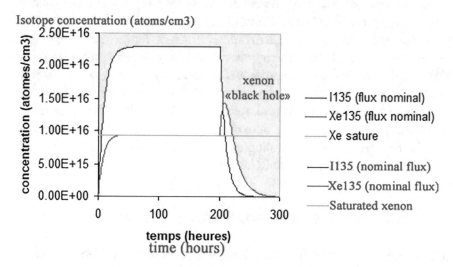

Fig. 4.27 Production of a Xenon pike during a reactor shutdown. Evolution of the concentration of iodine 135 and xenon 135 with a shutdown placed at 200 h of operation. The zone of the peak is sometimes called "black hole" because of the very strong neutron absorption of Xenon 135 (Marguet 2017, p. 1143)

level of control, not "in general", but only in a limited area adjacent to the 13–33 fuel channel. Outside this area, the core remained poisoned (by xenon 135)." The rapid power excursion of this local area resulted in overheating and massive destruction of the fuel element cladding. Abakumov recalls that when the radioactivity alarm went off, *"The chief engineer's reaction was immediate: He shouted, "Shut down the reactor!""*—And the reactor was shut down by pressing the red emergency shutdown (manual scram) button. Operating at full power with all rods removed was a violation of the Technical Operating Specification. Violations of regulations are never welcome. *"But at the same time, they were not perceived as dangerous at the time. Therefore, violations of the regulated lower limit of the operational reactivity reserve value were common practice in Leningrad and were tacitly perceived as evidence of a particular mastery of the supervisor engineer,"* writes Abakumov. This accidental sequence follows exactly the same logic as what happened at Chernobyl, with the fundamental difference that the Lenigrad-1 reactor is much less advanced in the cycle and its average burn-up rate was much lower, making the neutronic weight of the rods stronger, which allowed to stop the power excursion at the time of the scram, while the power excursion in Chernobyl-4 could not be stopped (the quotations of Abakumov are taken (and explained technically) from an article of Lina Zernova in https://bellona.ru/2016/04/04/laes75/. Bellona is a Russian ecological foundation founded in 1986, but it seems to lack specialists in these very technical issues).

Estimates of releases into the environment are confused and poorly referenced: between 137,000 and 1.5 million Curies were released into the environment according to different sources. Tons of liquid radioactive waste without any particular treatment were certainly discharged into the Baltic Sea. Afterwards, the defective channel was replaced by a repair team. After the analysis of the accident, an order was issued to introduce an additional local automatic reactor power control system on all RBMK-1000 reactors, and some technical improvements were made to the design, namely: water tanks for the emergency reactor cooling system were installed, check valves were used on the water transfer manifolds…

Sosnovy Bor (Leningrad, Russia, 1992)

On March 24, 1992, at 4:30 a.m. (French time), a severe accident occurred in the reactor-3 (Photo 4.20), resulting in the release of radioactive xenon and krypton gas, as well as iodine gas, via a radioactive steam leak. The safeguard system of the plant was triggered automatically. These releases were quite

Photo 4.20   Service desk and reactor loading face (photo RIA Novosti)

important *a priori* even if the Russian government tends to sweeten the consequences of this incident (especially only 6 years after the Chernobyl disaster) by explaining that there was no need to evacuate the population. The
Russians affirmed that they were cooling the core and that they had the situation under control. According to them, the gas releases into the atmosphere
were *"in conformity with the norms"* and there was *"no evacuation of people
either in the plant or near it."* The accidental release of radioactive products
places this accident at least in INES level-3 and the Russians declared that
they would only evacuate from level-4. According to the information provided, with great speed it is true, by the Russians to the International Atomic
Energy Agency (IAEA) in Vienna, this leakage would be the result of a cladding rupture in the reactor core. This incident would have led to a sudden
increase in radioactivity in the primary circuit of the plant due to the immediate release of radioactive gases contained in the cladding. These gases, xenon,
krypton and especially iodine, were evacuated through the iodine filters and
then the chimney of the installation, which led to the alert. The actual cause
of these cladding integrity losses is rather unclear. One cause put forward by
some sources is that one of the pressure seals failed due to a failure on a water
supply valve leading to a loss of pressure in the primary circuit and sub-
cooling. The Russian Ministry of Atomic Energy stated that the primary

reason was this failed valve. The failure of the valve caused the flow of fluid to be disrupted, one or more of the 1660 pressure tubes to overheat and fuel cladding to leak. Gaseous fission products were released into the affected pressure tubes, contaminating the outside via the leak. Yuri Rogozhin, the spokesman for the Russian nuclear inspection service (*Gosatomnadzor*), declared that "*the incident was serious, with possible consequences for the environment and the population.*"

When *Minatom* (the Russian Federation in charge of nuclear plants) reported the accident to the IAEA, the accident was rated at level-3 on the International Nuclear Event Scale (INES). According to the IAEA, level-3 means that there were radioactive releases, but that public health was not endangered. The accident nevertheless released radioactivity (via the ventilation system) of inert gases (4000 Curies) and iodine 131 (2.5 Curies).

There was then a media conflict between the Russians who tend to play down the accident, and the opponents who try to prove the toxic releases. At 3 p.m. the next day, measuring stations in Finland, Sweden, and the United Kingdom reported a slight increase in radioactivity readings. According to the spokesman of the Finnish Center for Radiation and Nuclear Safety in Helsinki, "*the increase is only detectable high in the sky.*" According to Minatom, the release of iodine-131 was less than 0.2 curies/day, while the maximum allowed was 0.05 curies/day. According to the ÖKO-Institut in Darmstadt, the information they received from Lovisa in Southern Finland indicated that on March 24, the level of ^{131}I was 4.4 millibecquerel/m^3. This is 1000 times higher than before the Sosnovy-Bor accident. The institute also reports that such a level of contamination probably indicates damage to the core and the fuel rods inside it. According to Greenpeace, the radioactive releases are estimated at 3000 Curies. If this figure is correct, it would make this accident the fifth worst accident in the history of nuclear power (with Chernobyl, Three-Mile Island, Windscale and Fukushima). Note that the 1975 accident at Reactor-1 is well above the 1992 accident. The Swedish specialists had inspected the Sosnovy Bor plant in January 1992 (Photos 4.21 and 4.22), just before the accident. They had expressed their concern that the plant was in a deplorable state. The conclusions were alarming: the risks of accident in this plant were estimated to be 1000 times higher than in Sweden.

Photo 4.21   A recent view of the Leningrad plant

Photo 4.22 Assembly handling system. The very large size of the system is due to the height of the assemblies (almost 7 m!)

5

The Fukushima Accident

Abstract The Fukushima accident in 2011 in Japan is the latest large-scale accident to date. A major tsunami led to the drowning of the plant and the destruction of four reactors by a succession of adverse circumstances, a historically unprecedented situation. This accident confirmed that even the most industrialized countries can cope with a large-scale nuclear crisis.

On March 11, 2011, at 5:46:23 UTC (Universal Time), or 14:46:23 local time, an earthquake of unprecedented power, 8.9 on the Richter scale, slightly less than the most powerful ever recorded, 9.5 in Chile in 1960, occurred off the east coast of Japan. Its epicenter is located 130 km east of Sendai, capital of Miyagi prefecture, in the Tōhoku region, located about 300 km northeast of Tokyo. While the reactors of the Fukushima Dai-ichi plant[1] (Photo 5.1) resisted the earthquake rather well, it was the induced tsunami that drowned all 13 emergency generators and destroyed the external power supplies (Photo 5.2). Three reactors heated up irreparably, and the fourth was destroyed by an induced hydrogen explosion. All four were finally destroyed, marking the greatest civil nuclear disaster to date.

At the confluence of tectonic plates, Japan is known for its high seismicity with about 20 significant earthquakes per year. Since the beginning of time, devastating earthquakes have been known. Thus, an earthquake followed by a

---

[1] Dai-ichi means number 1 in reference to the site n°1 which includes the 6 incriminated reactors. Site n°2 (Dai-ini) includes 4 other more recent reactors.

© The Author(s), under exclusive license to Springer Nature Switzerland AG 2022
S. Marguet, *A Brief History of Nuclear Reactor Accidents*, Springer Praxis Books,
https://doi.org/10.1007/978-3-031-10500-5_5

Photo 5.1 The Fukushima Dai-ichi plant: in the foreground, unit 4 appears as a cube located between the two red and white towers, then reactors 3, 2, and finally 1 (Photo Tepco). A large gap separates Reactor-1 and Reactors-5 and -6 sharing a ventilation stack painted in white and red. The building located between reactor-1 and reactor-5 along the coast is the Dry Cask Storage Facility common to all reactors

tsunami nicknamed the "*Big One*" (7.9 magnitude) killed 143,000 people on September 1, 1923, in the Tokyo area, especially because of the huge fire that followed. On July 16, 2007, at 10:13 am, the Chu-Etsu-Oki earthquake in Niigata province, about 250 km from Tokyo, reached a magnitude of 6.6 on the Richter scale (11 dead, a thousand injured). It took place about 10 km off the Kashiwazaki-Kariwa plant. (also operated by TEPCO) which includes 7 BWRs for a total power of 8212 MWe. During the earthquake, which proved to be 2.5 times more powerful than the one taken into account for the design of the reactors, the reactors in operation were shut down correctly on the signal of the accelerometers[2] detecting the earthquake. Afterwards, it was noted

---

[2] The accelerations experienced by the reactor are measured to trigger the shutdown systems from a level equivalent to an earthquake of intensity V on the Japan Meteorological Agency scale.

Photo 5.2   Fukushima plant drowned by tsunami—March 11, 2011 (photo Tepco)

that buildings of earthquake-resistant design were indeed resistant, while older buildings were sometimes heavily impacted. The consequences of this earthquake on the plant were relatively low: only the spent fuel storage pool, common to several reactors, overflowed due to the movement of water, causing a release of slightly radioactive water. The reactor cores were not damaged. The reactors were shut down for inspection for almost 2 years before being restarted in May 2009.

Returning to Fukushima, the earthquake caused an automatic shutdown of the operating reactors by triggering the accelerometers, the accidental loss of power, and the tripping of the generators. Exactly fifty-one minutes later, a tsunami caused by the earthquake hit the east coast of Japan. The wave reached an estimated height of more than 30 m in some places, but about 15 m in the plant, traveling up to 10 km inland and devastating nearly 600 km of coastline (Photo 5.3, Fig. 5.1). The height of the wave depends on the configuration and depth of the seabed near the coast. This wave will partially or totally destroy many cities and harbor areas. The Fukushima Dai-ichi nuclear plant is located 160 km from the epicenter. It has six reactors: the reactor-1 has a gross electrical power of 460 MWe, Reactors 2–5 had a power of 784 MWe, and reactor-6 had a power of 1100 MWe. Reactors -1, -2, and -3 were in operation at the time of the earthquake and were running at full power. Reactors -4, -5, and -6 were shut down for maintenance. Reactor 4 had all its

Photo 5.3 The tsunami wave completely submerges the protective dikes (left) and rushes into the waterfront facilities (right) (photos TEPCO)

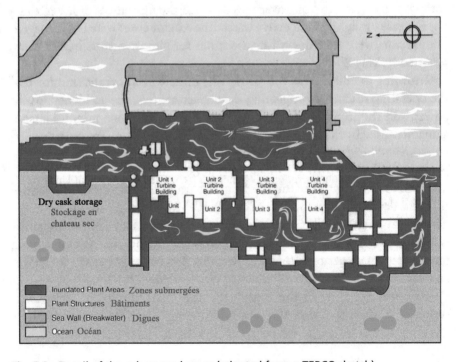

Fig. 5.1 Detail of the submerged areas (adapted from a TEPCO sketch)

fuel unloaded into the fuel pool located inside the conventional building containing the reactor block. However, the conventional building of reactor-4 will also explode.

The four damaged reactors (1, 2, 3, 4) are boiling water reactors (BWRs) of General Electric design, including the Oyster Creek plant, built in 1967 in the USA, which is the reference model, in particular the vessel, the core, and the fuel assemblies (Figs. 5.2 and 5.3).

Unlike the Pressurized Water Reactors that we have in France, boiling water reactors produce steam directly in the core (Fig. 5.4), which avoids a possible secondary circuit since the steam is sent by two loops evacuating the steam directly to the turbine , which makes it simple to build, but with the disadvantage that any possible release of fission products from the fuel contaminates the turbine.

The economic gain due to the absence of a secondary circuit is appreciable, but it is necessary to be very rigorous about the quality of the water (thanks to demineralizing filters that treat all the condensates leaving the condenser) to avoid contaminating the turbine, since there is no longer any physical separation by a secondary circuit between the primary circuit and the turbine as in the case of Pressurized Water Reactors. The liquid water that comes out of the active core is recirculated via pumps inside the dry cavity and placed against the vessel. Only a few external loops are needed to evacuate the steam, which limits the risk of LOCA. The vessel containment has a very particular bulbous shape. Finally, a cylindrical torus in the lower part of the reactor building contains the pressure suppression pool. This pool contains cold water in which the primary circuit can be depressurized to condense the steam. These two features are representative of the so-called *Mark*-I of the BWR's containment (Fig. 5.5), which will evolve thereafter towards a suppression of the torus (chamber of suppression of the pressure) in favor of a "wet" pit containing water in the concepts Mark-II then -III (Table 5.1). It should be noted that the Fukushima reactors do not have the same high-pressure cooling systems (Table 5.1).

The reactor vessel, which is much higher than a PWR vessel because it contains the steam separators, is 16 cm thick of steel. It is housed in a hermetic containment called a "*dry pit*," capable of holding an overpressure of 4 bars. The dry pit is a steel cavity (30 mm thick) in the shape of a bulb that hugs the 2-m thick reinforced concrete cavity (Fig. 5.5). The only way to enter this cavity is through a double-door access hatch for personnel or a bolted main hatch. A bolted cover closes the upper part of the dry pit. Finally, a concrete biological cap closes the pit to protect the personnel working on the service desk. The drywell is connected to a toroidal pressure

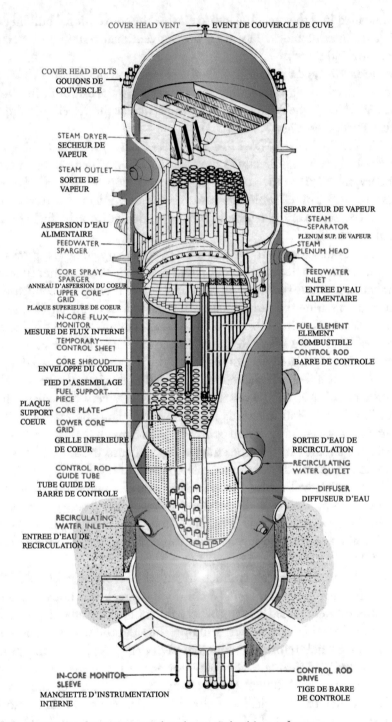

Fig. 5.2 Oyster Creek reactor vessel and core, Fukushima reference

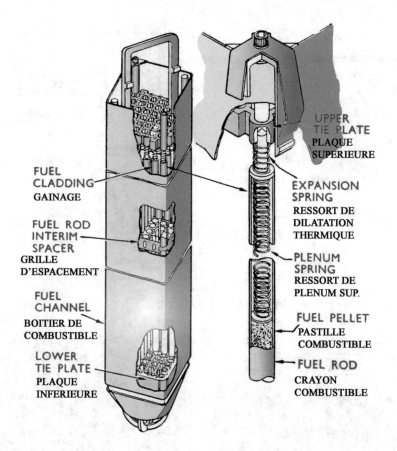

Fig. 5.3 Oyster Creek canister fuel assembly, Fukushima reference

suppression pool located below and around the reactor. This cold water pool, also known as the "*wet pit*", allows the steam to be bubbled up to condense it if necessary (if the reactor vessel is to be depressurized, for example). This pool is cooled by a heat exchanger and serves as a water tank for the safeguard injection systems when the tank of the condenser at the turbine outlet can no longer supply water. The spent fuel storage pool is located in the upper parts of the reactor building, near the dry pit. The transfer of the fuel is done under water by filling the reactor pool in the same way as for PWRs.

The Fukushima-1 fuel is composed of assemblies with 7×7 fuel rods of 14.5 mm diameter surrounded by a 0.9 mm thick cladding separated by a 0.28 mm thick pellet/cladding gap. The maximum linear power is 492 W/cm for a maximum heat flux of 108 W/cm^2. The number of spent fuel assemblies

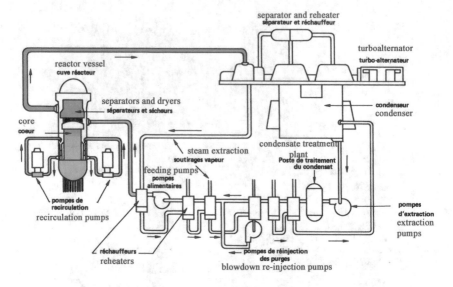

Fig. 5.4 Classic primary circuit of a BWR (x2 loops)

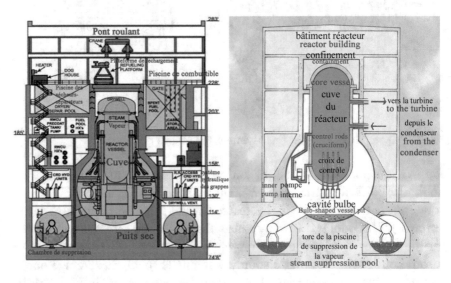

Fig. 5.5 MARK-I geometry of the Fukushima containment-1 to -5

in the core, irradiated in the pool as well as the number of fresh fuel assemblies in the pool, are provided in Table 5.1.

Several safeguard systems have been designed in case of loss of cooling by the primary pumps. The residual power evacuation circuit (*Residual*

Table 5.1 Type of reactors at the Fukushima plant. IC: isolation condenser (presence of a containment condenser of approximately 100 m^3 of water. In case of turbine isolation, the steam is sent to tubes embedded in this condenser located above the vessel. The condensed water then returns to the core by gravity. The steam produced by the condenser is evacuated to the atmosphere. Once the condenser is empty, it must be refilled with water, the IC system is passive, but valves must be opened (manually or by DC servo-motor). HPCI: high-pressure coolant injection, this system can passively supply (turbopump) about 800 tons/h to more than 70 bars, RCIC: reactor core isolation cooling. The steam is condensed in the pressure suppression chamber (also called wet well) (about 84 tons/h), HPCS: high-pressure core spray

Fukushima	Power (MWe)	Reactor configuration	Containment configuration High-pressure cooling mode	Number of assemblies Core/Pool/Fresh
Unit-1 (March 1971)	439	BWR-3	MARK-I IC + HPCI	400/292/100
Unit-2 (July 1974)	760	BWR-4	MARK-I RCIC+HPCI	548/587/28
Unit-3 (March 1976)	760	BWR-4	MARK-I RCIC+HPCI	548/514/52
Unit-4 (October 1978)	760	BWR-4	MARK I RCIC+HPCI	0/1331/204
Unit 5 (April 1978)	760	BWR-4	MARK I RCIC+HPCI	548/946/48
Unit-6 (October 1979)	1067	BWR-5	MARK-II RCIC+HPCSI	764/876/64

Heat Removal System, RHRS) allows the water in the vessel to circulate through a heat exchanger by means of an electric pump. Figure 5.6 shows the residual power to be evacuated over time for a BWR-4 reactor and situates the cooling issues. A recirculation system (on reactors -2 and -3) in an "isolated core" situation makes it possible to draw water from the core to the vessel thanks to a turbopump fed by the steam produced in the core. On the older Reactor-1, an isolation condenser allows the cooling of the water in the vessel if a cold source is available. For loss of coolant accidents, there is a high-pressure safety injection, as well as a low-pressure injection in the form of a spray that floods the core. Finally, a borication system guarantees the subcriticality of the reactor by injecting boric acid diluted in water.

Residual heat (MWth) Puissance résiduelle (MWth)

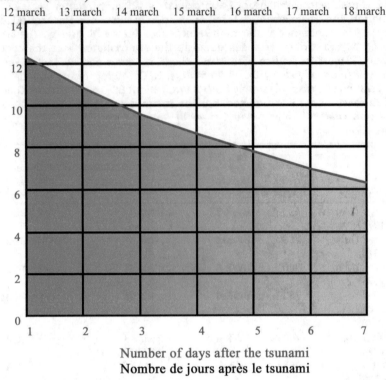

Number of days after the tsunami
Nombre de jours après le tsunami

Fig. 5.6 Residual power over time of a 700 MWe BWR-4 reactor operating at nominal power before the Automatic Reactor Shutdown. Knowing that the latent heat of vaporization of water is about 2 MJ/kg, only some 7 kg/s (about 25 m³/h) of liquid water is needed to remove the residual power. This table corner calculation shows that even this very modest issue could not be won in the cataclysmic situation of the site

Technical Insert: "Fukushima Reactor Safeguard Systems"

This technical insert aims to present the safeguard systems of the Fukushima reactors for the technologically curious reader (Figs. 5.7, 5.8, 5.9, 5.10, and 5.11). This more technical part can be skipped by the reader who wants to get to the point. However, I thought it would be useful to provide an accessible technical explanation to understand how the situation degenerated to the point of losing four reactors in a similar chain of events, but still with different system effects. Reactor-1 stands out from the others because it is the oldest (1971) and of BWR-3 model. On BWRs in general, there is the possibility of discharging the steam produced in the core into a pressure suppression torus located in the lower part of the reactor. This discharge is necessary when the steam is no longer required to be sent to the turbine (this is called bypass). This bypass will send the steam either to the turbine condenser if it is still available, or to the toroidal pressure suppression condenser if it is still operating, or even to the atmosphere when this

(continued)

(continued)

is not the case. In the toroidal chamber, the steam percolates into cold water through submerged distribution nozzles (in cold water that will nevertheless become saturated over time, making cooling more and more irrelevant), which allows for efficient condensation at least initially. In the more recent BWR-4 model, a connection has been established between the torus and the atmospheric discharge valves in case of steam overpressure in the torus, the objective being to "disgorge" it in pressure to avoid a rupture of the torus (Fig. 5.7). When a reactor is shutdown in a normal situation, the residual power due to the radioactivity of the fission products must be evacuated by the SHC (Reactor Shutdown Cooling) system. As long as alternating current is available, the SHC pumps can be operated, which send the reactor water to a heat exchanger which, after several intermediaries that we will not describe here for simplicity's sake, evacuates this heat to the cold source (in this case the sea) if the water intakes in the sea are still able to ensure their service.

The cooling of the core, after the introduction from below of the control rods, is performed by an isolation condenser (IC, heat exchanger) in the case of Fukushima-1, which requires the use of electric recirculation pumps, or a feeding pump from the vessel, in the more recent BWR-4 concept (Fig. 5.9). The isolation condenser is a simple heat exchanger that cools the core by isolating it from the outside environment. The advantage of the RCIC turbopump is that it uses the steam produced in the core to turn a small steam turbine, the axis of rotation of which drives a coupled positive displacement pump that pumps (cold) water into a condensate water reserve tank. This water is injected at the top of the core to cool it down. When the residual power in the core decreases over time, there will not be enough steam produced

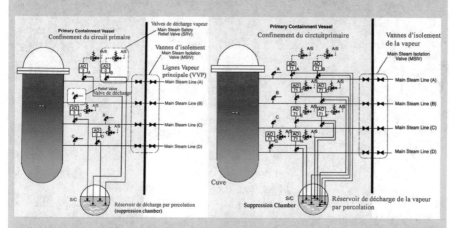

Fig. 5.7 Main steam discharge system of Fukushima-1 (BWR-3, left) compared to Fukushima-2 or -3 (BWR-4, right). There is a greater redundancy of the atmospheric discharge valves (equivalent to the GCTa system on PWRs), as well as a possibility to discharge the depressurization torus to the atmosphere in case of saturation of the water-producing steam in the torus. The system is called *Main Steam Relief Valve* (MSRV) (The illustrations in this insert are adapted from diagrams provided by TEPCO)

(continued)

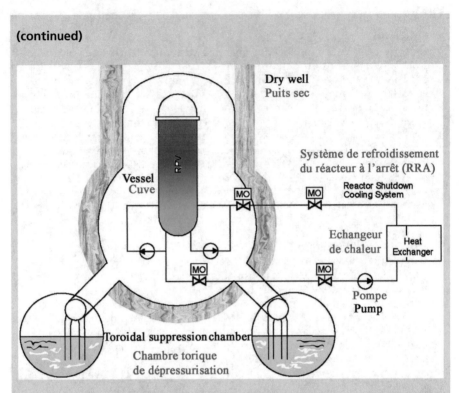

Fig. 5.8 Shutdown reactor cooling system. This system uses pumps running with alternating current to circulate water from the shutdown reactor (the thermal residual heat is lower) to cool it by circulating it through an exchanger to evacuate the residual power to the cold source. It should be noted that the loss of electrical power renders the pumps inoperative. This system is called SHC (for *SHutdown Cooling*). The equivalent of this system on PWRs is the RRA circuit which is almost identical

to drive the turbopump and electric pumps will have to be activated to replace the turbopump. If the water balance in the core is threatened by the draining of the water reserve tank, it is imperative to bring back water by some means.

A safeguard system in case of loss of primary coolant allows to reinject water in the vessel even at pressure (about 70 bars) (Fig. 5.10). This system called High-Pressure Coolant Injection (HPCI) is very similar to the equivalent system on PWRs: the RIS-HP, with the difference that it is a turbopump that supplies the vessel, whereas in PWRs it is electric pumps. Remember that the nominal pressure in the vessel of PWRs is 155 bars, more than twice as high as in BWRs. The HPCI system operates as long as there is sufficient steam production in the core and as long as the reserve water tank is not empty. These two conditions will deteriorate as the accident progresses.

On the newer BWR-4 model, the designers have increased the power supply redundancies (Fig. 5.11), This was insufficient during the accident as all external power sources were lost, as well as the emergency generators. This means that when the batteries were drained, it was impossible to recharge them despite desperate attempts to line the batteries of buses and cars present on the site.

(continued)

(continued)

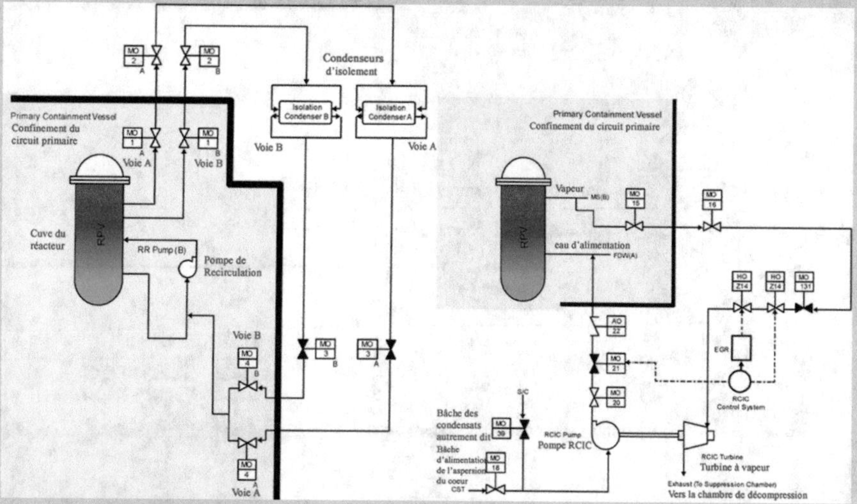

Fig. 5.9 Isolation cooling of the Fukushima-1 core (IC, left) compared to Fukushima-2 and -3 (RCIC, right). This system differs significantly between BWR-3 and BWR-4. In particular, the BWR-4 has a feeding pump that uses the steam produced in the core to turn an RCIC pump that draws water from a condensate storage tank to cool the core by flooding (spraying into the vessel). This tank is called the Condensate Storage Tank (CST) and collects the liquid condensate from the primary circuit, which comes from the condenser of the main turbine. The RCIC pump can also pump water from the toroidal pressure suppression chamber. The RCIC is designed to maintain the water level in the vessel above the active core, but care must be taken to ensure that the water level is not too high so that water can flow into the steam hot leg and drown the RCIC turbine pump, which would slow it down or even degrade it immediately. In the case of PWRs, this tactic of using turbopumps also exists on the normal supply system (ARE) of the steam generators (SGs) and the emergency supply system of the SGs (ASG). The ASG turbopump also draws its water from an ASG tank whose volume is sized to reach a safe state of the PWR plant by connection to the RRA shutdown cooling system. The BWR-3 model does not benefit from the turbopump and must therefore guarantee an AC power supply for the recirculation pumps

(continued)

(continued)

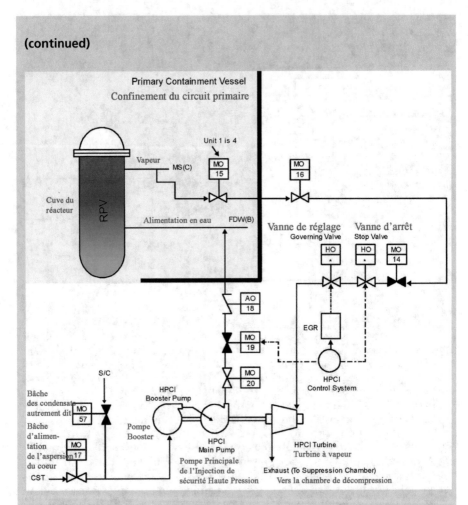

Fig. 5.10 High-pressure safety injection (HPCI). This circuit is identical on all the Fukushima reactors. It is based on the principle of a turbopump fed by the steam produced in the core. The axis of the turbine turns a pump called "booster" and a main pump. The booster pump pumps water from the CST feed tank and raises it to an intermediate pressure. The booster pump discharges the water by feeding the HCPI main pump. The purpose of this tactic is to stagger the water pressure to avoid cavitation of the main pump. This allows water to be injected into the vessel at a pressure of over 70 bars. This feeding system is found on the PWR steam generators at about the same pressure, with the difference that the water at the exit of the turbopump is heated by two stages of reheaters (R5 and R6) before entering the SGs

(continued)

(continued)

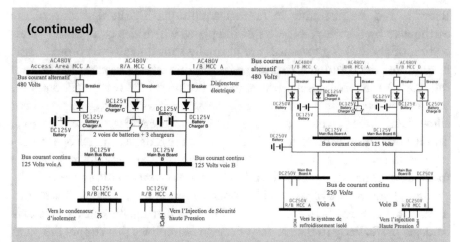

Fig. 5.11 Diagram of the power supplies of Fukushima-1 (left) compared to Fukushima-2 and -3 (right). We notice the presence of a 250 V direct current stage (more powerful) on the BWR-4 concept. The redundancy of the electric batteries is also more important

On March 9, 2011, an earthquake of magnitude 7.3 was recorded off the coast of Japan, in the subduction zone between the oceanic plate and the Eurasian plate, which did not cause any damage or victims at the time. But it is the first signs of a gigantic earthquake that occurred on March 11 at 2:46 p.m. local time, and of magnitude 8.9, which occurred at a depth of 24.4 km below the Pacific Ocean, about 100 km off the coast of Miyagi Prefecture on the east coast of Japan. Japan, located on a plate that moves about 8 cm per year, moved 2.40 m in one go! Three minutes later, the Japanese Meteorological Agency issued a maximum tsunami warning,[3] the giant wave that sometimes accompanies marine earthquakes. An initial wave of about 20 m high progresses concentrically from the epicenter and hits the coast in less than 50 min, barely damped due to the proximity of the point of origin (estimated to be more than 10 m at impact). The Fukushima-Dai-ichi site (managed by the Tokyo Electric Power Company—TEPCO), which includes six boiling water reactors, is located 160 km southwest of the epicenter and 240 km north of Tokyo. These reactors were commissioned between 1970 and 1979 and five of them have an old Mark I containment. The vessel of reactor-3 was changed at the end of the 1990s during a renovation. Reactor-3 was certified to receive MOX fuel (mixed oxide of uranium and plutonium), which was loaded in February 2011. Of the 6 reactors in the plant, reactors -4, -5, and -6 are in shut down condition, with fuel unloaded.

Reactors 1–3 are operating in normal conditions. To the detection of the earthquake by the accelerometers of the plant, which detect a ground acceleration of about 0.5 g,[4] reactors are normally shut down by oleo-pneumatic insertion of the shutdown control rods. On this type of reactor, the control rods and shutdown rods enter through the bottom of the reactor (Photo 5.4), because the size of the upper part of the vessel with the steam dryers does not allow them to be placed on top as in the case of Pressurized Water Reactors. This is a definite advantage for PWRs, because the rods can then drop by gravity even in the case of a total absence of electrical power, which is not the case for Boiling Water Reactors, which depend on oleo-pneumatic controls.

The general course of what will happen to reactors -1, -2, and -3 is shown in Figs. 5.12, 5.13, 5.14, 5.15, and 5.16. The earthquake caused extensive damage, particularly to the six power transmission lines that supply the plant and to the transformer, which will prevent the plant from being re-supplied when power is restored to the grid within an hour. All the reactors were left without an external power source, but 13 of the 14 diesels started up normally to make up for the loss of the electric grid.[5] When the tsunami hit the plant an hour later, a wave of more than 14 m advanced inland, destroying almost everything in its path, including the emergency diesels and their seawater cooling pumps[6] (Photo 5.5).

The reactors are deprived of their cold source because the water intakes were completely blocked by debris, and above all the only remaining electrical sources were batteries with power and duration limited to about 8 h, which prevents any active circulation of water in the reactors over a long period to be able to evacuate the residual power. The dimensioning of the plant with respect to a wave (Higher Safety Level) was only 5.70 m. The reactor buildings of the three reactors were voluntarily isolated by closing the valves not necessary for safety, to avoid an early release of radioactivity. Reactor-1 was then cooled by its isolation condenser, while reactors -2 and -3 were cooled by the RCIC turbopump (Fig. 5.9). In the absence of an external cold source, this system can only be effective as long as the water in the torus is not saturated (when the water boils). At 3:27 p.m., a wave of more than 14 m hits the plant and destroys all the diesels and their fuel tanks, as well as the emergency cooling auxiliaries (water tanks of the residual power

---

[3] The term tsunami means in Japanese *tsu* = port and *nami* = wave, literally "*wave in a port.*"

[4] Accelerations are commonly expressed in units of the earth's acceleration at ground level, which is 9.81 m/s², which corresponds to 1 g.

[5] The 14th diesel (on plant-4) was in shutdown condition for inspection.

Photo 5.4 Operators inspecting vessel bottom penetrations by control rods on a Fukushima Dai-ini plant. The reactor core, shutdown in this photo because one does not go into the reactor pit during operation, is thus located above the operators. The system is identical on Fukushima Dai-ichi (TEPCO)

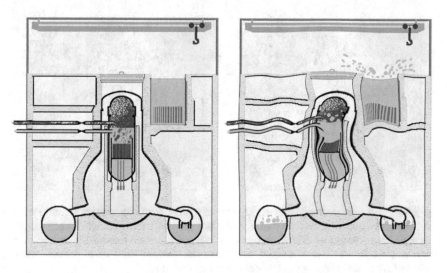

Fig. 5.12 The Fukushima plants -1, -2, and -3. March 11, 2001, 2:46 p.m. Reactors -1, -2, and -3 of Fukushima-Dai-ichi are in operation, Reactor-4 is discharged at shutdown. The structures of the reactors vibrate but resist

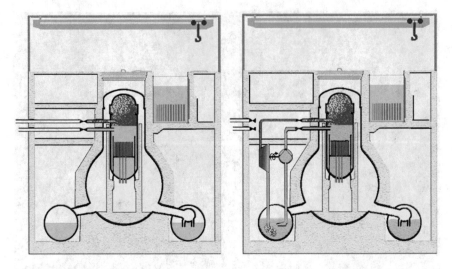

Fig. 5.13 Emergency shutdown: the rods are inserted correctly and stop the chain reaction. But the electric grid is lost. The emergency diesels take over. 3:41 p.m.: the ensuing tidal wave drowned the emergency diesels. Only batteries with limited autonomy (a few hours) remained as an electrical source. All the core cooling systems were out of order, except for the isolation turbopump, which operates on a turbine fed by the core steam (in the compartment on the left of the drawing on the right)

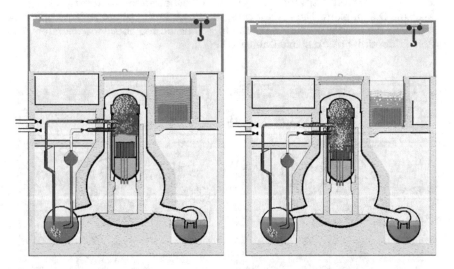

Fig. 5.14 The Fukushima plants -1, -2, and -3 (continued). Temporary cooling of the core: the turbopump circulated water in the core, but the water in the steam suppression torus must be below 100 °C, and power was still required from the batteries. Since there was no longer a cold source available, the temperature of all compartments increased. The water in the deactivation pool evaporated

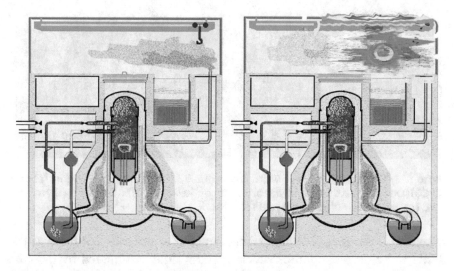

Fig. 5.15 Loss of the isolation pump and degradation of the core: the batteries are emptied on March 11, at 4:36 p.m. for unit-1, on March 13, at 2:44 a.m. for unit-3. In unit-2, the pump was lost. The steam was discharged from the core to the steam suppression torus via the discharge valves. The water level in the core decreased. The assemblies were uncovered, and the cladding oxidized producing hydrogen. The rods swelled and cladding ruptures released volatile fission products. Hydrogen spilled into the dry pit of the core. More than 300 kg of hydrogen filled the reactor containment

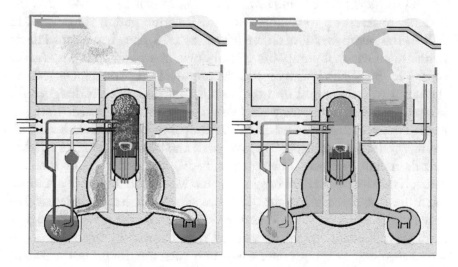

Fig. 5.16 A hydrogen explosion blows the reactor building. About half of the core is depleted. The temperature exceeds 2500 °C and the core melts. The hydrogen accumulates in the upper parts of the building. There was a hydrogen explosion in the buildings of units-1 (12 March, 3:36 p.m.) and -3 (14 March, 11:00 a.m.). In unit-2, there was an explosion (March 15, 6:10 a.m.) probably in the containment at the level of the torus. Helicopters and then the fire brigade sent water from outside the building

Photo 5.5 Damage on the seawater pumps: unit-3 (left) and unit-5 (right) (photos TEPCO). One can see the immediate proximity of these pumps to the sea by their function. The dam was ineffective to protect them

system). All the core cooling systems were out of order except for one: the RCIC isolation turbopump, which operates on a turbine fed by the core steam. The emergency batteries of Unit-1 were also flooded, unlike those of reactors -2 and -3. Unit-1 had lost all its electrical sources. Units 2 and 3 were running on batteries designed to last 8 h. As there was no longer a cold source for Reactor 4, the temperature of all compartments increased. The water in the deactivation pool was slowly evaporating. At 3.40 p.m., with no battery, the isolation valve of the condenser of Reactor-1 closed, which shut down the cooling of the core. On March 13, at 2:44 a.m., the batteries in reactor 3 were empty and the isolation pumps shut down. In unit-2, the pump was lost on 14 March at 1:18 p.m. It was possible to switch to an external pump, but it shut down at around 5:00 p.m. As the cores-1 to -3 were no longer cooled, the pressure in the vessel increased and the steam was discharged into the toroidal wet pit, which is designed for this purpose. As a result, the water level in the reactors decreased. The cores were dewatered, and the zirconium cladding of the fuel rods oxidized with steam, producing hydrogen. The rods swell and cladding ruptures released volatile fission products. Hydrogen spilled into the dry pit of the core through the wet pit vacuum breaker system. More than 300 kg of hydrogen filled the reactor containment. The core was about half dewatered. The temperature exceeded 2500 °C and the core melted. The pressure increases in the containment, which is designed to withstand a pressure of about 5 bars absolute. As the pressure raised to 8 bars, the dry pit in the reactor hall is depressurized (on 12 March at 4:00 a.m. for unit-1, on 13 March at

---

[6] Even if the diesels had not been drowned, the destruction of their cooling pump would have prevented their operation.

midnight for unit-2, on 13 March at 8:41 a.m. for unit-3). The hydrogen produced by the oxidation of the zirconium in the cladding at high temperature accumulated by stratification in the upper parts of the building, above the service desk. There was a hydrogen explosion above the service desk in the buildings of units -1 (March 12, 3:36 p.m.) and -3 (March 14, 11:00 a.m. In unit-2, there was also an explosion (March 15, 6:10 a.m.), but probably at the level of the torus in the containment, which created a crack in the building that leaked contaminated water. This last explosion, less spectacular than those of the other two reactors, was the one that will pose the most problems because of the release of highly radioactive water.

At 7:46 p.m. on March 11, the Japanese government reported a cooling problem in the shutdown reactors. The situation deteriorated during the night and on the morning of March 12, the authorities decided to evacuate the 20,000 inhabitants of the region. On March 12 at 3:36 a.m., an explosion blew the outer building of the reactor-1, due to a hydrogen explosion. TEPCO decided to inject borated water into the reactor caisson (dry pit). On March 13, the Japanese Nuclear Safety Agency provisionally classified the event at 4 on the INES scale, but the worst was yet to come. On March 14, a double explosion destroyed the roof of the reactor-3 building. The explosion was filmed live and broadcast on Japanese television. On March 15, a new explosion took place in reactor 2, followed by a fire in reactor 4. In the case of Reactor-2, it is believed that the explosion damaged the pressure suppression pool, causing an increase in dose rate of 1 mSv/h[7] up to 8 mSv/h. In the case of Reactor-4, a fire broke out, but the low rate of oxidation of the assemblies in the pool proves that the pool was never mostly dewatered. The measurement of certain isotopes of iodine at low half-lives, which could only come from nuclear fissions, could even suggest a criticality outbreak, but their small quantity showed that they were rather spontaneous fissions, a rare form of radioactivity of certain fissile nuclei. In fact, the dose rate rose to 400 mSv/h, which can only be explained by a release of fission products from the assemblies in the pool. However, the pool in question being in the open air following the fire, on 16 March, an attempt was made to drown the pool of Reactor-4 by dropping water through the gutted roof from a heavy-lift helicopter (7 tons of water per rotation), a very difficult and desperate maneuver. At the

---

[7] The milli-Sievert is a unit of dose, i.e., energy deposited in matter, which characterizes the damage produced by the absorption of radiation. The milli-Sievert per hour is therefore a dose rate.

initiative of the director Masao Yoshida[8] and without referring to the Tokyo hierarchy, which was totally overwhelmed by the events, the situation of the three reactors was stabilized by injecting directly available sea water at first, then fresh water for fear of corrosion and salt crystallization (we talk about a salt deposit[9] from 1 to 2 m ! i.e., 25 tons of salt in reactor 1). The vessel pit

---

[8] Masao Yoshida (1955–2013) is a Japanese nuclear engineer. He graduated from the University of Tokyo in nuclear engineering and joined TEPCO in 1979. He was appointed Deputy Head of Reactors 5 and 6 from 1986 to 1988, then Head of the Maintenance Department for Reactors -1 and -2 from 1993 to 1995, and became Unit Manager for Reactors -1 to -4 from 2005 to 2007. He alternated between positions of responsibility at the Dai-ichi and Dai-ini sites and at TEPCO headquarters in Tokyo. He was appointed Director of the entire Dai-ichi site in 2010. During the accident, he will remain at the head of the "*Fukushima 50*", a group of volunteers who stayed after March 15 to try to manage the disaster. Wakamatsu Tetsuro's film in 2020 pays tribute to them. Yoshida would not leave TEPCO until the announcement of his esophageal cancer in late 2011. Having always defended the point of view of his hierarchy to which he remained faithful, this leader described as sometimes angry but respected by his employees, will nevertheless take the decision to inject seawater into the reactors (by a fire truck) without referring to his management, whose operational assistance had proved disastrous since the beginning of the crisis. These events are particularly well described in the comic strip: "*Fukushima: Chronique d'un accident sans fin*" of Bertrand Galic and Roger Vidal (2021) which I recommend reading. It shows the desperate attempts of the operators to cool the reactors, under the direction of Yoshida. The timing is well respected and the realism very credible.

Masao Yoshida

The poster of the film "Fukushima 50" and a synthetic tsunami scene from the film.

Photo 5.6 Reactor-3: The disturbingly persistent black smoke from Reactor 3 probably indicates an ongoing corium-concrete interaction in the bottom of the dry pit and possibly on the building foundation (TEPCO photo)

was thus drowned. This did not prevent the corium from relocating to the vessel bottom of reactor-1, by far the most degraded. It is known since TMI-2 that corium can progress even under water. Reactor-3 has also seen a breakthrough in its vessel and the black fumaroles seen coming out of the remains of the building seem to indicate the appearance of an interaction between the molten corium and the concrete of the dry pit (Photo 5.6).

Tables 5.2, 5.3, and 5.4 present the chronology of events of reactors -1, -2, and -3, respectively.

Reactor-4 has been shut down since November 30, 2010, for a major maintenance operation on the core barrel. In fact, reactor-4 has been completely emptied of its fuel, which is now located in its fuel pool. However, hydrogen produced by oxidation of the cladding in reactor-3 will enter building-4 through a common connection to the gas evacuation chimney of plants -3 and -4, Photo 5.9).

The explosion in building-4 initially raises many questions (Fig. 5.18, Photo 5.10). Since the core was empty, the most logical explanation would be that the storage pool located in the building was dewatered, producing an oxidation of the cladding in the air by a rise in temperature due to the residual power,

---

[9] The salt mixed in the water remains in the liquid phase in case of evaporation (therefore the water in the clouds created by evaporation from the oceans is fresh) and the salt accumulates until it crystallizes and settles in the lower parts of the reactor. At 0 °C, 1 l of water cannot dissolve more than 357 g of salt. Beyond that, the salt will settle at the bottom of the vessel. Note that the solubility increases with the temperature.

Table 5.2 Chronology of events related to the Fukushima-1 reactor

Day/month/year hour/minutes	Duration since the origin of the accident (hours)	Event
03/11/2011 14 h 46	0.00	Triggering of the earthquake detectable by accelerometers
03/11/2011 14 h 46	0.00	Scram signal and automatic reactor shutdown by insertion of the shutdown control rods
03/11/2011 14 h 47	0.02	Emergency diesels start-up on loss of off-site power sources
03/11/2011 14 h 52	0.10	The Isolation Condenser (IC) is automatically activated
03/11/2011 15 h 03	0.28	The operators manually shutdown the Isolation Condenser. The IC train B is manually shutdown because the core cooling was too fast compared to the Technical Operating Specifications. Only the A-train is operating.
03/11/2011 15 h 10	0.40	IC (train A) re-engaged
03/11/2011 15 h 19	0.55	IC (train A) re-stopped
03/11/2011 15 h 24	0.63	IC (train A) re-engaged
03/11/2011 15 h 24	0.67	IC (train A) re-stopped
03/11/2011 15 h 27	0.68	**The first wave of the tsunami hits the coast**
03/11/2011 15 h 32	0.77	IC (train A) re-engaged
03/11/2011 15 h 34	0.80	IC (train A) re-stopped
03/11/2011 15 h 35	0.83	**The second wave of the tsunami hits the coast**
03/11/2011 15 h 37	0.85	Loss of all AC sources
		Loss of all DC power sources
		IC control valve status is lost in the Control Room
		High-Pressure Safety Injection (HCPI) appears to be inoperative
		The vessel SLC control and control rods (CRD) pressurization by loss of AC power
		Containment Cooling (CCS), Seawater Closed Circuit Cooling (CCSW), Water Makeup (MUWC), Fire Protection and Fuel Pool Cooling (SFP) are inoperative due to loss of AC power

(continued)

Table 5.2 (continued)

Day/month/ year hour/ minutes	Duration since the origin of the accident (hours)	Event
03/11/2011 18 h 18	3.53	Partial return of the DC current. The isolation condenser (train A) is put into operation and steam is observed. The Control Room partially recovers power. Heat transfer in the IC condenser (train A) seems to be degraded (presence of hydrogen?)
03/11/2011 18 h 25	3.65	The isolation condenser (train A) is shutdown
03/11/2011 21 h 30	6.73	The isolation condenser (train A) is put back into service and steam is observed. It was at about this time that a very clear peak in overpressure in the reactor, which rose from 5 bars (the reactor had been depressurized after 6 h) to 35 bars, certainly indicates a relocation of the corium in the vessel bottom. The in-core instrumentation tubes must have been degraded, causing a leak of radioactive products into the containment
03/11/2011 21 h 30	11.03	The isolation condenser (train A) is shutdown (the diesel-driven fire pumps have failed. Seawater from the fire system can be sent to the core spray with difficulty because the pressure in the core and in the drywell is twice the containment design pressure. The initial pressure in the dry pit is 100 kPa, which will rise to 850 kPa. The plug of the dry pit will rise at approximately 12 h following this overpressure and probably did not close properly. This overpressure will last for a day, preventing this water injection from being truly effective. It is almost certain that the vessel broke through, letting corium flow to the bottom of the dry pit and initiating a corium-concrete interaction
03/12/2011 15 h 36	23.83	Hydrogen explosion blows out the conventional containment building (Photo 5.7)

resulting in the production of hydrogen, and then an explosion. This scenario will be particularly considered insofar as the fire broke out in the spent fuel pool located in the upper part of the containment. But the visual inspection by camera after the accident showed that the assemblies in the water-filled pool were in rather good condition (zirconium oxide is easy to spot on the surface of the fuel rods because of its whitish color, while the metallic zirconium has a tinge close to steel). The camera shows quite clearly that the assemblies in the pool are only covered by rubble from the explosion (Photo 5.11).

Table 5.3 Chronology of events related to the Fukushima-2 reactor

Day/month/ year hour/ minutes	Duration since the origin of the accident (hours)	Event
03/11/2011 14 h 46	0.00	Triggering of the earthquake detectable by accelerometers
03/11/2011 14 h 47	0.02	Scram signal and automatic shutdown of the reactor by insertion of the shutdown control rods
		The emergency diesels start when the electrical sources outside the site are lost (at the same time as reactor-1)
03/11/2011 14 h 50	0.07	The RCIC system is engaged by the operators
03/11/2011 14 h 51	0.08	The RCIC system automatically trips because the water level in the vessel is too high, and the water may drown the RCIC turbine
03/11/2011 15 h 02	0.27	The RCIC system is re-engaged by the operators
03/11/2011 15 h 27	0.68	**The first wave of the tsunami hits reactor-2**
03/11/2011 15 h 28	0.70	The RCIC system is engaged by the operators. Contrary to reactor-1, the cooling of reactor-2 will not be uncooled after the loss of electrical sources thanks to the RCIC
03/11/2011 15 h 35	0.82	**The second wave of the tsunami hits reactor-2**
03/11/2011 15 h 41	0.92	Loss of AC power sources
		Loss of DC power sources, therefore no more control of the valves and no more control of the RCIC injection rate
		Unknown RCIC government
		HPCI not functional
		Water level control and pressurization of the control rods (CRD) not functional due to loss of AC power
03/11/2011 21 h 02	6.27	RCIC state unknown, no water level measurement in vessel
03/12/2011 00 h 30	9.73	RCIC system apparently operational on auditory indications
03/12/2011 02 h 66	12.15	RCIC system apparently operational based on high RCIC discharge pressure measurement
03/12/2011 05 h 00	14.23	The RCIC system water source is switched from the condensate storage tank to the toroidal pressure suppression chamber
03/12/2011 21 h 00	30.23	RCIC system apparently operational
03/13/2011 10 h 40	43.90	RCIC system apparently operational

(continued)

Table 5.3 (continued)

Day/month/ year hour/ minutes	Duration since the origin of the accident (hours)	Event
03/13/2011 13 h 50	47.07	RCIC system apparently operational
03/14/2011 13 h 25	70.65	Assumed loss of the RCIC system due to very low water level in the vessel. The pressure increase in the containment can no longer be controlled. The overpressure in the dry pit (twice the design pressure), causes the dry pit plug to rupture at about 80 h. After 80 h, a sharp increase in dose around Reactor 1 was observed, clearly indicating a rupture of the drywell plug. The containment depressurized by itself after 90 h (rupture?). The presence of fission products indicates a degradation of the core, which may have remained localized in the vessel
03/15/2011 06 h 10	87.40	Hydrogen explosion in the toroidal suppression chamber (Fig. 5.17)

Table 5.4 Chronology of events related to the Fukushima-3 reactor

Day/month/ year hour/ minutes	Duration since the origin of the accident (hours)	Event
03/11/2011 14 h 46	0.00	Triggering of the earthquake detectable by the accelerometers
03/11/2011 14 h 47	0.02	Scram signal and automatic shutdown of the reactor by insertion of the shutdown control rods
03/11/2011 14 h 48	0.03	The emergency diesels start-up on loss of electrical sources outside the site (at the same time as reactor-1)
03/11/2011 15 h 05	0.32	The RCIC system is engaged by the operators
03/11/2011 14 h 25	0.65	The RCIC system automatically triggers because the water level in the vessel is too high
03/11/2011 15 h 27	0.68	**The first wave of the tsunami hits reactor-3**
03/11/2011 15 h 35	0.82	**The second wave of the tsunami hits reactor-3**

(continued)

Table 5.4 (continued)

Day/month/ year hour/ minutes	Duration since the origin of the accident (hours)	Event
03/11/2011 15 h 38	0.87	Loss of alternating current sources
		Partial loss of DC sources. The RCIC valves can still be controlled
		The RCIC remains controllable for about 20 h after the scram. RCIC is pumping water from CSE
		HPCI not working
		Packed water level control and control rod pressurization (CRD) not functional due to loss of AC power
		Low-pressure core spray (CS), residual power removal RHR, residual power removal by cold source (sea water) RHRS not functional due to loss of AC power
		Core water makeup, fire protection pumps and containment cooling are non-functional following the loss of AC power
03/11/2011 16 h 03	1.28	The RCIC system is re-engaged by the operators
03/12/2011 11 h 36	20.83	The RCIC system is permanently activated (shutdown) following a mechanical (not electrical) problem
03/12/2011 12 h 35	21.82	The High-Pressure Injection System (HPCI) is automatically activated by a low water level signal in the vessel. The HPCI uses steam to drive a steam turbopump. The pump draws water from either the CST or the suppression chamber. The HPCI removes a lot of steam from the core, which causes the pressure to drop rapidly, decreasing the efficiency of the HCPI turbine
03/13011 17 h 30	26.73	The HPCI system is still operational but injects less water because the turbine is fed at a lower pressure
03/13011 02 h 42	35.93	The HPCI system shuts down by emptying the available water tanks. The system cannot be restarted due to lack of voltage to the batteries. The same goes for the RCIC system. The operators tried to control the pressure of the containment by venting, but at 67 h, the hydrogen in the containment exploded. The explosion was more powerful than that of Reactor-1. The corium pierced the vessel and relocate in the dry pit, attacking the concrete
14/03/2011 11 h 01	68.25	Hydrogen explosion in the containment (Photo 5.8)

Since reactor 4 is in communication with reactor-3 through common buildings connected by pipes, it is conceivable that hydrogen was released with some delay. In any case, the explosion of reactor-3 took place 19 h before the explosion of reactor 4 which exploded on March 15, 2011, at 09:38 a.m.

Photo 5.7   A picture of the explosion of reactor-1 taken on TV (NHK). On the left, reactors -5 and -6 and their shared chimney. On the right, reactors -2, -3, and -4 whose buildings are still intact

Photo 5.8   Hydrogen explosion in the building of reactor-3 taken on television (NHK). On the left reactor-2, on the right reactor-4

The explosion of reactor 3 must have destroyed the connecting pipe. If there was indeed a hydrogen explosion in reactor-4 and that this hydrogen does not come from the pool of reactor-4, one can imagine a process of dis-inerting by condensation of steam, creating a pocket of explosive hydrogen (initially coming from reactor-3), and/or a slow stratification of the hydrogen. This hydrogen will concentrate in the upper part of the reactor by gravitational effect (hydrogen is less heavy than air), creating zones richer in hydrogen in the upper parts of the building. The fire will be shut down by the supply of water by helicopters (Photo 5.12) and trucks. No particular attention had been paid to reactor-4 by the operators, since no one could have imagined that a reactor empty of fuel could explode. A simple ventilation of the containment could probably have saved it. This human error, which is finally quite understandable, exacerbates the catastrophe.

Reactor-5 has been shut down since January 3, 2011 and the fuel has been loaded into the core awaiting restart. The residual power is very low. The situation

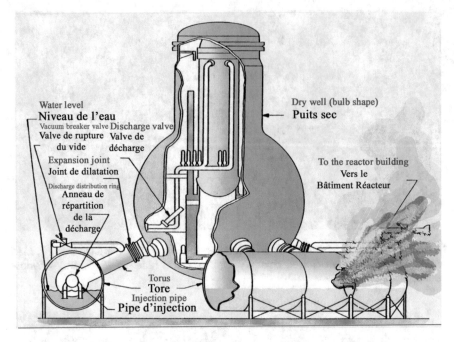

Water level
Niveau de l'eau
Vacuum breaker valve Discharge valve
Valve de rupture Valve de
du vide   décharge

Expansion joint
Joint de dilatation

Discharge distribution ring
Anneau de
répartition
de la
décharge

Dry well (bulb shape)
Puits sec

To the reactor building
Vers le
Bâtiment Réacteur

Torus
Tore
Injection pipe
Pipe d'injection

Fig. 5.17 Postulated Explosion of the torus of the Fukushima reactor n°2

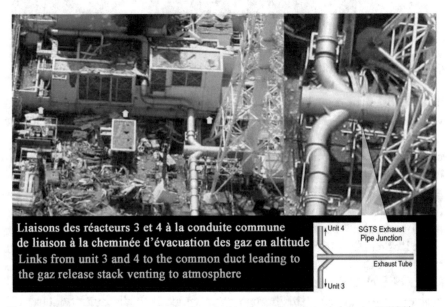

Liaisons des réacteurs 3 et 4 à la conduite commune
de liaison à la cheminée d'évacuation des gaz en altitude
Links from unit 3 and 4 to the common duct leading to
the gaz release stack venting to atmosphere

Unit 4   SGTS Exhaust
              Pipe Junction

Exhaust Tube

Unit 3

Photo 5.9 Connections from units -3 and -4 to the exhaust stack (photo TEPCO)

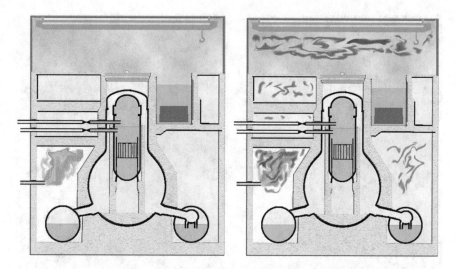

Fig. 5.18 Unit-4 is a special case because the reactor was shut down for heavy mainte-nance of the core barrel. It was therefore completely unloaded at the time of the earthquake. However, the deactivation pool was full and contains the equivalent of two cores. Hydrogen from reactor-3 will enter building-4 through a common pipe. This hydrogen will concentrate by buoyancy (buoyancy of a gas lighter than air) in the upper parts of the building, then explode 19 h after the explosion of building-3

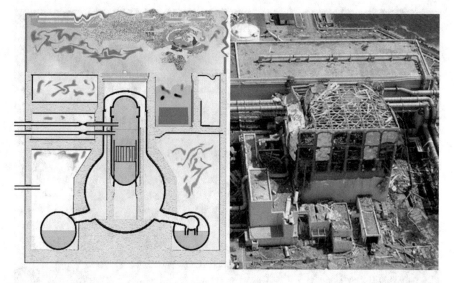

Photo 5.10 Plant-4 of Fukushima—11–15 March 2011. Hydrogen explosion in plant-4: the scenario of the other reactors is repeated: the hydrogen produced by the oxidation of the cladding of the twin plant-3 exploded. It appears after the accident that the pool seems little degraded on inspection by cameras and that the fuel cladding is not oxidized. As a precaution, an attempt was made to fill the pool from the outside (by dropping water from a helicopter and then a fire hose) to cool the fuel (photo TEPCO)

Photo 5.11 Reactor-4 pool. This view of poor quality taken from a camera still shows that the fuel has not degraded dry in case the pool was dewatered. On the contrary, the assemblies seem intact and little oxidized, supporting the idea of an external origin of the hydrogen (from reactor-3) (TEPCO)

Photo 5.12 A helicopter on its way to drop a water supply on the building of reactor-4 (photo taken on NHK TV)

is identical for reactor-6. At the time of the tsunami, an air-cooled emergency diesel generator in reactor-6 survived. This diesel will be used to rescue reactor-5 afterwards. Reactors -5 and -6 emerged relatively unscathed from the disaster.

Technical Insert: "The Hydrogen Explosion"

Hydrogen gas is mainly produced in severe accident situations by the oxidation of metals, mainly zirconium in fuel cladding. The oxidation of metals by water steam at high temperature is a phenomenon that is exacerbated by temperature. In a severe accident situation, the Zircaloy-4 cladding (a zirconium alloy) of the fuel rods, the grids (nearly 20 tons in all of zirconium for a 900 MWe CPY) and the steel of the core structures can oxidize considerably. Hydrogen is produced during the oxidation reaction of zircaloy by water by an equation of the type:

$$Zr_{metal} + 2H_2O_{steam} \rightarrow ZrO_2 + 2H_{2\,gaseous} + \approx 600\,kJ$$

This reaction is extremely exothermic due to the energy of the reaction, which varies from 576 kJ/mol of Zr at 2000 K, up to 599 kJ/mol of Zr at the melting point of zirconia ZrO_2.[10] Note that the increase in the energy of the reaction is mainly due to the melting of zirconium. From a chemical point of view, the oxidation of Zircaloy by water can be modeled as follows: first a dissociation of water at the oxide-water surface by the reactions:

$$H_2O + 2e^- \rightarrow O^{2-} + H_2$$

$$2H_2O + 2e^- \rightarrow H_2 + 2OH^-$$

Then a diffusion of species through the layer: O^{2-} and OH^- inward, e^- to the outside and the formation of the oxide at the metal-oxide interface:

$$Zr + 2O^{2-} \rightarrow ZrO_2 + 4e^-$$

$$Zr + 2OH^- \rightarrow ZrO_2 + H_2 + 2e^-$$

The most common assumption is that oxygen diffusion alone governs the oxidation reaction.

---

[10] D.F. Fletcher, B.D. Turland, S.P.A. Lawrence: *A review of hydrogen production during melt/water interaction in LWR's,* Nuclear safety Vol. 33, n°4, pp 514–534 (1982).

(continued)

(continued)

Hydrogen is a very flammable gas. We all have in mind the images of the gigantic fire of the German airship Hindenburg (Photo 5.13) inflated with hydrogen, at its arrival in the USA in 1937.[11]

The complete combustion of hydrogen is strongly exothermic and generates water in the form of vapor according to the reaction:

$$H_2 + \frac{1}{2}O_2 \rightarrow H_2O + Q$$

The energy Q of this reaction is 141.79 MJ/kg of hydrogen, almost three times that of butane! In the containment, the combustion of hydrogen releases thermal energy, which leads to an increase in enthalpy[12] of the gases after combustion, resulting in a sudden increase in pressure which can threaten the integrity

Photo 5.13

---

[11] After a smooth two-day journey of the Atlantic Ocean from Frankfurt, Germany, the Hindenburg, pride of the Nazi regime and flagship of the German airship fleet, arrived on May 6, 1937, at the US Air Force Base in Lakehurst, New Jersey. The weather conditions of arrival were uncertain and stormy, the aircraft was late, but the American public came in numbers to salute the exploit. Suddenly, in a matter of seconds, the shell caught fire when it touched the mooring mast, probably as a result of an electrostatic discharge. It took only a few minutes for the fire to spread to the 245-m long sides, filled with 200,000 m³ of hydrogen. In the gondola, there was panic, and people threw themselves out of the windows. Of the 97 people, passengers and crew, 35 perished in atrocious conditions under the eye of the cameras mobilized for the event. The Zeppelin LZ 129 Hindenburg was on its 63rd commercial flight and its twentieth Atlantic journey. Nowadays, the non-flammable helium has advantageously replaced the hydrogen in the mounted balloons.

[12] The enthalpy of a gas is the energy it stores.

(continued)

(continued)

of the containment, and which blew up the conventional buildings of the Fukushima reactors. The term explosion is used if the combustion generates a significant pressure peak (more than 2 bars). The flammability of hydrogen in the context of a LOCA accident was first studied by Shapiro[13] (Photo 5.14) and Mofette[14] in 1957, and the name Shapiro diagram is often associated with the curve of flammability of hydrogen as a function of the hydrogen content in the presence of water steam (Fig. 5.19).

The flammability limit is defined as the ability of a mixture of hydrogen and air to maintain a self-sustaining flame. The Shapiro diagram distinguishes a lower flammability limit for a hydrogen content of about 5%, and an upper limit for a content of about 80%, beyond which the atmosphere is too rich in hydrogen. From 550 °C onwards, there is a zone of self-ignition by volumetric oxidation of the hydrogen. This oxidation always takes place, but we can distinguish relatively arbitrarily a zone where a strong overpressure is observed. Note that self-ignition does not require an initiator such as a spark.

Photo 5.14 Zalman Shapiro

---

[13] Zalman Shapiro (1920–2016). American chemical engineer. After a thesis at Johns Hopkins University in 1948, he joined the Westinghouse research teams at the Bettis Naval Nuclear Power Laboratory where he became a specialist in zirconium chemistry. He developed the process of purifying zirconium by iodide vapor deposition and was one of the architects of the first US atomic submarine reactor: the USS Nautilus. He then worked on the production of fuel cladding for the Shippingport reactor. He developed the method of continuous fabrication of uranium oxide powder, then plutonium, used to produce high-density sintered fuel pellets. At 89 years old, he filed another patent on the mass production of industrial-grade synthetic diamond. In 1957, he created NUMEC, a company specializing in nuclear materials, and returned to Westinghouse in 1971. At the end of the 1960s, he was at the center of a controversy over the possibility that he had stolen uranium (nearly 300 kg!) for Israel, because of his proven Zionist sympathies, without ever being questioned, because no decisive proof was ever brought forward.

[14] Z.M. Shapiro, R.T. Mofette: *Hydrogen flammability data and application to PWR loss-of-coolant accident*, WAPD-SC-545, Westinghouse Electric Corporation (1957).

(continued)

(continued)

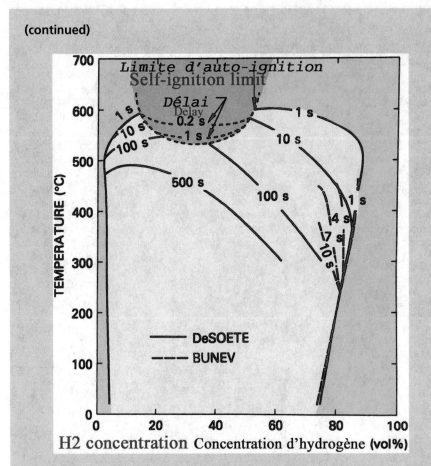

Fig. 5.19 Flammability diagram of hydrogen in air and self-ignition range (according to Douglas W. Stamps, Marshall Berman: *High-temperature hydrogen combustion in reactor safety applications*, Nuclear Science and Engineering Vol. 109, n°1, pp 39–48 (1991))

After drowning the reactors with seawater, the situation gradually stabilized, especially from March 20 when electricity was restored in reactor-2. From March 25, fresh water was reinjected because of fears of corrosion and the risk of clogging by salt deposits. Concerning reactor-2, the very high radioactivity of the water escaping from the building through a 20 cm break suggests that the containment was cracked, either because of the hydrogen explosion or because of penetration by corium. A leak of about 7 tons/h let escape very radioactive water towards the sea. After several days of effort, this leak was sealed by injecting sodium silicate around the break, which has the property of solidifying in contact with water to form a glassy structure. Very large volumes of radioactive water were stored in the reactor buildings, which must be treated to facilitate the entry of operators. A purification station proposed by AREVA (1200 tons of water/day in theory) began to solve the problem of the 100,000 tons of accumulated radioactive water on June 17. The overall release of radioactivity due to the accident is estimated at 10% of that of Chernobyl and the favorable winds have pushed it towards the Pacific Ocean. The accident will finally be classified at the highest level-7, on the INES scale, The four reactors of the plant have been definitively destroyed and the future will show what the residual state of the fuel-laden cores of reactors -1 to -3 is. It is certain that the corium has pierced the vessel of reactor-1 and that a corium-concrete interaction has started to attack the concrete of the dry pit of reactor-3. It is estimated by calculation that 60 cm may have been eroded. Only reactor-2, which suffered an explosion in the lower parts, retains its roof. The presence of MOX fuel in reactor-3 has caused much public concern because of the very high radiotoxicity of plutonium. As for the storage pools, water analyses have shown that the fuel they contained was not significantly damaged.

While the mainstream press focuses on the distress of the population (more than 20,000 deaths due to the tsunami) and the visible effects ($300 million in damage), using "*the shock of the pictures,*" many opinion papers questioned the foundations of nuclear power and the possibility of abandoning it. The economic press analyzed the credibility of abandoning nuclear power in France (*Alternatives économiques* n°301 of April 2011) and the environmental magazine Terra-eco did not hesitate to speak of "*the end of a World*" (Photo 5.15). In any case, it is certain that Fukushima will slow down the nuclear revival in the world, with some countries, such as Germany, Belgium, and Switzerland, announcing their withdrawal from nuclear power and others freezing their investments. This choice will lead these countries into a situation of energy dependence, the limits of which can be seen in the 2022 war between Russia and Ukraine, with President Putin using Russian gas to put pressure on the West.

In Europe, the European Commission called for "stress tests" for all plants in the European Union as of March 15, 2011. The additional safety assessments were conducted voluntarily by the various governments. In France, the French Nuclear Safety Authority (ASN, *Autorité de Sûreté Nucléaire*) asked EDF on May 5, 2011, to reassess the safety of its reactors considering the Fukushima events, in particular the risks of earthquakes, flooding, loss of electrical sources and/or cold sources, and fuel management. However, the

Photo 5.15   A panel of the French press about Fukushima at the time of the accident. Many are announcing the end of nuclear power. In any case, the nuclear industry will have to rethink the total loss of electrical sources concomitant with the loss of the cold source. "*A change of model is required!*" in Alternatives Economiques. "*End of the World*" in Terraeco. "*Japan, terror and survival*" in Paris Match

need to secure the power supplies can be emphasized. If only one coolable generator had been raised at Fukushima, the plants could have shared a power source. The absence of a Mark-I containment reactor building is an unfavorable element in the mitigation of the accident. The successive Mark models had moreover evolved favorably for a reinforced containment (Fig. 5.20). TEPCO's management of the crisis also raises questions. The operator took a long time to re-establish an emergency power supply and was probably understaffed, which suggests the creation of a national emergency task force (which France did), or even a supra-national one (not easy?).

The future of Fukushima now lies in the management of the nuclear waste produced by the accident, the decontamination of polluted areas and the dismantling of the 4 reactors. A very high contamination of the marine environment by runoff occurred during and just after the accident at least until April 8, 2011. The dilution factor of the Pacific Ocean mitigates the long-term impact of this contamination, which was only measurable on local fisheries for a limited time. The contaminated water (presence of tritium) from the reactors is now stored in tanks built on the site. In November 2020, there were 1040 tanks for a total of 1,234,000 m^3 which corresponds to 10 years of systematic storage. A very gradual release into the sea will probably be chosen within the limits corresponding to the release authorizations of an operating nuclear plant. Unfortunately, the wind at the time of the accident pushed the radioactive cloud in a north-easten direction, which contaminated the

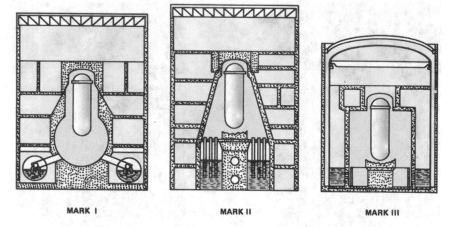

MARK I   MARK II   MARK III

Fig. 5.20 Evolution of the BWRs containments: Mark-I to -III. Mark-I (1963) is nick-named "torus and bulbous": the bulb in question being the containment in the shape of a bulb and the torus the steam suppression pool or pit/wet enclosure. Communication between the dry and wet containment is provided by sloping pipes and dip tubes. This is the Fukushima containment model. The Mark-II (1967) is truncated cone with a cylindrical wet containment. Communication is ensured by vertical dip tubes. The Mark-III (1972) dry pit is cylindrical with a cylindrical wet containment around the dry containment with horizontal vent holes

territories in this direction by deposits. An exclusion zone of 30 km with evacuation of the population was established in the most affected areas (Fig. 5.21), but a voluntary action of the Japanese Authorities to reclaim the soil has limited the exposure of the population. This zone was later reduced to 20 km for some municipalities. Very large quantities of waste (nearly 20 million m³) were concentrated in storage sites known as *kariokiba*, waste that will gradually be sorted, incinerated, or reused (civil engineering foundation materials, for example) depending on their activity.

The dismantling of the reactors involves removing the radioactive corium spread in the lower structures of the buildings (Photo 5.16). This task will require the use of teleoperated tools and/or robots, when radiation decreases.

And France?

Even if the tsunami phenomenon is considered unrealistic in metropolitan France, but more serious in some overseas departments or territories (Caribbean, New Caledonia, Polynesia, where there are no reactors), the occurrence of this type of event has been studied by the *Bureau de Recherches*

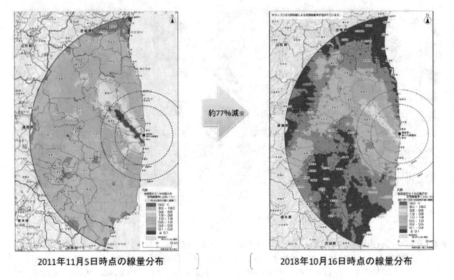

2011年11月5日時点の線量分布   2018年10月16日時点の線量分布

Fig. 5.21 Comparison of the activities measured in dates of 11/05/2011 and 10/16/2016 by the Japanese authorities. The color scale is comparable, and the temporal "cooling" is easily noticeable. A detailed treatment of the soil can accelerate a return to a certain normality. Time does its work, and radioactive products with half-lives of less than 1 year have already almost completely disappeared. It is usual to consider that we have a good approximation of the almost complete disappearance of an isotope after ten times its radioactive half-life (division of the concentration and activity in *Bq* of a coefficient $2^{10} = 1024$)

Photo 5.16 Reactor-2 vessel pit. One can see the vessel bottom torn open by the molten corium and the solidified corium, which has spread in the dry pit (photo TEPCO). One should be wary of the pasty aspect that the cooled corium takes because a metallic corium in meltdown can be more fluid than water

Géologiques et Minières.[15] Important works of historical bibliography have shown that there was no reference of events of intensity higher than 3 in the scale of tsunamis in metropolitan France. This scale of tsunamis is not the Richter scale. The number 3 meaning *"Wave strong enough, generally noticed. Flooding of the coasts on gentle slopes, light boats stranded, light constructions near the coasts slightly damaged. In estuaries, reversal of the watercourses to a certain distance upstream."* As a curiosity, let's mention an event classified 2 in the bay of Flamanville, site of the EPR, where the sea would have presented in 1725 a retreat and then a wave of 1.5 m amplitude (classified 2, *light wave*). The reasonable conclusion that can be drawn from this work is that the risk of tsunamis seems very low on the metropolitan coast. It is more the risk of earthquakes, especially for the plants in the Rhone Valley, along the Rhine, and the experimental reactors of Cadarache, which seems to be feared.

The "Blayais" Case

However, the risk of flooding is not totally absent in France. During the night of December 27–28, 1999, occurred a storm combined with an exceptional tidal coefficient to swell the waters of the Gironde. The Blayais plant, located 50 km from Bordeaux (France), is built in a low, marshy area along the Gironde estuary, 4.5 m above sea level. The plant is theoretically protected from the millennial flood[16] by a dike made of earth and stones which will however prove to be insufficient to face the storm. The plant is also cut off from the electric grid because high voltage lines have fallen. The low-voltage grid is also cut. As a result of the loss of the grid, the operating reactors were shut down and the diesels were started to produce the power needed for the residual heat removal (RRA) system. At 10:00 p.m., the tide overflowed the dike, and the water rushed into the technical galleries, then into the basements, where the safety injection RIS pumps found themselves *"feet in the water."* The Internal Emergency Plan was declared in the early morning. The situation, which combined the loss of electrical power and safety injections (which were not

---

[15] J. Lambert, P. Daniels: *Inventaire des tsunamis historiques en France*, Rapport final, BRGM/RP-55132-FR (2006). Le BRGM est un organisme public français dont la fonction est l'étude des sciences de la Terre.

necessary at the time as long as the RRA shutdown cooling circuit was operating), could have deteriorated very quickly if the cold source had been lost.

For the dimensioning of the elevation of the structures with respect to floods, a safety margin (CMS, *Cote Majorée de Sécurité*) is calculated, which depends on the location of the plant, on the riverbank, on the seashore or in the estuary. At the river's edge, the CMS depends on the millennial flood, the 100-year flood and the flood caused by the removal (i.e., the loss) of the most constraining dam upstream. At the seaside, the CMS depends on the millennial flood and the millennial marine surge. Finally, for sites located in estuaries, the CMS depends on all the previous criteria. The Blayais site is located in the Gironde estuary. Nevertheless, the estuary being very open, the maritime effects are preponderant on the fluvial effects. The CMS is therefore calculated according to the criteria of the seashore. The level of flood retained is 5.02 m NGF,[17] It corresponds to the maximum tide (with a tidal coefficient of 120) increased by a surcharge linked to local meteorological and topographical conditions. The plant is protected by a dike of 5.2 m in front of the Gironde and 4.75 m on the lateral sides. Studies conducted by EDF in 1998 reevaluated the CMS at 5.46 m. Work was planned to raise the dike to 5.70 m in 2000 but was postponed until 2002. After the storm, studies showed that the water had crossed obstacles located between 5.00 and 5.30 m. On December 27, 1999, reactors -1, -2, and -4 were operating at nominal power, reactor-3 was shut down on RRA cooling, but loaded and in the restart phase. In the early evening of the 27th, the plant was confronted with problems with the electric grid due to the storm. At 7:30 pm, the site lost its 225 kV auxiliary power supply on the four plants, as well as the main 400 kV grid on plants -2 and -4 (for safety reasons, one plant is always connected to two different 400 kV sources). The reactors are automatically shut down without any particular problem by switching to the emergency diesels, which start up correctly. It appears that house-load operation, an action dreaded by the operators and which consists of re-supplying the plant with its own production, was not

---

[16] A millennial flood is a flood that has a one in a thousand chance of occurring within a year. It is therefore a frequency of occurrence, which should not be confused with the millennium flood, which is the most important flood of the last thousand years.

[17] NGF: *Nivellement Général de la France*, this corresponds to the altitude in relation to sea level 0.

possible, the main 400 kV grid having been lost, which prevents this maneuver from being carried out.[18]

The situation worsens when the storm pushes waves up the Gironde, associated with a high tidal coefficient and a significant depression related to the storm. The waves went over the dike and submerged part of the site. At 9 p.m., the access road to the site was also blocked by the flooding, which prevented the teams from taking over. The plants -1 and -2 located in the northwest corner of the site were badly affected by the water inflow. The higher plants -3 and -4 are much less affected. Water entered the underground technical gallery through several trains, mainly through handling holes, as well as through gaps in deformed sheets. The flow of water that entered this gallery is estimated to be between 20,000 and 40,000 m³/h. Several rooms in units -1 and -2 were flooded, in particular those containing the pumps for the raw water backup system (SEC), the base of which is located at −10.75 m, resulting in the loss of the cold-source SEC pumps in track A of unit-1. Water was entering part of the fuel building (BK). The water rendered inoperative the low-pressure safeguard injection system (ISBP) and the containment spray system (EAS), whose pumps are based at −10.50 m.

EDF mobilized its crisis teams around 3:00 a.m. and informed the Safety Authority. The level-1 (non-radiological) Internal Emergency Plan (PUI, *Plan d'Urgence Interne*) was activated. Water pumping operations were started but the water level did not drop. At 5:45 a.m., EDF called on the national crisis organization and the IPSN. The crisis team was operational by 7:00 a.m. The situation worsened around 8:00 a.m. with the flooding of the SEC pump rooms in plant-1 and the loss of SEC train-A. The state of train B was uncertain. The PUI level-2 (internal radiological) was then triggered. The situation, without being catastrophic, was worrying. Reactor-1 has lost one of its two trains in the river cooling system as well as the RIS and the EAS, which were unavailable. These two circuits are only activated during an accident, which is not the case. It is only from 10:40 a.m. that the pumping of the water works. However, it took more than a week to restore the safeguard systems. In this case, the loss of the second cold source (SEC) train, for example by a blockage of the intake by debris and mud from the storm, would have deprived the plant of any cold source, a textbook case close to the Fukushima situation, but credible in the context. The loss of intermediate cooling RRI would have led to the loss of injection at the primary pump seals that ensure the tightness of

---

[18] It is only at 10:20 p.m. that the main grid is available again. The auxiliary grid will finally be recovered only on December 28th at 23 h 30.

the primary circuit, resulting in a small water leak, a scenario with very slow kinetics and easily countered. Water could have been injected from the PTR tank, which can be fed by two plants (twinned plants). In the event of a total loss of power, the turbopumps and the Emergency Power Unit (a kerosene-fired turbine) or a conventional diesel would have been available. By lowering the pressure to less than 32 bars, it was possible to connect the RRA, fed by the PTR tank, and to cool the core by the SGs using the PTR tank and the ASG turbopump. This turbopump could operate as long as the SGs have sufficient temperature to produce steam to power the turbopump (i.e., at least 120–140 °C). Once empty, the ASG tank could be refilled with raw water (pumped from the cold source or fire truck).

After the incident, EDF was criticized for the late declaration of a level-1 emergency response plan, which could have been launched earlier as soon as the high-level alert for the Garonne was detected by plant-4. This would have saved several hours in the course of operations to secure the site. The attitude of EDF, which postponed raising the dike despite repeated requests from the Safety Authority since 1996, was widely criticized in the press and poorly regarded by public opinion. The dike was subsequently raised. The feedback from this real case of external aggression has made it possible to reflect on the concomitance of two situations: the loss of the cold source and the loss of electrical power, and in particular on the interest of a level-2 Probabilistic Safety Assessment (we analyze failures up to core meltdown and the induced radioactive releases) of such an occurrence. The actual incident at Le Blayais was finally classified as level-2 on the INES scale, although no release occurred. Following this incident, a certain number of recommendations were issued by the Parliamentary Office for the Evaluation of Scientific and Technological Choices for the safety and security of civil nuclear installations, the main one concerning access to the plant by road under all circumstances.

The Consequences of Fukushima in France

After the Fukushima accident in March 2011, EDF promptly conducted an assessment of the robustness of its facilities to potential natural hazards. EDF submitted the supplementary safety assessment reports (RECS) to the French Nuclear Safety Authority (ASN) on September 15, 2011, for the reactors in operation and under construction. The ASN authorized continued operation of the nuclear facilities on the basis of the results of the stress tests performed by EDF on all the Fleet's plants and considered that continued operation required increasing their robustness in the face of extreme situations as soon as possible, beyond the safety margins they already have. Following the submission of these reports, on June 26, 2012, the ASN published regulatory technical requirements for EDF reactors (Decision n°2012-DC-0276). These initial requirements were supplemented by the ASN in January 2014 with decisions setting out additional requirements to be met by structures, systems, and components to satisfy the "*hard core*" (as it is called) of Post-Fukushima corpus (Decision n°2014-DC-0396). The supplementary safety assessment reports on reactors undergoing decommissioning were submitted to the ASN on September 15, 2012. EDF has already embarked on a vast program over several years, which consists in particular of verifying that the facilities are properly sized to cope with natural hazards; equipping all French nuclear power plants with new means, first mobile (phase 1) and then fixed (phase 2), to increase their autonomy in terms of water and electricity (Fig. 5.22); to equip the Fleet with a Nuclear Rapid Action Force (FARN) that can intervene within 24 h on a site with six reactors (operational since 2015). This force of 300 regularly trained specialists has its own resources pre-positioned at several sites in France to intervene on all sites in France, such as helicopters, barges to cross water cuts, all-terrain vehicles, mobile electricity production resources, pumping resources, etc.

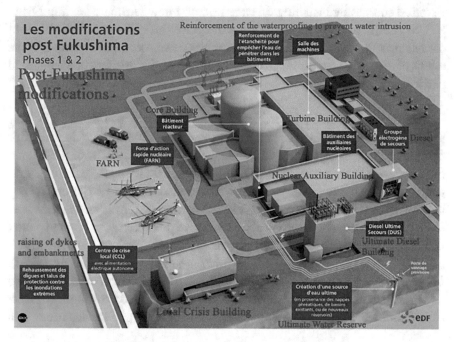

Fig. 5.22 Summary of Post-Fukushima corpus of modifications (EDF iconography)

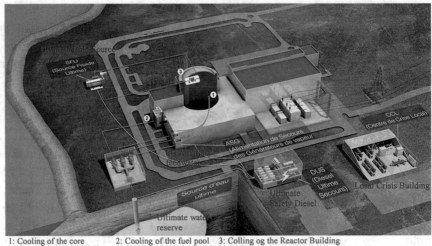

1: Cooling of the core	2: Cooling of the fuel pool	3: Colling og the Reactor Building
1 : refroidissement du réacteur	2 : refroidissement de la piscine	3 : refroidissement du bâtiment réacteur

Photo 5.17   The Local Crisis Center (LCC) and the new *"Hard Core"* facilities (according to ASN). (1) Cooling of the reactor, (2) Cooling of the fuel pool, (3) Cooling of the reactor containment, SPU: ultimate cold source, ASG: steam generator auxiliary feed water, CCL: local crisis center, DUS: ultimate safety diesel

It was also decided to reinforce the robustness to situations of total loss of electrical sources by installing on each reactor a new Ultimate Backup Diesel (DUS) robust to extreme hazards; to integrate the situation of total loss of cold source on the whole Fleet in the safety demonstration; to improve the safety of the fuel assemblies storage; to improve the crisis management, in particular by setting up new Local Crisis Centers (CCL) (Photos 5.17 and 5.18).

In addition, we will reinforce and train the driving teams on shift. This program initially consisted of implementing several short-term measures. This first phase was completed in 2015 and allowed the deployment of the following resources: Emergency generator set (complementary to the existing emergency turbo generator) to ensure the electrical re-supply of the emergency lighting of the control room, the minimum control system as well as the water level measurement of the spent fuel storage pool; Backup borated water makeup during maintenance shutdown (mobile pump) on the 900 MWe reactors (the 1300 and 1450 MWe reactors are already equipped with it); Implementation of taps to connect mobile water, air (in particular compressed air for controlled valves) and electricity supplies; Increased autonomy of electrical batteries; Reliability of pressurizer valve opening; Mobile means and

Photo 5.18 The first Local Crisis Center in Flamanville (France) went into service on January 10, 2020, the site that will host the first French EPR. The strongly defended and particularly austere building (blockhouse without windows!) contains 2500 m² of premises. It has 72 h of electrical autonomy and has three floors located more than 20 m above sea level to avoid submergence. The means of communication have been particularly studied. This unique building, built in 5 years, allows 3 days of autonomy to the teams who would use it. All French nuclear sites will progressively be equipped with a Local Crisis Center (photo EDF)

their storage (pumps, hoses, portable lighting, etc.); Earthquake reinforcement of the reactor building); Reinforcement of the crisis management premises for the earthquake; New crisis telecommunication means (satellite telephones).

The Ultimate Safety Diesel (DUS)

Following the post-Fukushima measures where all 13 emergency diesels had been drowned by the tsunami, it was decided to install from 2017, at the request of the Safety Authority, an Ultimate Emergency Diesel (DUS, Fig. 5.23) par plants. This new building is placed on a 5-m high concrete platform, with a 1.40-m thick reinforced raft, with a 72-h autonomy without

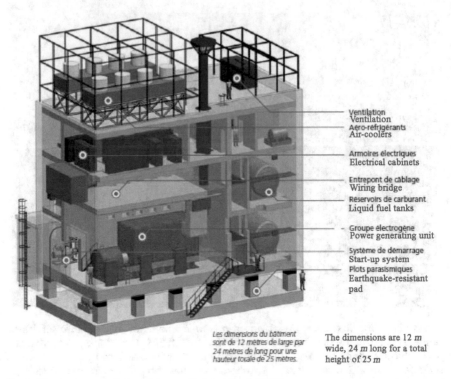

Ventilation
Ventilation

Aéro-réfrigérants
Air-coolers

Armoires électriques
Electrical cabinets

Entrepont de câblage
Wiring bridge

Réservoirs de carburant
Liquid fuel tanks

Groupe électrogène
Power generating unit

Système de démarrage
Start-up system

Plots parasismiques
Earthquake-resistant
pad

Les dimensions du bâtiment
sont de 12 mètres de large par
24 mètres de long pour une
hauteur totale de 25 mètres.

The dimensions are 12 m
wide, 24 m long for a total
height of 25 m

Fig. 5.23 Structure of the Ultimate Safety Siesel (DUS)

human action, and robust to aggressions of levels much higher than the design standards. The building is highly recognizable with its blue openwork hood housing the ventilation systems placed on the roof of the building. This new building, which is part of the Hard Core of post-Fukushima provisions, has a high level of resistance to hazards (earthquakes, storms, floods). Its design led to the conception of a "bunkerized" building (L: 24 m, W: 12 m, H: 25 m) with special provisions such as earthquake-resistant pads, a low floor located above the reference flood level, a wall thickness of 50 cm or a structure and exterior equipment resistant to the "Hard Core" reference tornado. Each diesel has two 12 ton fuel oil vessels of 63 m^3 of fuel oil each, giving the plant concerned 72 h of electrical autonomy. Each DUS produces 3.5 MWe of power, sufficient for the safeguard systems. In addition to the generator set, the DUS includes lubrication pumps, batteries, an electric switchboard, and a fire detection system. The generator set starts automatically in case of a power failure. The first two DUS were operational at the end of 2018 at the Saint-Laurent-des-Eaux site. The entire Fleet has been fully equipped with DUS since February 2021.

The SEU System: The Ultimate Cold Source

Within the framework of the post-Fukushima hard core measures and the creation of the Nuclear Rapid Action Force (FARN, Photos 5.20 and 5.21), EDF has been led to consider an ultimate water supplement coming from a diversified cold source: drilling in deep water tables, water intake in existing basins/ponds (Photo 5.19) or new reservoirs. This backup system is designed to ensure 72 h of cooling water autonomy in the event of a natural event of an intensity exceeding the unit's usual safety standards. This new system, called SEU supplies the steam generators (via the ASG emergency supply to the steam generators), the reactor pool and the fuel pool (BK).

A new system of ultimate heat removal from the reactor containment by means of an EASU heat exchanger has also been introduced. which will be connected to the cold source (possibly the EAS containment spray system if the conventional cold source is lost) by the Nuclear Rapid Action Force (FARN) within 24 h of the hazard. The EASU system intervenes in case of loss of the safety injection systems (RIS) and the EAS containment spray. EASU injects borated water contained in the PTR borated water tank in

Photo 5.19 Fresh water pumping test by FARN (Bugey). Taps were placed in 2013 to be able to re-supply the backup circuits with water

reserve into the containment. When the tank is empty, the system draws water from the sumps and cools it by the EASU heat exchanger. While waiting for the EASU to be connected, the instructions are to refill the PTR tank and to inject borated water into the reactor as quickly as possible if possible (Photos 5.20 and 5.21).

Photo 5.20   FARN: Transport of heavy water replenishment equipment by helicopter

Photo 5.21   A FARN team welcomes EDF President Jean-Bernard Levy, named FARN Honorary Team Member (November 18, 2021, Bugey site)

Conclusions and Perspectives

Civil nuclear power suffers from a lack of public perception because it is technologically complex and often wrongly associated with military nuclear power, from which it has inherited a certain culture of secrecy. Moreover, impalpable radioactivity is a much more frightening danger (an insidious danger?) than the risk of taking your car to go on an errand. However, car accidents undoubtedly kill many more people each year than nuclear power! Why are we more afraid of a shark attack than of a bee sting when the risk of dying from a bee sting (about 400 people per year in the world and about 15 deaths per year in France alone) is out of all proportion to the lethal attack of a shark (8 deaths in the world in 2016, including 3 in Australia, a record), and what can we say about the deaths from road accidents (1.35 million deaths per year).

Is nuclear power dangerous? This is a question that is difficult to answer with a binary and reassuring "*yes or no*" answer. It is undeniable that past accidents should teach us humility because nuclear power has potential for harm at the level of a region, or even a country, that is rarely equaled by other human technologies.[1] No answer can seriously be given if we do not put the profits that humanity gets from it in the equation. In an energy-intensive society, the choices of replacements are all problematic. Man will have consumed in barely two centuries all his non-renewable oil resources, and the greenhouse effect,

---

[1] We will not try to clear the responsibilities of nuclear power by citing the Bhopal disaster in India in 1984, where the explosion of a chemical plant killed 3500 people in one night due to a release of methyl isocyanate. In another register, the Seveso disaster (Italy) gave its name to a directive classifying chemical industries at risk.

© The Author(s), under exclusive license to Springer Nature Switzerland AG 2022
S. Marguet, *A Brief History of Nuclear Reactor Accidents*, Springer Praxis Books,
https://doi.org/10.1007/978-3-031-10500-5

undoubtedly due to the human production of carbon dioxide, is beginning to show all its perverse effects. What about the production of green gasoline that monopolizes huge areas of arable land? There is little doubt that the next wars will be more about access to drinking water, food, and energy than about ideologies. The societal debate is beyond the scope of this book. However, it is reasonable to think that nuclear power is safer today than it was yesterday, and that it will become even safer tomorrow, thanks to the acquisition of knowledge and the use of feedback. History shows us, however, that the most serious accidents (INES-7) are the most recent, perhaps because the gains in safety have only succeeded in eliminating the least serious accidents, without succeeding in countering the most violent. It also shows us that all industrialized countries have been affected, which proves that we should not let our guard down and shift on the side of self-confidence.

This is why on existing reactor designs, engineers are improving safety with new devices, such as passive auto-catalytic recombiners that recombine hydrogen (produced during an accident by oxidation of zirconium and metals) and oxygen into harmless water steam, reducing the risk of hydrogen explosion that was so harmful at Fukushima (Fig. 1). These recombiners have been introduced in the reactor buildings of the French nuclear fleet.

So-called ultimate procedures have been put in place to take into account the meltdown of the reactor. To manage severe accident situations beyond the design basis, and as part of the defense-in-depth approach to protect the population and the environment, an ultimate procedure known as U5 of decompression-filtration of the containment has been implemented.[2] The objective of this procedure is to avoid the loss of containment that could result from a slow pressurization of the Reactor Building (BR) atmosphere, leading to exceeding the design basis pressure, following a core meltdown. This pressurization would be the consequence of the production of hot gases coming from the attack of the concrete raft by corium (a mixture of molten fuel, oxidized or metallic cladding, and structural materials of the core internals), flowing from the vessel after it has been breached, but also from the water steam flowing from a possible breach in the primary circuit The U5 procedure aims at depressurizing the enclosure via a sand filter (Fig. 2), before channeling the filtered releases to the TEG chimney. This should be seen as a last-chance procedure before irreparable loss of the containment, in an extreme situation. The function of the sand filter is to retain aerosols in particular (purification by a factor of about 10), whereas iodine in gaseous form (I_2), for example, would not be retained. The filter contains sand, glass wool, and expanded clay (Figs. 3 and 4, Photo 1). The filter box consists of a cylindrical

---

[2] A. L'Homme ; G. Servière: *Les filtres à sable*, Revue Générale Nucléaire n°2, Mars 1988, pp. 159–160.

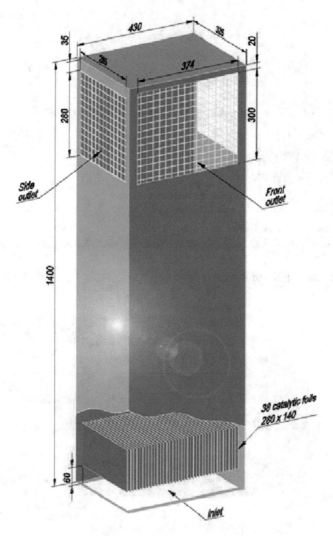

Fig. 1 A passive autocatalytic recombiner (model AREVA/SIEMENS FR-380, dimensions in mm, adapted from (Antoni Rożeń: *Simulation of start-up behaviour of a passive autocatalytic hydrogen recombiner*, Nukleonika - Original Edition - 63(2), July 2018)). Heterogeneous catalytic recombination of hydrogen with oxygen is one of the effective methods used to remove hydrogen from the containment of a nuclear reactor. Inside a recombiner, hydrogen and oxygen molecules are adsorbed at the active points of the catalyzer (including platinum) deposited on parallel plates and recombine to form water. The heat released by this exothermic reaction creates a natural convection of gas in the spaces between the catalytic elements. The hot, moist gas rises into the recombiner stack, while the fresh, hydrogen-rich gas enters the recombiner from below. Catalytic recombination should ideally start spontaneously at room temperature and low hydrogen concentration. As soon as the gas inside the recombiner absorbs enough heat to become lighter than the gas outside the recombiner, it starts to flow upward. Be aware that the temperature of the plates can rise sharply (close to 1000 °C). The French fleet is entirely equipped with recombiners

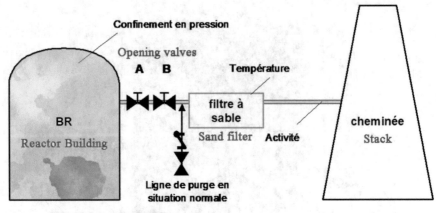

Fig. 2   The sand filter of the U5 procedure

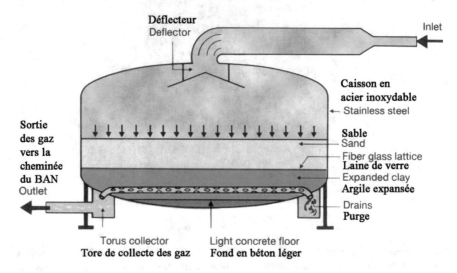

Fig. 3   The U5 sand filter

casing with a diameter of 7 m, closed by curved bottoms. The casing is made of stainless steel and is about 4 m high. It contains a sand bed of 0.8 m thickness, and the gas circulation is downward with an arrival by the top of the filter. Upstream, the filter is connected by piping to the containment, and the feedthrough to the containment wall is an existing feedthrough by design. The feedthrough has two manual isolation valves external to the containment, normally closed. A diaphragm downstream of the isolation valves allows the

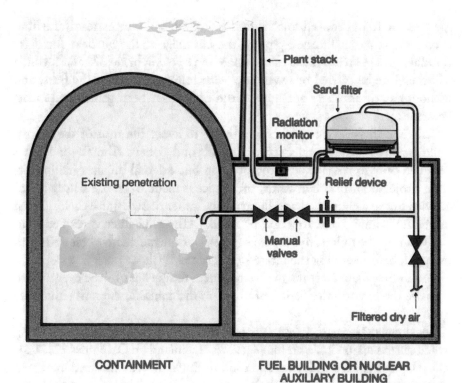

Fig. 4 Location of the sand filter

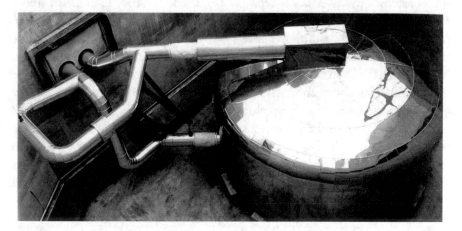

Photo 1 Principle and photo of a U5 filter of a P4 unit (photo EDF). The aerosols arrive at the top of the filter and the non-retained volatile gases are evacuated through the lower pipe toward the TEG stack. The filter can be bypassed in case of severe blockage

gas pressure to be reduced to a value close to one bar. Downstream, the filter is connected by an internal pipe to the chimney of the Nuclear Auxiliary Building. There are no plans to automate the opening of the U5 filter, which will therefore be carried out manually with full knowledge of the facts, and without risk of untimely operation. Sand filters have been generalized in the French fleet.

This "sacrificial" tactic makes it possible to avoid the ruin of the reactor building (BR) by internal overpressure beyond 5 bars. The BR is in fact designed only to resist violent compression but external impacts (falling aircraft, projectiles). In other words, the choice is made to release a little radioactivity, but in a controlled and filtered way, rather than definitively losing the important containment function of the BR. The sand filter is very efficient and will trap the radioactive aerosols (99.9% of the radioactive cesium is thus trapped), before sending the filtered gases to the stack.

For boiling water reactors, the containment design has evolved over time to improve the retention of fission products in the containment and reduce the hydrogen risk.

The Franco-German design of the European Pressurized Reactor (EPR), which started up in China on June 6, 2018, in Finland on December 21, 2021, and is being finalized in France and soon in England, is an illustration of this. Severe accidents involving core meltdowns have been taken into account in the design of the reactor. The translation into probabilistic terms for the EPR has set as reactor objectives a frequency of occurrence of core meltdowns under non-aggressive power conditions of 10^{-6} events per reactor per year, an even lower occurrence for shutdown states, and a frequency of occurrence of core meltdowns due to internal events, associated with an early loss of containment, lower than 10^{-7} events per reactor per year.[3] An emblematic device of severe accident fuel management is the core catcher , sometimes called an ashtray (Fig. 3). In the event of a core meltdown with a hole in the vessel bottom, the corium is collected in the vessel pit, where it comes into contact with a fusible plug. The role of this plug is to retain the corium long enough for it to rise in temperature and be very fluid. When the fusible plug fails, the corium flows through the transfer tube in a gentle slope into a spreading chamber of about 180 m². The more the corium spreads out, the easier it will be to cool. Here we illustrate an efficient way to avoid the Chinese syndrome (Fig. 5).

The EPR has 4 emergency trains (instead of 2 at present), which means that the reactor has 4 redundant safeguard systems. This design increases both

---

[3] To relativize its very weak occurrence, the age of the universe being estimated at 13.7 billion years, we can consider that the occurrence of appearance of the universe is at least 10^{-10} events per year!

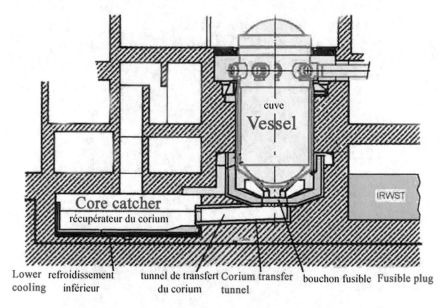

Lower refroidissement   tunnel de transfert Corium transfer bouchon fusible Fusible plug
cooling   inférieur   du corium   tunnel

Fig. 5   Schematic diagram of the core catcher of the European Pressurized Reactor (EPR)

safety and operational availability, as personnel can enter the reactor building during operation, thanks to a judicious "*bunkerization*" of the reactor, to carry out repairs…

New concepts of passive reactors aim at proposing "*forgiving*" reactors with respect to the loss of active systems. A concrete example is the American concept AP600 and then AP1000 from Westinghouse. In these concepts, the aim is to passively cool a steel containment vessel with air, using an ingenious system of air recirculation in contact with the vessel (Fig. 6). The air enters through openings in the reactor building, licks the containment vessel by chimney effect, and is evacuated through an opening placed above the reactor building. The residual power of the core is evacuated in natural circulation by an exchanger called PRHR-HX (*Passive Residual Heat Removal Exchanger*, Fig. 6). This exchanger evacuates the power to a pool inside the containment: the IRWST (*In-containment Refueling Water Storage Tank*) which serves as a heat sink (cold source). The borated water tanks of the safety injection circuit are located above the reactor to always benefit from gravity ("*water cask*" effect). It is the IRWST that ensures the long-term water injections. This concept makes it possible to manage the total loss of electrical power but does not allow a significant increase in the thermal power of the reactor (Fig. 7).

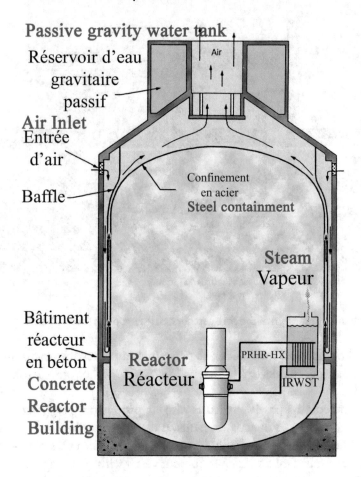

Fig. 6 Passive cooling of the AP600 (Advanced Passive 600 MWe)

Zaporizhzhia or the Madness of Men

The recent events of the 2022 war in Ukraine, following the attack on Russia at the initiative of its President Vladimir Putin, stunned the world. If this war, for the moment conventional, escapes the scope of this book, the attack on the Zaporizhzhia (Ukrainian: Запоріжжя) plant, formerly named Aleksandrovsk, cruelly highlights what is called "*external aggression*" in nuclear language, usually flooding, loss of cold source, or external power. Aggression if ever there was one, since the plant described as the largest in Europe, was attacked by Russian forces from the Crimea on March 4, 2022. The plant of Zaporizhzhia (Photos 2 and 3) is cooled by the Dnieper River via the Kakhovka pond. It is located 71 km south of Dnipro and 445 km southeast of Kiev.

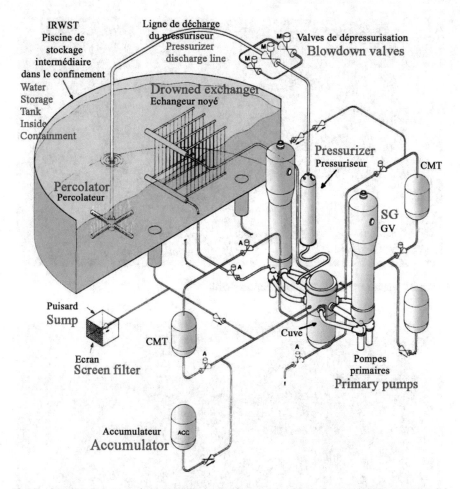

Fig. 7 Passive systems of the AP600 (according to (A. Hall, C.A. Sherbine: *PWRs with passive safety systems*, Nuclear Energy Vol. 30, n°2, pp. 95–103 (1991).))

Located near the town of Enerhodar, it consists of 6 WWER 1000/320 reactors of 950 MWe each, whose construction by Energoatom started in April 1980 and whose reactors were commissioned on December 25, 1985 (Unit 1), February 15, 1986 (Unit 2), March 5, 1987 (Unit 3), April 14, 1988 (Unit 4), October 27, 1989 (Unit 5), and September 17, 1996 (Unit 6). On the same site, there is a conventional coal-fired power plant with very high chimneys, which are easily visible in the photos. The plant supplies one-fifth of the energy of the whole Ukraine and half of the country's nuclear energy.

After the Russian invasion of Ukraine began on February 24, 2022, Energoatom shut down units -5 and -6 to reduce risk and kept units -1 through -4 operating until March 3, 2022. On March 3, 2022, artillery fire

Photo 2   View of the 6 WWERs in Zaporizhzhia

Photo 3   Aerial view of the Zaporizhzhia plant

from Russian forces damaged some of the plant's non-essential buildings (Photo 3). A fire broke out near unit-1 of the reactor, but the nuclear equipment was not damaged. The Russians, who built the plant, appear to have carried out non-strategic scare shots to deter on-site resistance, allowing automatic reactor shutdowns (ARS). Reactor-3 was shutdown at 2:26 a.m.,

Photo 4 A flare falls on the parking lot in front of the line of six reactors in the left-hand follower of the shot. Russian armored vehicles are seen at the shutdown at the entrance of the site on the right. The chimneys of the conventional thermal power plant can be seen in the background, at least one of which seems to be in operation. This inconsequential event was repeated on television

leaving only Reactor-4 operating (Photo 4). Russian troops captured the plant after confirming that radiation levels had not changed. The Ukrainian foreign minister confirmed this information on March 4 at 2:30 a.m., stating on Twitter that the Russian military was *"firing from all sides at the Zaporizhzhia nuclear plant, the largest nuclear plant in Europe. A fire has already broken out."* He called for an immediate ceasefire to allow firefighters to bring the fire under control. At 04:20 a.m. UTC on March 4, the IAEA also reported that the fire, which was in a training building (presumably the training simulator building), had been extinguished (Photo 5). It did not impact the safety of the reactor or any critical equipment. The plant lost 1.3 GWe of capacity, which was then compensated by 9 additional power units (thermal plants) in the nearby. How ironic to try to build safer reactors only to have them attacked by military forces!

Photo 5 Video of one of the control rooms of the plant of Zaporizhzhia. A loud-speaker broadcasts a warning message in Russian to the attackers, subtitled in English on the video (DR). As none of the control panel lights are on, I conclude that this is the control room of one of the shutdown reactors. The video shows non-active operators on the control consoles. A precise shot at the Auxiliary Power System could endanger the cooling of a recent shutdown reactor from which the residual power could no longer be evacuated, leading in the medium term to a meltdown of the shutdown reactor itself. A missile exploding in the spent fuel storage area could also have caused considerable damage, leading to radioactive releases...

It should be noted that the IAEA indicated on March 8, 2022, that a nuclear research facility producing radio-isotopes for medical and industrial applications, the Kharkiv Institute of Physics and Technology (Photo 6), was damaged by bombing in the city of Kharkiv, in northeastern Ukraine, says the International Atomic Energy Agency (IAEA). The incident had not resulted in an increase in radioactivity at the site. "*As the nuclear material it contains is*

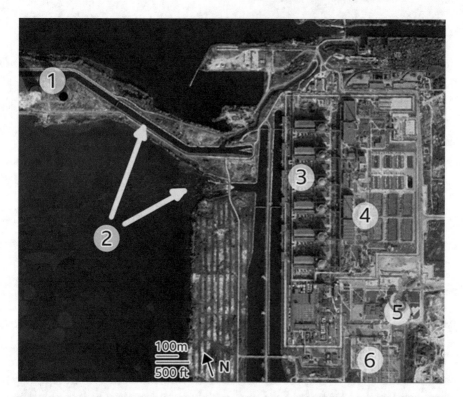

Photo 6   A satellite view of the site (source IAEA). 1: Air coolers (x2); 2: Channels for the supply and discharge of cooling water (cold source); the channels run along the 6 plants; 3: The 6 WWERs, the rectangular buildings with red roofs perpendicular to the water supply channel for the condensers, correspond to the turbine buildings (engine room); the reactors are in the cylindrical buildings with red domes; 4: Spent fuel and radioactive waste storage area; 5: Training building where the fire started; 6: Power evacuation electrical lines station

still subcritical and its stockpile of radioactive material is very small, the IAEA confirmed after its assessment that the reported damage could not have any radiological consequences" said IAEA Director General Rafael Grossi (Photo 7).

What Future for Nuclear Energy?

In conclusion, nuclear power is a technology that requires constant financial and intellectual efforts, as well as absolute rigor. There is a real risk of a "*cheap*" nuclear power plant if financial interests take precedence or in the context of a sector in sharp decline (loss of competence, aging equipment, obsolete

Photo 7   Kharkiv Institute of Physics and Technology

design, etc.). The renewal of the French fleet is a unique opportunity to improve the safety of nuclear reactors and to "*raise the level.*" We have a duty to remember the accidents of the past, which should prevent us from making the same mistakes.

The recent war in Ukraine, whatever its outcome, has created an unprecedented tension on the price of energy as several European countries are highly dependent on Russian gas and oil. When the war began, Germany imported 55% of its gas from Russia. Moreover, just before the war, it had chosen to quickly get out of nuclear power by closing all its plants (except 3), a very unwise choice in the circumstances. As for Italy, 95% of the gas consumed is imported, and Italy is one of the European countries most dependent on Russian gas. About 45% of the gas imported by the peninsula comes from Russia. Italy has also chosen to leave nuclear power "*definitively*" and has not had any operational reactors for a long time. The effort required to return to it would be considerable, given that all the last Italian nuclear specialists have left the country (lots of them work in France) and the fact that the industrial background has totally disintegrated. Even in France, a country with a strong nuclear industry, there are voices calling for the reactivation of the two reactors of the Fessenheim plant shut down in February and June 2020. This is without knowing that the turbines and elements of the secondary circuit have already been *"cannibalized"* for the benefit of other plants of the CP0 type elsewhere in France. It is easy to understand that shutdown is easy, but to resume is much more difficult! The war in Ukraine and its aftermath will, in any case, teach a bitter lesson in industrial strategy to the *"cicadas"* of Monsieur de La Fontaine.[4] Something to ponder on the future of civil nuclear power.

In another register, perhaps more distant, Man will have to leave Earth with nuclear reactors during his conquest of Space because it is today the best power/weight ratio at our disposal. What a way to revive the nuclear dream?

---

[4] Jean de La Fontaine (1621-1695) is an immense French poet who certainly never thought he would be quoted in a book on Nuclear. So, let's give thanks to him here: *"The Cicada, having sung all summer, found herself very deprived when the breeze came"*. This immortal fable is the first of the first collection (124 fables, divided into 6 books) published in March 1668. This collection is dedicated to the *Dauphin*, the son of Louis XIV, king of France.

Appendix A: The INES Scale

The need to compare incidents and accidents in order to learn from them has led to the establishment of an international severity scale: the INES (International Nuclear Event Scale) since 1987. This scale is essentially a communication tool that does not constitute a safety assessment tool as such, and especially not usable to compare the safety level from one country to another, everything being based on the voluntary declaration of the operator and on the level of expertise and independence of the local Safety Authority. The scale has 8 levels (from 0 to 7) in a progression of severity according to a logic of frequency of occurrence. Level-3 is limited by the annual public dose limit (decree n 88-521 of April 18, 1988) of 5 mSv per year.

EDF adopted the INES scale on March 1, 1994. Previously, the use of the French scale led to the declaration of level-1 incidents, which had no real impact on safety. Since 2008, many IAEA countries have applied the INES scale for radiation protection events, taking into account the problems of radioactive sources and materials and their transport.

INES scale of severity of nuclear events

Level INES	Designation	Impact on the site	Off-site impact	Comment
0	Difference	None	None	Anomaly of no safety significance. One thousand situations in France per year.

(continued)

© The Author(s), under exclusive license to Springer Nature Switzerland AG 2022
S. Marguet, *A Brief History of Nuclear Reactor Accidents*, Springer Praxis Books,
https://doi.org/10.1007/978-3-031-10500-5

(continued)

Level INES	Designation	Impact on the site	Off-site impact	Comment
1	Anomaly	None	None	Exit from the authorized operating regime. About a hundred cases in France per year.
2	Incident	Low local and targeted contamination	None	A few cases in France per year.
3	Severe incident	Important targeted contamination	Low release due to loss of defense lines	A few cases in 10 years, including Gravelines-1 in 1989: the use of solid rather than hollow screws in the SEBIM valves for protection of the primary circuit in the event of overpressure rendered them inoperative by inhibiting a useful gap in their opening and by modifying their opening kinetics. The defect was only detected after 1 year. Although there was no contamination at all, the incident was classified 3 by the Safety Authority because of the potential risk to the installation

(continued)

(continued)

Level INES	Designation	Impact on the site	Off-site impact	Comment
4	Accident without off-site consequences	Localized reactor damage or lethal exposure on site	Minor release within legal limits	1980: Meltdown of a Natural Uranium Graphite Gas (UNGG) fuel cartridge at Saint-Laurent-des-Eaux-A2 (France). This event, too old to be classified by INES, which did not exist at the time, would most likely have warranted a classification of 4 1999: Tokaimura criticality accident (dissolved fissile material in a tank, Japan)
5	Accident with off-site consequences	Severe damage to the reactor, loss of biological barriers	Loss of defense-in-depth and severe off-site contamination	1957: Windscale fire (England) 1979: TMI-2 (USA)
6	Severe accident: Large release	Partial destruction of the installation	Application of countermeasures	1957: Reprocessing plant in Kyshtym (USSR)
7	Major accident: massive release	Total destruction of the reactor	Major release and death of many people. Considerable effects on the environment	1986: Chernobyl-4 (Ukraine) 2011: Fukushima-1-2-3-4 (Japan)

Appendix B: The Deterministic Approach to Barrier Design: Applications to Pressurized Water Reactors

Incident/Accident Classes

Following the analysis of the frequency of occurrence of accidents, 4 categories of situations are defined. The first category covers normal operation. In this context, the releases from the plant are limited to those authorized. The second category covers incidents of moderate frequency (about once a year, between one and one event per plant per year), the consequences of which are limited to authorized releases. The third category covers accidents of low frequency (less than once in the life of the plant, between 10^{-4} and 10^{-2} event per plant per year) where releases must be limited to 0.5 rem or 0.005 Sievert (Sv) for the whole body[1] and where the situation must be controlled. The fourth category includes extremely hypothetical events between 10^{-6} and 10^{-4} event per plant per year) where releases must remain below 15 rem or 0.15 Sv for the whole body[2] to 2 hours at the site boundary. A complementary domain, which was not explicitly foreseen in the definition of the categories, makes it possible to cover even lower frequency accidents corresponding to accumulated failures: the so-called H procedures. This domain attempts to respond to scenarios outside the initial design basis of the plants and covers the combinations of events linked to the short-term loss of redundant safety systems frequently called upon, or to the medium or long-term loss of systems intervening in LOCA scenarios, where residual power must be evacuated for several

---

[1] With 0.015 *Sievert* maximum for the thyroid.

[2] With 0.45 *Sievert* maximum for the thyroid.

S. Marguet, *A Brief History of Nuclear Reactor Accidents*, Springer Praxis Books,
https://doi.org/10.1007/978-3-031-10500-5

months. Four complementary areas have been studied with specific procedures associated:

H1 corresponds to the total loss of the cold source (RRI/SEC). If the primary circuit is closed (vessel not open), if the pressure is higher than 32 bars, RRA not connected, we fall back to the conditions to connect the RRA. If the RRA is connected, the pressure and temperature rise due to the residual power and it is evacuated by the steam generators fed by the ASG. This can work as long as the ASG tank can be externally recharged. If the primary circuit is open, there is a priori a low residual power; it is then sufficient to compensate the free boiling of the primary circuit by RCV which draws from the PTR tank..

H2 corresponds to the total loss of feed water from the steam generators (ARE+ASG). Without operator action, the temperature rises in the primary circuit and the SGs dry up. This triggers an emergency shutdown on low water level in the SGs. The pressure in the pressurizer increases and the pressurizer valves are activated. Without the action of the safety injection, the pressurized core will melt. The H2 line then consists in shutting down the primary pumps which inject power into the water and cooling the primary circuit by the safety injections (SI). The pressurizer valves are voluntarily opened to depressurize (which facilitates the work of the SI), and the water from the sumps is reinjected into the core via the PTR tank also known as feed and bleed operation). The problem then lies in the possible loss of the SI to the recirculation of water in the sumps (clogging of the sumps?).

H3 corresponds to the total loss of internal and external power supplies. It is the simultaneous failure of the two external power sources of the grid, the failure of the house-load operation and the two emergency generators on train A and B. In this case, the injection at the pump seals is lost, which will lead to a degradation of the seals and a small primary circuit leak. We also lose the safety injection. A turbo-alternator set LLS will be used, which produces current from the steam of the SGs and re-supplies the control unit and the injection at the pump seals. The recharge of the ASG tank ensures the supply of the SGs.

H4 corresponds to the loss of the ISBP or EAS pumps. A mutual backup of the ISBP low-pressure safety injection and the EAS containment spray is then established by a mobile alignment, possibly filled by mobile exchangers (U3 procedure), in case of failure of one of the circuits (long-term LOCA situation).

Some anticipated transient without scrams (ATWS) , which could have been classified as H5, although the terminology does not exist), are covered by specific measures (diversification of turbine trip signals and ASG pump start-up). On the other hand, the accumulation of situations such as a steam line break (SLB)+nSGTR (multiple steam generator tube rupture), or the total loss of the ISMP medium pressure injection, or the emptying of 2 SGs, is also part of the additional situations covered by specific procedures.

Description of incidents/accidents of internal origin and safety requirements for PWRs

Class of incidents/ accidents (of internal origin)	Description
1 Normal operating situations Occurence frenquency ≥ 1 (event per year)	Normal operation in compliance with the Technical Operating Specifications (STE). The study of class 1 transients allows the sizing of the boiler control devices. Examples: House-load operation of the turbo-generator set Normal operating transients The fuel element must remain intact: perfect integrity of the 3 barriers
2 Moderate frequency incidents $10^{-2} \leq$ occurence frequency < 1	Uncontrolled removal of control rod clusters (subcritical or power reactor) Incorrect positioning, dropping of a control rod cluster or group of clusters Uncontrolled dilution of boric acid due to malfunction of the RCV (primary circuit charge and discharge) Partial loss of primary circuit flow Introduction of cold water by starting an inactive loop Load shedding (total pressure drops and turbine trip) Loss of normal feeding water Malfunction of the normal steam generator supply system (ARE) Loss of external power supplies (H2) Excessive load step-up Untimely opening of a pressurizer safety valve Untimely opening of a secondary safety valve Untimely start of the IS (Safety Injections) The fuel element must remain intact: perfect integrity of the 3 barriers A category 2 condition must not cause a higher category condition

(continued)

(continued)

Class of incidents/ accidents (of internal origin)	Description
3 Low-frequency incidents $10^{-4} \leq$ occurence frequency $< 10^{-2}$ Releases at site boundary (public): – Whole body <5 mSv – Thyroid alone <15 mSv	LOCA (Loss of Coolant Accident) small break on the primary circuit Small break in secondary piping (SLB small break) Total loss of primary circuit flow Incorrect positioning of an assembly in the core (loading error) Removal of a single power control rod cluster RCV tank rupture Rupture of the effluent gas storage tank TEG SGTR (Steam Generator Tube Rupture) 1 tube Limited damage of some fuel elements: perfect integrity of the last 2 barriers (primary circuit and containment) A category 3 condition must not be the cause of a higher category condition
4 Very low frequency incidents (even hypothetical) $10^{-6} \leq$ occurence frequency $< 10^{-4}$ Whole body < 150 mSv Thyroid alone < 450 mSv	Fuel handling accident Major steam line break in the secondary (SLB, Steam Line Break) Blocked rotor of a primary pump GMPP Ejection of a control rod cluster Steam generator tube rupture (SGTR) and a blocked SG valve open (containment bypass scenario) SGTR 2 tubes (for N4) Large primary circuit break Geometrical structure of the core allowing cooling, perfect integrity of the last barrier (containment), i.e., no additional damage to the primary circuit and the containment
Beyond design basis	Multiple SGTR+SLB Decapping of the vessel bottom

The deterministic approach consists in calculating the releases induced by these types of accidents to confirm the safety approach (definition of the source term). First of all, the most exhaustive list of events of internal origin to the plant likely to occur is established. These scenarios are classified in the previously defined categories. Then the conservative events that maximize the consequences are selected from each category. The design and sizing of buildings and systems must protect against the consequences of conservative accidents by applying the single failure criterion. These scenarios define the so-called design basis operating conditions. To these conditions, we add aggressions of internal origin to the plant (broken pipe whip aggravating the initial accident, internal projectiles, fire, etc.) and those of external origin to the plant (plane crash, flooding by the cold source, earthquake, etc.). The intervention of automatic systems is then taken into account by applying the single aggravating factor rule (in the fourth category, it is added to the External Voltage Failure MDTE for *Manque De Tension Externe*), by increasing the duration of the interventions of the automatic controls and by not taking into account the controls that have a beneficial effect by penalization. But this approach raises the question of the exhaustiveness of the scenarios retained, and of the understanding of the phenomena by the engineers, the famous *"expert judgment"* on conservatism. The probabilistic approach of probabilistic safety assessments (PSA) complements the deterministic approach for families of accidents with very low frequencies (per plant and per year), whose probability is sometimes so low that it is difficult to define. The probabilistic approach is applied in particular to external hazards (airplane crash, explosion), which are difficult to treat in a deterministic way. It also makes it possible to quantitatively justify the fallback times specified in the Technical Operating Specifications in the event of accidental unavailability of equipment or a system classified as safe that is not used in a normal situation (quantification of the increase in risk).

Deterministic Safety Criteria

During the 1970s, a set of deterministic criteria was established, based on experiments (fuel behavior during depressurization tests, reactivity insertion tests, SPERT and CABRI tests, etc.), and on calculations and expert judgment, which make it possible to conclude that an accident is harmless with respect to the three major safety functions. These criteria have undergone adjustments over time, both because of the progress of knowledge (new tests) and because of the appearance of new fuel management and use (extension of burn-up rates, grid monitoring, use of MOX). These criteria are a common basis accepted by the operator, who is responsible for the safety of his plant, and by the Safety Authority, which must check them. Verification of the criteria allows validation of a fuel loading pattern.

Safety criteria for PWRs by accident category

Class of incidents/ accidents (of internal origin)	Description
1, 2, and 3 Normal operation and moderate frequency incidents	**Non-fragilization of the cladding:** In class 1, zirconium oxide layer ≤100 μm, i.e., approximately 12% of the initial thickness of the cladding. Oxidation weakens the cladding because it is the metal layer that holds the material together
	In class-2 and -3, the temperature at the Zr/ZrO$_2$ interface must remain below 425 °C
	No boiling crisis: The maximum thermal flux is (largely) lower than the critical flux evaluated by the WRB1 critical flux correlation, verification of the critical thermal flux ratio (CTFR) criteria
	No fuel melting: No point of the fuel must reach the melting temperature of the uranium oxide, i.e.:
	– 2810 °C for fresh fuel
	– This temperature is lowered by 7.6 °C for each 10,000 MWd/t plant to take into account the appearance of fission products that lower the melting point of the spent fuel
	To avoid a complex calculation of the fuel temperature during the plant monitoring, we use a decoupling criterion based on the linear power: P_{lin}≤590 W/cm, which guarantees the default temperature criterion, and which allows to call the reactor protections
4 Very low frequency incidents (even hypothetical)	**Accidents of RIA (Reactivity Initiated Accident):** Limiting the number of rods in boiling crisis to less than 10% of the total number of rods
	Less than 10% by volume of molten fuel at the hot spot
	No runaway oxidation of zirconium: Average cladding temperature <1482 °C (2700 °F)
	No dispersion of liquid fuel in water by fuel explosion: Maximum fresh fuel enthalpy <225 cal/g, spent fuel <200 cal/g
	Additional criteria for high burn-up rates > 47,000 MWd/t:
	– Thickness of the zirconia ≤100 μm
	– Power pulse width ≥30 ms
	– Energy deposit of the pulse ≤57 cal/g
	– Maximum cladding temperature ≤700 °C
	LOCA accidents:
	– **Fuel resistance during reflooding:** Maximum cladding temperature ≤1204 °C(2200 °F) to prevent the oxidation reaction of the cladding from getting out of control
	– « *Equivalent*» oxidation[a] ≤17% of the thickness of the cladding
	In the long-term phase, avoid the crystallization of boron (by over-concentration) and the return to criticality (by under-concentration) and ensure the evacuation of the residual power, according to the fuel management:
	– High limit of boron concentration in the core
	– Low boron concentration limit in the sump for recirculation
	– Sufficient safety injection rate
	Coolable geometry

[a]The equivalent oxidation is calculated by assuming that all oxygen absorbed by the cladding is in the form of zirconia, ignoring the α-Zr layer (a partially oxidized metal layer). To calculate this oxidation rate, the oxidation thickness resulting from normal operation and the additional oxidation occurring during the accident are summed

Some criteria depend on the kinetics of the accident, as for example in the case of class-2 incidents. In order to simplify the monitoring of the plant, we have been led to look for decoupling criteria, applicable in operation, such as a limit value of the linear power, and which guarantee that the real criteria will be maintained in case of an accident. Indeed, it is difficult to access the maximum fuel temperature online, which would require a complete 3D calculation at the exact conditions of the core. The decoupling criteria affect the operability of the plant by limiting, for example, the rate of return to full power after prolonged operation at intermediate power. They are used to size the thresholds of the automatic protections of the reactor. We also note that certain historical criteria (such as the term "*coolable geometry*") are relatively vague and qualitative, which leaves trains of progress in the determination of criteria with very high stakes.

Appendix C: History of Significant Nuclear Reactor Accidents in the World

June 23, 1942: Leipzig (Germany) within the framework of the *Uranverein* program to develop an atomic weapon. Explosion of the L-IV heavy water pile following a heat-up of the uranium fanned by an air inlet. The vessel of the experimental reactor exploded and the ignited uranium, projected to the ceiling, caused a severe fire in the laboratory. This was the first reactor accident in history. Developed in a very secret setting under the Nazi regime, the information that has survived is very fragmentary (described in this book).

12 December 1952: Chalk River Research Centre (Canada). Power excursion on the NRX heavy water moderated reactor, stopped by draining heavy water. During the test phase, a vertical pressure tube was cooled with air; the other channels were cooled with light water. Due to a valve opening error, 3 or 4 control rods were ejected. Despite the closing of the valves, the shutdown control rods do not drop correctly. A misinterpretation of the rod position signals leads to the removal of control rod steps. The power rises to about 4 times the nominal power, i.e., about 90 MWth. Some calandria tubes burst, as well as several fuel rods. A hydrogen explosion of moderate intensity located in the heavy water calandria shut down the accident. This is the most severe accident in Canada and the first in a power reactor controlled by rods.

November 29, 1955: Argonne Research Center, Idaho (USA). Meltdown of almost half of the core of the fast neutron reactor EBR-1 (sodium).

October 10, 1957: Military reactor n °1 of the Natural Uranium Graphite Gas (UNGG) type, Windscale (England). Fire caused by the Wigner effect. Classified 5 on the INES scale (described in this book).

S. Marguet, *A Brief History of Nuclear Reactor Accidents*, Springer Praxis Books, https://doi.org/10.1007/978-3-031-10500-5

May 24, 1958: NRU heavy water reactor, Chalk River, Canada. Fuel element catches fire during unloading.

October 15, 1958: Heavy water moderated experimental reactor in Vinča (Yugoslavia). Power surge by exceeding the critical level of heavy water. An operator dies following his burn-up (described in this book).

July 13, 1959: Sodium-cooled graphite-moderated experimental reactor at Santa Susana Field (California, USA). Flow blockage leading to partial meltdown of 20% of the core. The accident was not even detected during the test! (described in this book).

January 3, 1961: SL-1 military boiling water reactor, Argonne Research Center, Idaho (USA). Power surge by rod withdrawal. The 3 operators are killed by the explosion (described in this book).

January 3, 1961: Soviet submarine K-19, in the Barents Sea (USSR). Loss of primary coolant accident on a PWR by rupture of a collector of the primary circuit. About twenty deaths by severe burn-up (described in this book).

December 30, 1965: Experimental reactor of Mol (Belgium); power excursion by rod movement. A serious injury.

October 5, 1966: Enrico Fermi fast neutron reactor using sodium coolant. Lagoona Beach (USA). Meltdown of two assemblies by cooling failure (described in this book).

May 11, 1967: Magnox reactor in Chapelcross (England). Magnox magnesium-aluminum cladding melting. Two shutdowns for cleaning (described in this book).

November 7, 1967: Siloé light water research pool reactor , Grenoble (France). Partial melting of a fuel element (described in this book)

January 21, 1969: Heavy water research reactor at Lucens (Switzerland) moderated with heavy water and cooled by carbon dioxide. Melting of the fuel following a depressurization. The reactor was placed in a cavern. The accident was classified 4 on the INES scale (described in this book).

October 17, 1969: Saint Laurent-A1 (France). Melting of 50 kg of fuel following a loading error. Classified 4 on the INES scale *a posteriori* (described in this book).

December 7, 1975: Lubmin WWER-440 plant (light water), Greifswald (Germany, former GDR). A fire destroyed the electrical cables supplying the pumps. An emergency pump re-powered by reactor n °2 saved the plant. Classified 4 on the INES scale.

November 28, 1975: RBMK type in Leningrad-1 (Russia). Fuel melting due to loss of cooling of a channel. Classified 3 on the INES scale (described in this book).

February 22, 1977: Bohunice A1 plant (heavy water, carbon dioxide), near Bratislava (Czechoslovakia). Two severe accidents occurred at the A1 plant.

The first occurred in 1976 killing two people by asphyxiation. The second severe accident occurred during refueling on February 22, 1977, and caused a fuel meltdown. The accident was classified 4 on the INES scale (described in this book).

March 28, 1979: Three-Mile-Island-2, Pennsylvania (USA). Core meltdown following an undetected LOCA small break. Rated 5 on the INES scale (described in this book).

March 13, 1980: Saint-Laurent-A2, (France). Meltdown of two combustible cartridges (described in this book).

September 23, 1983: Experimental reactor of Constituyentes, (Argentina). Power excursion: 1 dead (described in this book).

August 10, 1985: Reactor of the Soviet submarine K-431, Vladivostok (Soviet Union). During the replacement of the vessel cover of the reactor after its reloading in nuclear fuel (thus new), this one is badly repositioned. When it was put back on to correct the error, the vessel cover pulled the control rods too far out of the reactor, causing a criticality outbreak and a power excursion. A steam explosion caused a fire in the fuel compartment of the submarine and the rupture of the pressure shell. Ten people died as a result of the explosion and many firemen were burned up.

April 26, 1986: Reactor n °4 of Chernobyl, (Ukraine, ex-USSR). Power surge and steam explosion. Classified 7 on the INES scale (described in this book).

October 18, 1989: Reactor n °1 of Vandellos (Spain). Oil explosion in a turbine bearing. The fire spread to the electrical circuits affecting the safeguard systems of the plant. Classified 3 on the INES scale (described in this book).

March 24, 1992: Sosnovy Bor (near Leningrad), partial meltdown of the RBMK reactor type, of the same reactor type as Chernobyl (described in this book).

December 8, 1995: MONJU reactor (FBR), (Japan). Breakage of a thermometric probe due to vibrations in the secondary circuit. Leakage of 2 m³ of liquid sodium causing a fire (1500 °C!). An inappropriate air-conditioning cladding transported sodium vapors into a part of the reactor building. The reactor was shut down for 14 years following a safety reassessment. Incredibly, on August 26, 2010, barely 3 months after the restart, the fuel transfer machine (3.3 tons), inside the vessel, fell into the vessel (!) at the end of a fuel handling operation. The sodium being totally opaque, it took nearly a year to safely remove the machine from the reactor.

March 11, 2011: Reactor n °1 to 4 of Fukushima, (Japan). Meltdown by loss of cooling. Rated 7 on the INES scale (described in this book).

Bibliography

[*Arnold, 2007*] Lorna Arnold, *Windscale 1957: Anatomy of a nuclear accident*, Palgrave-MacMillan, Basingstoke, United kingdom, ISBN 978-0-230-57317-8, 2007, 236 pages. 3rd edition.

[*Ash, 1979*] Milton Ash, *Nuclear reactor kinetics*, 2nd edition, McGraw Hill, USA, ISBN 0-07-002380-8, 1979, 445 pages.

[*Barjon, 1993*] Robert Barjon, *Physique des réacteurs nucléaires*, autoédition, Grenoble, ISBN 2-7061-0508-9, 1993.

[*Bar-Zohar, 1965*] Michel Bar-Zohar, *La chasse aux savants allemands (1945-1960)*, Fayard, 1965, 285 pages.

[*Brown, 2021*] Kate Brown, *Tchernobyl par la preuve: vivre avec le désastre et après*, Actes Sud, ISBN 978-2-330-14413-5, 2021, 524 pages.

[*Chelet, 2006*] Yves Chelet, *La radioactivité: manuel d'initiation*, Nucléon, Paris, ISBN 2-84332-019-4, 2006, 557 pages.

[*Chouha et Reuss, 2011*] Michel Chouha, Paul Reuss, *Tchernobyl, 25 ans après... Fukushima: Quel avenir pour le nucléaire?* Lavoisier Tec et Doc. Cachan, France, ISBN 978-2-7430-1364-6, 2011, 215 pages.

[*Galic et Vidal, 2021*] Bertrand Galic, Roger Vidal, *Fukushima: Chronique d'un accident sans fin*, Glénat, Grenoble, France, ISBN 978-2-344-03437-8-001, 2021, 128 pages including a documentary booklet.

[*Glasstone et Sesonske, 1994*] Samuel Glasstone, Alexander Sesonske, *Nuclear reactor engineering tome 1 et 2*, Chapman-Hall, USA, ISBN 0-412-98521-7 et 0-412-98531-4, 1994, 841 pages in two volumes, 4th edition.

[*Hetrick, 1993*] David L. Hetrick, *Dynamics of nuclear reactors*, American Nuclear Society, La grange Park, USA, ISBN 0-89448-453-2, 1993, 542 pages.

[*Keepin, 1965*] G. Robert Keepin, *Physics of nuclear kinetics*, Addison-Wesley, Library of congress 64-20831, 1965, 435 pages.

© The Author(s), under exclusive license to Springer Nature Switzerland AG 2022
S. Marguet, *A Brief History of Nuclear Reactor Accidents*, Springer Praxis Books,
https://doi.org/10.1007/978-3-031-10500-5

[*Lamarsh et Barrata, 2001*] John Lamarsh, Anthony J. Barrata, *Introduction to nuclear engineering*, Prentice Hall, ISBN 0-201-82498-1, 3rd edition, 2001, 783 pages.

[*Marguet, 2017*] Serge Marguet, *The physics of Nuclear reactors*, Volume 1 and 2 Springer, Suisse, ISBN 978-3-319-59560-3, 2017, 1 445 pages.

[Marguet, 2019] Serge Marguet, *La technologie des réacteurs à eau pressurisée*, Editions EDP-Sciences, collection EDF/R&D, ISBN 978-2-7598-2360-4, 2019, 1 168 pages.

[*Métivier, 2006*] Henri Métivier (coordinateur), *Radioprotection et ingénierie nucléaire*, EDP Sciences-INSTN, Paris, ISBN 2-86883-769-7, 2006, 505 pages.

[*Réacteurs nucléaires à caloporteur gaz, 2006*] Collectif piloté par Pascal Anzieu, *Les réacteurs nucléaires à caloporteur gaz*, Monographie de la DEN, 2006, CEA, ISBN 2-281-11317-5, 2006. 162 pages.

[*Renoux et Bouland, 1998*] André Renoux, Denis Boulaud, *Les aérosols: Physique et Métrologie*, Lavoisier, collection Tec et Doc, Cachan, ISBN 2-7430-0231-X, 1998.

[*Reuss, 2003*] Paul Reuss, *Précis de neutronique*, EDP Sciences, collection INSTN, Paris, ISBN 2-86883-637-2, 2003.

[*Rozon, 1992*] Daniel Rozon, *Introduction à la cinétique des réacteurs*, Editions de l'école polytechnique de Montréal, Canada, ISBN 2-253-00223-8, 1992, 413 pages,

[Stacy, 2000] Susan M. Stacy, *Proving the principle: a history of the Idaho National Engineering and Environmental Laboratory*, 1949-1999, DOE, USA, ISBN 0-16-059185-6, 2000, 320 pages.

Printed in the United States
by Baker & Taylor Publisher Services